Essen

Essential heat transfer

CHRISTOPHER LONG

University of Sussex

LONGMAN

© Pearson Education Limited 1999

Pearson Education Limited
Edinburgh Gate
Harlow
Essex CM20 2JE
England

and Associated Companies throughout the World.

The programs in this book have been included for their instructional value. They have been tested with care but are not guaranteed for any particular purpose. The publisher does not offer any warranties or representations nor does it accept any liabilities with respect to the programs.

Many of the designations used by manufacturers and sellers to distinguish their products are claimed as trademarks. Pearson Education Limited has made every attempt to supply trademark information about manufacturers and their products mentioned in this book. Kaplon, Mylar and Teflon are registered trademarks of DuPont; Plexiglas is a trademark of Ato Haas Industries. The publishers are grateful to John Wiley & Sons Inc. for permission to reprint the tables shown in Appendices A and B, adapted from Incropera, F.P. and DeWitt, D.P. (1996) *Fundamentals of Heat and Mass Transfer* 4th edn.

Typeset in 10/12pt Times by 32

First published 1999

ISBN 0-582-29279-4

British Library Cataloguing-in-Publication Data
A catalogue record for this book is available from the British Library

Printed in Malaysia, PP

In memory of Ron Mansfield (1954–1999)
my friend, companion and brother.

Contents

Preface   x
Nomenclature   xii

1. Introduction   1
1.1   What is heat transfer?   1
1.2   Conduction   3
1.3   Convection   6
1.4   Radiation   10
1.5   Closing comments   12
1.6   End of chapter questions   13

2. Conduction – Analytical methods and applications   14
2.1   Introduction   14
2.2   Derivation of the general conduction equation   15
2.3   Boundary conditions and initial conditions   18
2.4   Dimensionless groups for conduction   20
2.5   One-dimensional steady-state conduction   22
      2.5.1   Simple plane walls   22
      2.5.2   Composite plane walls   25
      2.5.3   One-dimensional radial conduction   28
      2.5.4   One-dimensional radial conduction in composite systems   31
2.6   Thermal contact resistance   34
2.7   Temperature-dependent thermal conductivity   34
2.8   Fins and extended surfaces   36
      2.8.1   Analysis of steady-state heat transfer from extended surfaces   36
      2.8.2   Fin performance   36
2.9   Two-dimensional steady-state conduction   44
2.10 Transient conduction   50
      2.10.1 One-dimensional approximation   51
      2.10.2 Low Biot number or lumped mass approximation   55
      2.10.3 The semi-infinite solid   57
2.11 Closing comments   60
2.12 References   61
2.13 End of chapter questions   61

3. Conduction – Numerical methods and application   64
3.1 Introduction   64
3.2 Finite difference equations for two-dimensional steady-state conduction   67
     3.2.1 Interior points   68
     3.2.2 Boundary conditions   69
3.3 Application to two-dimensional steady-state conduction problems   71
3.4 Solution methods for problems in numerical conduction   76
     3.4.1 Iterative techniques   77
3.5 Numerical methods for transient conduction   80
     3.5.1 Explicit solutions   81
     3.5.2 Implicit and semi-implicit solutions   82
3.6 Closing comments   90
3.7 References   90
3.8 End of chapter questions   91

4. Principles of convection   94
4.1 Introduction   94
4.2 Continuity, momentum and energy equations   95
     4.2.1 Mass continuity equation   95
     4.2.2 Momentum equations   97
     4.2.3 Energy equation   100
4.3 Boundary layers   106
4.4 Dimensionless groups for convection   109
     4.4.1 Forced convection   110
     4.4.2 Free convection   112
     4.4.3 Average and local values   113
     4.4.4 Choice of reference temperature and temperature for evaluating fluid properties   114
4.5 Turbulence and the time-averaged equations   115
     4.5.1 The nature of turbulence   115
     4.5.2 Time-averaged equations   116
     4.5.3 Prandtl's mixing length model of turbulence   120
     4.5.4 Analogies between heat and momentum   121
4.6 Integral equations   123
4.7 Scale analysis   126
     4.7.1 Laminar forced convection   127
     4.7.2 Laminar free convection   128
4.8 Closing comments   130
4.9 References   131
4.10 End of chapter questions   131

5. Forced convection   135
5.1 Introduction   135
5.2 Parallel flow   135

5.3	Laminar flow in a pipe	139
5.4	Turbulent flow in a pipe	143
5.5	Laminar flow over a flat plate	149
5.6	Turbulent flow over a flat plate	155
5.7	Heat transfer from a rotating disc	159
5.8	Impinging jets, cylinders in crossflow and spheres	163
	5.8.1 Impinging jets	163
	5.8.2 Cylinders in crossflow	166
	5.8.3 Spheres	168
5.9	Closing comments	168
5.10	References	169
5.11	End of chapter questions	170

6. Free convection — 173

6.1	Introduction	173
6.2	Free convection from a vertical surface	175
6.3	Inclined and horizontal surfaces	182
6.4	Other external flows: cylinder, sphere and rotating disc	186
	6.4.1 Horizontal cylinder	186
	6.4.2 Sphere	186
	6.4.3 Rotating disc	188
6.5	Vertical channel flow	189
6.6	Free convection in enclosures	191
	6.6.1 Enclosure heated from the side	192
	6.6.2 Enclosure heated from below	195
6.7	Closing comments	200
6.8	References	201
6.9	End of chapter questions	202

7. Condensation and boiling heat transfer — 205

7.1	Introduction	205
7.2	Condensation	205
	7.2.1 Droplet condensation	206
	7.2.2 Laminar film condensation on a vertical surface	207
	7.2.3 Turbulent film condensation on a vertical surface	210
	7.2.4 Other surface configurations	214
	7.2.5 Forced flow condensation	216
7.3	Boiling	217
	7.3.1 Pool boiling	217
	7.3.2 Forced convection boiling	225
7.4	Closing comments	228
7.5	References	229
7.6	End of chapter questions	230

8. Radiative heat transfer — 232

8.1	Introduction	232

8.2 Radiative properties 239
  8.2.1 Emission, irradiation and radiosity (E, G and J) 239
  8.2.2 Absorptivity, reflectivity and transmissivity 239
  8.2.3 Emissivity 240
  8.2.4 Kirchoff's law of radiation 242
8.3 View factors 243
8.4 Radiation exchange between black bodies 249
8.5 Radiation exchange between grey bodies 251
  8.5.1 Grey body enclosures 251
  8.5.2 Radiative exchange between two grey bodies 257
8.6 Environmental radiation 259
  8.6.1 General principles 259
  8.6.2 Solar heating panels 262
8.7 Closing comments 264
8.8 References 265
8.9 End of chapter questions 265

9. Heat exchangers 268
9.1 Introduction 268
9.2 Classification 269
9.3 Overall heat transfer coefficient 272
9.4 Analysis of thermal performance: the logarithmic mean temperature difference 281
9.5 Analysis of thermal performance: effectiveness and NTU 289
9.6 Pressure losses 295
9.7 Closing comments 300
9.8 References 301
9.9 End of chapter questions 301

10. Heat transfer instrumentation 304
10.1 Introduction 304
10.2 Temperature measurement 305
  10.2.1 Liquid-in-glass thermometer 306
  10.2.2 Thermocouples 306
  10.2.3 Resistance thermometers 311
  10.2.4 Thermistors 311
  10.2.5 Solid-state devices 312
  10.2.6 Thermal paints and liquid crystals 312
  10.2.7 Radiation pyrometer 312
10.3 Measurement of heat flux 313
  10.3.1 Simple one-dimensional steady-state conduction solution 313
  10.3.2 Two-dimensional transient conduction solution 315
  10.3.3 Low Biot number method 317
  10.3.4 Semi-infinite solid method 320
  10.3.5 Heat flux gauges 321
10.4 Errors in temperature measurement 323

10.4.1 Dynamic error   323
10.4.2 Errors due to radiation effects   324
10.4.3 Errors due to conduction along lead wires   326
10.4.4 Embedding errors   327
10.5 Closing comments   328
10.6 References   329
10.7 End of chapter questions   329

Appendices   333
A. Thermophysical properties of matter   333
B. Mathematical relations and functions   368
C. Program listings   370

Index   381

Preface

Heat transfer is perceived by students and practising engineers as a difficult subject: highly mathematical and a minefield of complex relations and assumptions that do not reflect the 'real' world. Although they serve as useful and comprehensive works, the physical size of most of the currently available heat transfer textbooks reinforces this unfortunate perception. Intentionally, therefore, this is a relatively short book.

This book aims to provide students and professionals with the essentials of engineering heat transfer. The material I have chosen to include takes the reader through the key topics in this subject: analytical and numerical conduction, principles of convection, forced and free convection, heat transfer during boiling and condensation, radiative heat transfer, heat exchangers and finally heat transfer instrumentation. This should adequately cover the requirements of most undergraduate level courses in heat transfer. In preparing the text, I have assumed that the reader has a working knowledge and understanding of physics and mathematics to English A Level or equivalent. The chapters are structured so as to lead the reader into each topic by way of introductory remarks; the theory is developed and applications given by way of worked examples.

It was also my intention to make this book appeal to a wider audience than undergraduate students. The chapter on heat transfer instrumentation, the tables of data and the presentation of different aspects of convection will be valuable to the practising engineer. Likewise, the relative weighting given to the principles of convection, in particular the derivation of the momentum and energy equations, is probably beyond the scope of most undergraduate study, but will serve as useful reference material for the burgeoning research worker. This book does not attempt to be encyclopaedic and the reader is given references in the text to further information where necessary.

Engineering is neither pure science, pure art nor purely practical work, and many a good engineer will admit to relying on his or her intuition as well as calculated results. In my experience of teaching engineering, three elements need to be addressed by the teacher: theory, example and experience; and I have tried to balance these elements in this book.

Students need to gain experience of the subject by considering practical examples: At what speed is a Reynolds number of 10^6 representative of conditions over a car body? How does a heat flux of $3\,kW/m^2$ actually feel? What are the consequences of making a cooling fin out of steel rather than aluminium? And so on. Gaining this experience naturally takes a long time, and I have imparted some of my own experiences to the reader to assist in this process. In particular, many of the problems included in the book reflect the real world of engineering design. I have also chosen to add commentary after the worked examples to promote exploration of the result, the assumptions and the practical consequences.

Additional problems are included at the end of each chapter to test the reader's knowledge. Although a single numerical answer is given to each of these problems, lecturers who adopt this book may obtain a manual from the publishers containing fully worked solutions.

I have also included listings of a number of computer programs written in BASIC in Appendix C. Their filename and a brief description are given below; they are also referred to at appropriate points in the text.

ERF.BAS	Calculates error function
FDEXPXT.BAS	One-dimensional transient conduction, numerical explicit solution
FDRZ.BAS	Two-dimensional steady-state conduction in r–z coordinates, numerical solution
ONEDXT.BAS	One-dimensional transient conduction, analytical solution
PN(NR).BAS	Calculates roots of a transcendental equation using Newton–Raphson iteration
SS2D(NR).BAS	Two-dimensional steady-state conduction, analytical solution

If readers have any comments or suggestions about this book, I would be grateful to hear of them. (My email address is c.a.long@sussex.ac.uk.)

Finally, I would like to thank my friends, family and colleagues for their encouragement, discussion and inspiration.

Christopher Long
December 1998

Nomenclature

Symbol	Meaning	Units
A	area	m^2
A_c	cross-sectional area	m^2
A_s	surface area	m^2
A_n	constant depending on value of n	
b	outer radius of a disc	m
$Bi = hL/k$	Biot number	
B_n	constant depending on value of n	
c	velocity of electromagnetic radiation, 2.998×10^8	m/s
C	specific heat (of a solid)	J/kg K
$C = \dot{m}C_p$	heat capacity rate of a heat exchanger fluid	W/K
$C_f = \tau_s / \frac{1}{2}\rho U_\infty^2$	skin friction coefficient	
C_p	specific heat at constant pressure	J/kg K
$C_{s,f}$	empirical constant in pool boiling	
D	diameter	m
$D_h = 4A_c/P$	hydraulic diameter	m
e	specific internal energy	J/kg
e	mean roughness height	m
E	energy	J
E	emissive power	W/m^2
E_b	black body emissive power	W/m^2
E_λ	spectral emissive power at wavelength λ	W/m^3
$Ec = U^2/C_p\Delta T$	Eckert number	
$f = (\Delta p / \frac{1}{2}\rho U^2)D_h/L$	friction factor	
F	heat exchanger temperature ratio	
F_{ij}	radiation view factor	

Symbol	**Meaning**	**Units**
$Fo = \alpha t/L^2$	Fourier number	
$F(Re)$	impinging jet parameter	
F_x, F_y, F_z	body forces per unit volume	
	in the x, y and z directions	N/m^3
g	acceleration due to gravity	m/s^2
G	irradiation	W/m^2
G_s	solar irradiation	W/m^2
$G_{s,o}$	extraterrestrial solar irradiation	W/m^2
$G(r/D, H/D)$	impinging jet parameter	
$Gr_D = \rho^2 g\beta\Delta T D^3/\mu^2$	Grashof number based on a diameter D	
$Gr_x = \rho^2 g\beta\Delta T x^3/\mu^2$	Grashof number; also Gr_L, Gr_H when based on length scales L and H	
h	Planck's constant, 6.6256×10^{-34}	J s
$h = q/\Delta T$	heat transfer coefficient	W/m^{2}K
$h_{av} = q_{av}/\Delta T_{av}$	average heat transfer coefficient	W/m^{2}K
h_c, h_b	convective and boiling heat transfer coefficients	W/m^{2}K
h_f	enthalpy of a saturated liquid	J/kg
h_{fg}	latent heat of vaporisation	J/kg
h'_{fg}	augmented latent heat of vaporisation	J/kg
h_g	enthalpy of a saturated vapour	J/kg
H	enthalpy flow rate	W/m
H	height, or other dimension	H
i	specific enthalpy (Chapter 4)	J/kg
I	current	A
I	radiation intensity (Chapter 8)	W/m^{2}sr
$I_{\lambda,b}$	black body spectral radiation intensity	W/m^{3}sr
J	radiosity	W/m^2
$Ja = C_p(T_{sat} - T_w)/h_{fg}$	Jakob number	
k	thermal conductivity	W/m K
k	Boltzmann constant, 1.3806×10^{-23}	J/K
k_c, k_e	contraction and expansion loss coefficients	
L	thickness, length, characteristic length scale	m
$L = \kappa y$	mixing length	m
L_c	flow boiling parameter	
$L_{h,entry}$	hydrodynamic entry length	m
$L_{t,entry}$	thermal entry length	m

Symbol	**Meaning**	**Units**
$m = (hP/kA_c)^{1/2}$	fin parameter	m^{-1}
m	mass	kg
$\dot{m}$	mass flow rate	kg/s
$\text{NTU} = UA/C_{min}$	number of heat transfer units	
$Nu_{av} = h_{av}L/k$	average Nusselt number	
$Nu_D = hD/k$	Nusselt number based on pipe diameter D	
$Nu_x = hx/k$	local Nusselt number; also Nu_L, Nu_H when based on length scales L and H	
p	pressure	Pa
P	wetted perimeter	m
P	heat exchanger temperature ratio	
$Pr = \mu C_p/k$	Prandtl number	
$Pr_T = \varepsilon_M/\varepsilon_H$	turbulent Prandtl number	
P^*	flow boiling parameter	
q	heat flux	W/m^2
q_{av}	average heat flux	W/m^2
q_b	heat flux due to black body radiation	W/m^2
q_c, q_b	convective and boiling heat fluxes	W/m^2
q'_g	internal heat generation term	W/m^3
q_s, q_w	surface heat flux	W/m^2
q_{solar}	heat flux due to solar radiation	W/m^2
q^*	flow boiling parameter	
$\dot{Q}$	heat transfer rate	W
$\dot{Q}_b$	black body heat transfer rate	W
$\dot{Q}_g$	internal heat generation rate	W
r	radial coordinate	m
r_{crit}	critical insulation radius	m
r_i, r_o	inner and outer radii	m
R	radial dimension	m
R	electrical resistance	Ω
R	heat exchanger temperature ratio	
R	characteristic gas constant	J/kg K
R	residual (Chapter 3)	
$\mathcal{R}$	recovery factor	
$Ra = Gr \cdot Pr$	Rayleigh number	
Ra^*	alternative Rayleigh number	
$Re_D = \rho U_b D/\mu$	Reynolds number based on diameter D	
$Re_x = \rho U_\infty x/\mu$	local Reynolds number; also Re_L, Re_H when based on length scales L and H	
$Re_y = 4\Gamma/\mu_L$	film condensation Reynolds number	

Symbol	Meaning	Units
$Re_\Omega = \rho \Omega r^2 / \mu$	local rotational Reynolds number	
R_f	heat exchanger fouling resistance	$m^2 K/W$
s	gap distance	m
s	specific entropy	J/kg K
S	entropy	J/K
S_c	solar constant, 1353	W/m^2
t	time	s
t	thickness	m
T	temperature	K
T_b	bulk mean temperature	K
T_f	fluid temperature	K
T_m	mean film temperature	K
T_{ref}	reference temperature	K
T_s, T_w	surface temperature	K
T_{sat}	saturation temperature	K
$T_{s,ad}$	adiabatic surface temperature	K
$\mathbf{T}$	finite difference temperature matrix	K
u, v, w	velocity components in the x, y and z directions	m/s
U	overall heat transfer coefficient	$W/m^2 K$
U	velocity	m/s
U_b	bulk mean velocity	m/s
U_c	centreline velocity	m/s
$U^* = (\tau_s/\rho)^{1/2}$	friction velocity	m/s
V	voltage	V
V	volume	m^3
V	velocity	m/s
V_r, V_z, V_ϕ	radial, axial and circumferential velocity components	m/s
W	width	m
W	work	J
x	vapour quality (Chapter 7)	
x_0	start of thermal boundary layer	m
$x^* = x/(4\alpha t)^{1/2}$	dimensionless distance	
$X = x/L$	dimensionless coordinate	
$y^* = y/v(\tau_s/\rho)^{1/2}$	dimensionless wall coordinate	
$Y = y/L$	dimensionless coordinate	

Symbol	Meaning	Units
α	thermal diffusivity	m^2/s
α	absorptivity	
α_s	solar absorptivity	
α, β	temperature coefficient of resistance (Chapter 10)	K^{-1}
$\beta = 1/\rho(\partial\rho/\partial T)_p$	volume expansion coefficient	K^{-1}
Γ	mass flow rate per unit area	$kg/m^2\,s$
δ	film thickness	m
δ	velocity boundary layer thickness	m
δ_T	thermal boundary layer thickness	m
ΔT_{lm}	logarithmic mean temperature difference	K
ε	emissivity	
$\varepsilon = \dot{Q}/\dot{Q}_{max}$	heat exchanger effectiveness	
ε_{fin}	fin effectiveness	
$\varepsilon_M, \varepsilon_H$	eddy momentum and thermal diffusivities	m^2/s
$\zeta = \delta_T/\delta$	boundary layer parameter	
$\eta = y/\delta$	dimensionless thickness of a velocity boundary layer	
η	efficiency	
η_{fin}	fin efficiency	
$\eta_T = y/\delta_T$	dimensionless thickness of a thermal boundary layer	
θ, Θ	temperature difference or dimensionless temperature ratio	
θ	weighting parameter	
θ^*	angle of local minimum in Nu for an inclined enclosure heated from below	deg.
κ	mixing length constant	
$\lambda = A_s/mC$	lumped mass conduction parameter	m^2K/J
λ	wavelength	m
$\lambda = 2\pi\{\sigma/g(\rho_L - \rho_V)\}^{1/2}$	length scale in boiling (Chapter 7)	m
μ	dynamic viscosity	$kg/m\,s$
v	kinematic viscosity	m^2/s
ρ	density	kg/m^3
ρ	reflectivity (Chapter 8)	
σ	Stefan–Boltzmann constant, 56.7×10^{-9}	W/m^2K^4
σ	surface tension (Chapter 7)	N/m
σ	area ratio (Chapter 9)	
σ	normal stress (Chapter 4)	N/m^2
$\tau = \alpha\tau/L^2$	dimensionless time (Chapter 2)	
τ	shear stress	N/m^2
τ	time	s

Symbol	Meaning	Units
τ	transmissivity (Chapter 8)	
Φ	dissipation function (Chapter 4)	s^{-2}
Φ	flow boiling parameter (Chapter 7)	
ω	solid angle	sr
Ω	numerical acceleration parameter (Chapter 3)	
Ω	rotational speed	rad/s

Symbol	Meaning
Subscripts	
av	average
b	pertaining to the base of a tin
c	cold surface or fluid
crit	denotes onset of free convection
f, u	with and without fins
h	hot surface or fluid
i, o	inner and outer surface values
in, out	inlet and outlet values
lam, turb	laminar and turbulent values
L	property of the liquid phase
m	mean value
max, min	maximum and minimum values
n	direction of the outward normal
n, s, e, w, p	pertaining to finite difference grid points
net	net value
o	outer surface, value at a reference state
s	value at the surface
tot	total value
trans	transition from laminar to turbulent flow
V	property of the vapour phase
w	wall value
xy, etc.	x direction in a plane normal to the y direction
x, y, z	x, y, z components
∞	value at some distance from the surface

Superscripts	
$\overline{}$	time average
$'$	fluctuating component
(n)	nth iteration or time step

Chapter one

Introduction

1.1 ● What is heat transfer?

To define heat transfer requires two important groups of words. Heat is a form of energy, and in this context we are concerned with the transport of heat energy from one body to another. The motivation for this transfer is provided by a difference in temperature. A definition of heat transfer would therefore include the words: a transport of energy due to a temperature difference.

Prior to the eighteenth century, heat was thought to be conveyed by an invisible substance known as 'caloric', or 'heat fluid'. Indeed, this is where we derive our expression of heat 'flowing' as if in a stream. Developments in the early nineteenth century recognised heat and work as forms of energy, or to be more precise, processes of energy exchange between two systems or a system and its surroundings. Work is the name given to an exchange which is neither accompanied by a transfer of mass, nor motivated by a temperature difference. The physical laws that govern the processes of work and heat exchange are known as the first and second laws of thermodynamics. The first law is a statement of energy conservation, that 'energy can neither be created nor destroyed', and with particular reference to the interchange of heat and work, we must add 'but can be converted from one form to another'. The second law clarifies the direction of the exchange, and arises from the observation that heat cannot of its own accord pass from a cold body to a hot body.

Although the process of heat transfer must concur with the laws of thermodynamics, it requires a separate science, that of heat transfer, to quantify the process. The reason for this is that classical thermodynamics concerns itself with equilibrium states, and clearly heat transfer occurs in a state of non-equilibrium. In a thermodynamic analysis, heat is merely a prescribed quantity, or the difference between an energy change and the work done; the mechanism of heat transfer is unimportant, as is the time taken to attain equilibrium conditions.

Today, we recognise that heat can be transferred by one, or by a combination, of three separate modes known as conduction, convection and radiation. Conduction occurs in a stationary medium; convection requires a moving fluid; radiation occurs in the absence of any medium, distinguishing it as part of the electromagnetic spectrum. Although we think of these as distinct processes, they can (and often do) occur together. The heat generated during combustion in a Diesel engine, for example, is transferred from the combusted gas to the steel cylinder walls by the combined action of radiation and convection. Heat flows through the cylinder walls by conduction. In turn, the outer surface of the wall is cooled by convection, and to some extent radiation, owing to water circulating in the cooling passages. The physical processes that govern conduction, convection and radiation are quite different, leading us to adopt different approaches for the purpose of analysis.

A heat transfer analysis usually involves quantifying the heat transfer rate for some known temperature difference; or the converse, from knowledge of the heat transfer rate, ascertaining the temperature difference. We may also want to know the time taken to reach a certain temperature or to transfer a known quantity of heat. Referring back to the Diesel engine, the physical reality is undoubtedly complex; the cyclic combustion of air and fuel takes place in a three-dimensional enclosure whose volume varies cyclically, and so may the thermal properties of the burnt mixture and the cylinder wall. The transformation from this to a suitable mathematical model, an ideal gas in a regular cylinder or pipe and with constant thermal properties, requires good understanding of the physical processes to assess what is and what is not significant, together with a fair grounding in mathematics to derive and solve equations. This is in essence, the 'art of engineering'. The degree of sophistication which we may build into our heat transfer model depends on the accuracy to which the physical variables are known, and also the accuracy to which we require our answers. Thus, in calculating the heat transfer rate from a 1.6 litre, 70 kW Diesel engine to the cooling water, an accuracy of ±100 W is probably adequate. In the overall process of heat transfer from the gas through the walls to the water, convection to the water offers the least resistance and is therefore the least sensitive to our assumptions. Consequently, we could be justified in simplifying the complex geometry of the passages to a single equivalent passage of circular cross-section. A further simplification without sacrificing accuracy would be to evaluate the relevant thermal properties – viscosity, thermal conductivity, density – at a suitable cycle average temperature. Also, in view of the rapid frequency at which the induction, compression, power and exhaust cycle takes place, and the large thermal capacity of the metal and water surrounding the cylinder, we could neglect the cyclic process altogether and attempt to find a cycle average value of the heat flux.

The past thirty years have seen gigantic improvements in computing facilities, from the days of there being one mainframe serving an entire university or company to a personal computer on every desk, which could well have the equivalent computing power of all the mainframes in the world of just a few decades ago. It is possible to obtain software to carry out a heat transfer analysis, and to the initiated and enlightened user, this provides a powerful tool. It is tempting to think that since the answers from computer software are correct, then they correctly portray the physical reality. In truth, the answers portray the model set up by the user. The above comments on understanding of physical processes, ability in mathematics and an awareness of accuracy are therefore as important today as they were when heat transfer calculations were done using a slide rule or a set of logarithm tables.

The following sections of this chapter introduce the separate processes of conduction, convection and radiation. The next two chapters look in more detail at conduction; convection is covered in Chapters 4, 5 and 6. Heat transfer during the phase change processes of condensation and boiling is addressed in Chapter 7, and thermal radiation in Chapter 8. The final two chapters of the book are concerned with heat exchangers (a general term for a common device for exchanging heat from one fluid stream to another) and heat transfer instrumentation. Self-contained discussions of the latter are notably absent from textbooks in this field, and it is for this reason and the author's experimental background that a brief discussion has been included here.

1.2 ● Conduction

Conduction occurs in a stationary medium. It is most likely to be of concern in solids, although conduction will be present, but to the engineer may not be significant, in liquids and gases. On the microscopic scale, there are differences between the physical mechanisms in fluids and those in solids that are responsible for the phenomenon of conduction.

The kinetic theory of gases tells us that the temperature of a molecule is related to its kinetic energy; the molecules in a high-temperature gas have greater mean velocities than those in a low-temperature gas. Thus a difference in temperature implies a difference in kinetic energy. So, when a fast-moving molecule from a region of high energy (or temperature) collides with slower-moving molecules from a region of lower temperature, an energy transfer takes place. Low-energy molecules gain temperature and the temperature of the high-energy molecule is reduced. In gases, therefore, heat is conducted by molecular collisions. The thermal conductivity of gases increases with absolute temperature. They become better conductors because the molecules are moving faster and therefore able to transport more energy. For gases, the distance between molecules is related to the pressure. However, unless the pressure is very large, this intermolecular distance will be large in comparison with the molecules' diameter, and so the thermal conductivity is relatively insensitive to pressure.

The mechanism in liquids is similar; however, the overall behaviour is much more complex. Most liquids become poorer conductors at higher temperatures, although water is a notable exception to this rule.

In solids, the mechanism is due to both vibration of the atomic lattice and the motion of free electrons, the latter generally being a more powerful effect. Thus non-metals are poor conductors because they have few free electrons, and metallic solids are good conductors because of the contribution made by available free electrons.

Conduction is governed by Fourier's law, which is based entirely upon observation and may be stated as 'the rate of heat conduction is proportional to the area measured normal to the direction of the heat flow and to the temperature gradient in the direction of the heat flow'. The constant of proportionality is termed the *thermal conductivity*, denoted by the symbol k, and has SI units of W/m K. Some representative values of thermal conductivity for various gases, liquids and solids are given in Table 1.1. It should be stressed that these are only representative values since, as previously noted, k varies with temperature, and for solids the exact composition (note the difference between mild steel, $k \approx 45$ W/m K and stainless steel, $k \approx 19$ W/m K). A more comprehensive list for various solids, liquids and gases is given in Appendix A.

Table 1.1 ●
Representative values of thermal conductivity for some common materials

	k (W/m K)	
Gases		
Argon (at 300 K and 1 bar)	0.018	
Air (at 300 K and 1 bar)	0.026	
Air (at 400 K and 1 bar)	0.034	
Hydrogen (at 300 K and 1 bar)	0.180	
Liquids		
Engine oil (at 20°C)	0.145	
Engine oil (at 80°C)	0.138	
Water (at 20°C)	0.603	
Water (at 80°C)	0.670	
Mercury (at 27°C)	8.540	
Solids		
Glass wool insulation	0.035	
Expanded polystyrene	0.040	
Wood	0.1–0.3	(depending on type, e.g. oak, pine, plywood)
Brick	0.3–6.0	(depending on composition)
Glass	0.8	
Stainless steel	19	(typical, but varies with composition)
Mild steel	45	(typical, but varies with carbon content)
Aluminium (pure)	204	
Copper (pure)	386	
Silver (pure)	410	

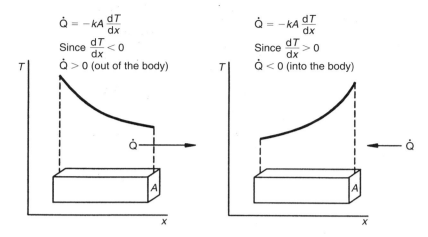

Fig. 1.1 ●
Conduction in a
solid

A mathematical statement of Fourier's law, for the x direction is thus

$$\dot{Q}_x = -kA\frac{\mathrm{d}T}{\mathrm{d}x} \qquad \text{(watts)}$$

$$q_x = \frac{\dot{Q}_x}{A} = -k\frac{\mathrm{d}T}{\mathrm{d}x} \qquad \text{(W/m}^2\text{)}$$

(1.1)

This is illustrated graphically in Figure 1.1, where for a heat flow in the positive x direction, the gradient $\mathrm{d}T/\mathrm{d}x$ must be negative, and vice versa for heat flow in the negative x direction. Conduction can occur in all three dimensions; we may also be interested in the time taken to achieve thermal equilibrium. Two-dimensional and transient conduction will be addressed in Chapters 2 and 3. For the purpose of this introduction, we will consider a simple problem in one-dimensional, steady-state conduction.

Example 1.1

Calculate the rate of heat flow through a 0.5 m wide, 0.3 m high and 3 mm thick steel plate, having a thermal conductivity of 45 W/m K when the temperature of the surface at $x = 0$ is maintained at a constant temperature of 198.0°C and its temperature at $x = 3$ mm is 199.7°C

Solution From Equation (1.1)

$$\dot{Q}_x = -kA\frac{\mathrm{d}T}{\mathrm{d}x}$$

$$= -45 \times (0.5 \times 0.3) \times \frac{(199.7 - 198.0)}{0.003}$$

$$= -3.825 \,\text{kW}$$

Alternatively this may be expressed as a heat flux, q_x,

$$q_x = \frac{\dot{Q}_x}{A}$$

$$= \frac{3825}{(0.5 \times 0.3)}$$

$$= -25.5 \, \text{kW/m}^2$$

Comment The negative sign preceding this value indicates the direction of the heat flow as being in the direction of decreasing x, i.e. into the surface at $x = 3\,\text{mm}$, and since the heat flow is one-dimensional and steady state, this 'flows' out of the surface at $x = 0$.

Note also the relatively large heat flow (as an aid, visualise the heat emitted from a domestic gas fire – typically $4\,\text{kW}$), sustained by a relatively small temperature difference. This is owing to the relatively high thermal conductivity of steel. For the same motivating temperature difference, aluminium would conduct 4.5 times this amount and stainless steel less than half.

1.3 ● Convection

The word 'convection' is derived from the Latin 'convehere' meaning to bring in or to carry. Convection occurs because of bulk fluid motion, as opposed to molecular motion within a fluid. When moving fluid encounters a solid surface, heat is convected either from it or to it, depending on the sign of the surface-to-fluid temperature difference. This is illustrated in Figure 1.2, with Figure 1.2b showing the temperature profile in the solid (due to conduction) and in the fluid (due to convection).

In convective heat transfer, the heat transfer rate, $\dot{Q}$, from a solid surface of area A and temperature T_s to an adjacent moving fluid stream of temperature T_f, is given by the simple relationship

$$\dot{Q} = hA(T_s - T_f) \quad \text{(watts)}$$

and the convective heat flux, q, by

$$q = h(T_s - T_f) \quad (\text{W/m}^2) \tag{1.2}$$

where h is the convective heat transfer coefficient (SI units, $\text{W/m}^2\text{K}$). The stunning simplicity of Equation (1.2) hides the complexity of the quantity h. The convective heat transfer coefficient represents the thermal resistance of the adjacent fluid stream. As is shown in Chapters 4, 5, 6 and 7, h is not a thermal property such as thermal conductivity. It depends, amongst other things, on the velocity of the fluid, the presence of turbulence,

Fig. 1.2 ●
Heat transfer by
convection due to
a moving fluid:
(a) nomenclature;
(b) temperature
distribution

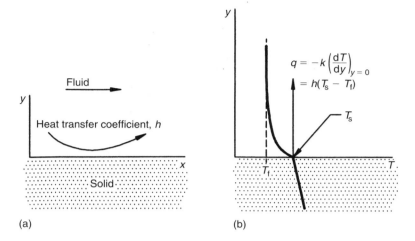

(a)   (b)

surface geometry, and fluid properties such as dynamic viscosity μ, density ρ, and thermal conductivity, k.

For any moving fluid that comes into contact with a stationary surface, at the surface itself, or the 'wall' as it is often termed, there is no relative motion of the fluid. This is known as the *no-slip condition* and, as is explained in Chapter 4, has considerable significance in the study of convective heat transfer. For the purpose of this introductory section, it is adequate to note the following. Since at the wall ($y = 0$) there is no relative fluid motion, heat transfer occurs through a process of conduction. Naturally, as soon as we move further away from the wall, heat is carried by the process of convection. If we examine an energy balance based on this statement, we obtain, from Equations (1.1) and (1.2)

$$h = \frac{-k\left(\dfrac{\partial T}{\partial y}\right)_{y=0}}{T_s - T_f} \qquad (\text{W/m}^2\text{K}) \tag{1.3}$$

Equation (1.3) tells us that the convective heat transfer coefficient is directly related to the temperature distribution in the fluid. This in turn is influenced by the velocity distribution, so to calculate or compute a value of h, we must solve the appropriate equations that govern the distribution of velocity and temperature in a fluid. Alternatively, or additionally, it is possible to measure the convective heat transfer coefficient, and some methods for achieving this are outlined in Chapter 10. Note the use of partial derivatives in Equation (1.3), since in general the fluid temperature depends on x, z and possibly time, t, as well as distance from the wall, y.

Convective heat transfer may be further divided into various regimes; these are explained at greater length in Section 4.1. One distinction of fundamental importance is the source of the fluid motion. The fluid may be forced to move by some external means such as a pump, a fan or by motion of the surface itself. This is referred to as *forced convection*. The other possibility is that there is no mechanism to 'force' the flow but the presence of a temperature difference creates density variations that then result in fluid motion. This is known as *free* or *natural convection*, and as one would expect, it results in a weaker effect. An important consequence of the distinction between free and forced convection is the dependence of the heat transfer coefficient on the surface-to-fluid temperature difference. For free convection, we would expect the heat transfer coefficient to increase with the 'strength' of the buoyancy-induced motion, which in turn we would expect to increase with the temperature difference that causes it. So, in free convection, unlike forced convection, the heat transfer coefficients also depend on the magnitude of the surface-to-fluid temperature difference.

Chapters 5 and 6 deal with forced and free convection, respectively, for fluids (a gas or liquid) where there is no change of phase during the heat transfer process. Convection will also occur during a change of phase, such as in boiling a liquid or the condensation of a vapour. This in itself constitutes a large branch of heat transfer, and the subject of convection during boiling and condensation is discussed in Chapter 7.

As noted above, the value of the convective heat transfer coefficient depends upon many factors, and a representative range of values for air and water is given in Table 1.2.

Note in Table 1.2 the difference between the values for free and forced convection. As is to be expected, forcing the flow, rather then relying on natural buoyancy, results in larger values of the heat transfer coefficient. This is made use of in, for example, the cooling of electronic devices such as computers where a large number of circuit boards are packed into a relatively small space and require forced ventilation by a fan to keep them at a safe operating temperature. The ability of a fluid to convect is affected by physical properties such as density, specific heat capacity, thermal conductivity and viscosity. This is reflected in the difference between the values cited for air and water. The processes of boiling and condensation are both accompanied by very large values of heat transfer coefficient. For boiling, this is associated with the complex interactions and motion of vapour bubbles. The condensation of discrete drops creates a thermal resistance that is virtually zero, so that the heat transfer coefficient is again very large.

Table 1.2 ● Representative values of h	h $(W/m^2 K)$
Air (1 bar, free convection)	6–30
Air (1 bar, forced convection)	10–200
Water (free convection)	500–1000
Water (forced convection)	600–8000
Boiling water	2500–100 000
Condensing steam	2500–70 000

Example 1.2

A refrigerator stands in a room where the air temperature is 20°C. The surface temperature on the outside of the refrigerator is 16°C, the sides are 30 mm thick and have an equivalent thermal conductivity of 0.1 W/mK. The heat transfer coefficient on the outside is 10 W/m²K. Assuming one-dimensional conduction through the sides, calculate the net heat flow and the surface temperature on the inside.

Solution From Equation (1.2), and with respect to Figure 1.3, the convective heat flux to the surface is

$$q = \frac{\dot{Q}}{A}$$

$$= h(T_{s,o} - T_f)$$

$$= 10 \times (16 - 20)$$

$$= -40 \,\text{W/m}^2$$

Since this must be equal to the heat conducted through the sides, and from Equation (1.1)

$$q = \frac{\dot{Q}}{A}$$

$$= -k\frac{dT}{dx}$$

$$= -\frac{k}{L}(T_{s,o} - T_{s,i})$$

$$T_{s,i} = \left(\frac{qL}{k}\right) + T_{s,o}$$

$$= -\left(\frac{40 \times 0.03}{0.1}\right) + 16$$

$$= 4°C$$

Comment This simple example demonstrates the combination of conduction and convection relations to establish the two desired quantities. It also demonstrates the concept of thermal resistance. Temperature can be considered a potential, and heat flow a rate parameter; writing Equations (1.1) and (1.2) in the general form, as for example Ohm's law ($V = I \times R$)

Potential = (Rate parameter) × (A resistance)

it can be seen that the resistance offered by convection is $1/h$, and that by conduction is L/k. Combining Equations (1.1) and (1.2) we see an overall

Fig. 1.3 ●
Heat transfer to
a refrigerator
(Example 1.2):
a cross-section
through the
refrigerator wall

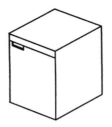

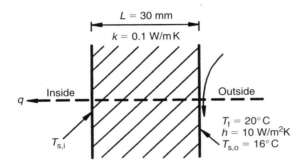

$L = 30$ mm

$k = 0.1$ W/m K

q Inside

$T_{s,i}$

Outside

$T_f = 20°$C
$h = 10$ W/m^{2}K
$T_{s,o} = 16°$C

resistance of $(1/h + L/k)$. In this example they are in the respective ratios of 1 to 3. Without insulation, assuming a steel plate side of, say, 1 mm thickness, the thermal resistance to conduction would be greatly reduced, as would the overall thermal resistance. However, just as in the case of the series resistance circuit, the overall resistance is dominated by the highest value, and here it is due to convection.

1.4 ● Radiation

Anything with a temperature above absolute zero $(-273.15°$C$)$, will experience molecular and atomic vibrations, and as a result will emit some energy as thermal radiation. Thermal radiation is an electromagnetic phenomenon and occurs in the region of wavelengths $10^{-7} < \lambda < 10^{-4}$ m. Since thermal radiation forms part of the electromagnetic spectrum, it is possible to exchange heat between bodies by this mechanism without the presence of any intervening medium. Indeed, this is confirmed by our everyday experience of being warmed by the sun, which is separated from the earth by approximately 1.5×10^{11} m of empty space. Thermal radiation can of course be transported through any 'transparent' medium such as air. Some gases, however, are particularly good absorbers of thermal radiation (notably carbon dioxide, which is transparent at shorter wavelengths but becomes virtually opaque at longer wavelengths).

As is discussed in Chapter 8, the heat transfer by radiation from an object at an absolute temperature T_1 has an upper limit known as the black body radiation. For a body of surface area A, this is given by the relation

$$\dot{Q}_b = \sigma A T_1^4 \quad \text{(watts)} \tag{1.4}$$

where σ is known as the Stefan–Boltzmann constant, having the numerical value of $\sigma = 56.7 \times 10^{-9} \, \text{W/m}^2\text{K}^4$.

For an object at temperature T_1 and area A_1, completely surrounded by an environment at temperature T_2, the net black body radiative heat transfer is the difference between that radiated from object 1 and that radiated from object 2 and received by area A_1:

$$\begin{aligned}
\dot{Q}_{b,1-2} &= \sigma A_1 T_1^4 - \sigma A_1 T_2^4 \\
&= \sigma A_1 (T_1^4 - T_2^4) \quad \text{(watts)} \\
q_{b,1-2} &= \frac{\dot{Q}_{b,1-2}}{A_1} \\
&= \sigma(T_1^4 - T_2^4) \quad \text{(W/m}^2\text{)}
\end{aligned} \tag{1.5}$$

The case of black body radiation corresponds to a surface which is an ideal emitter and absorber of thermal radiation. In practice, however, surfaces are not ideal. An allowance is made for this by the introduction of the thermal radiative properties, known as absorptivity, transmissivity, reflectivity and emissivity. In addition, the geometric relation of one surface to another will modify the net interchange (for two close parallel plates one surface 'sees' the entirety of the other, but this is not true for two flat surfaces at right angles to each other). Geometry is handled by way of the view factor or radiation configuration factor.

It is also instructive to rewrite Equation (1.5) as

$$\begin{aligned}
q_{b,1-2} &= \sigma(T_1^4 - T_2^4) \\
&= \sigma(T_1^2 - T_2^2)(T_1^2 + T_2^2) \\
&= \{\sigma(T_1 + T_2)(T_1^2 + T_2^2)\}(T_1 - T_2) \\
&= h_{\text{rad}}(T_1 - T_2) \quad \text{by comparison with Equation (1.2)}
\end{aligned}$$

where, for black body radiation

$$h_{\text{rad}} = \{\sigma(T_1 + T_2)(T_1^2 + T_2^2)\} \quad \text{(W/m}^2\text{K)} \tag{1.6}$$

is known as a 'radiation heat transfer coefficient'.

Example 1.3

The surface temperature of a central heating radiator is 60°C. What is the net black body radiative heat transfer between the radiator and surroundings, which are at 20°C?

Solution   From Equation (1.5)

$$q_{b,1-2} = \sigma(T_1^4 - T_2^4) \text{ W/m}^2$$
$$T_1 = (60 + 273.15) = 333.15 \text{ K}$$
$$T_2 = (20 + 273.15) = 293.15 \text{ K}$$
$$q_{b,1-2} = 56.7 \times 10^{-9}(333.15^4 - 293.15^4)$$
$$= 280 \text{ W/m}^2$$

Comment   This figure represents an upper bound to the radiative heat transfer, since for a real surface the emissivity will be less than unity. However, surface coatings on heating radiators are chosen with this in mind (which is why manufacturers warn against painting with metallic paints), so the figure is a reasonably good approximation.

Heating 'radiators' also convect heat by free convection. In fact there are roughly equal contributions from convection and radiation (see Question 8.3).

Note the use of thermodynamic temperature (in kelvins). In conduction and convection problems we deal with linear temperature differences, e.g. $(T_1 - T_2)^n$, and any consistent temperature scale may be used. In radiation the temperature differences are non-linear and the phenomenon itself is related to absolute temperature (it ceases at 0 K), hence the appropriate unit in the SI system is the kelvin.

1.5 ● Closing comments

We have seen that the science of heat transfer has evolved from thermodynamics as a means to quantify heat exchange and temperature. Heat transfer occurs by three separate modes: conduction, convection and radiation, which can, and often do, act together. Conduction takes place in a stationary medium, and although present in fluids, it is usually more of a concern in solids. Convection requires a moving medium; the motion may be generated by buoyancy forces (free convection) or through external means (forced convection). Thermal, or infrared radiation is part of the electromagnetic spectrum; all bodies with a temperature above absolute zero emit thermal radiation. Whereas the thermal conductivity is a thermophysical property of matter, the heat transfer coefficient is not; it depends on, for example, the velocity of the fluid, the presence of turbulence, surface roughness, geometry and the various thermophysical properties. A prerequisite for any heat transfer analysis is to identify the separate modes of heat transfer and, in particular, the relative contributions of convection and radiation.

1.6 ● End of chapter questions

1.1 Sketch and identify the modes of heat transfer in the following examples.

(a) Heating a room using a central heating radiator.
(b) Heating a room using an electric fan heater.
(c) Soldering an electric circuit board.
(d) Gas welding two sheets of steel plate.
(e) The dissipation of electrical energy into heat by an electronics component mounted on a heat sink.
(f) A car disc brake during braking.
(g) An exhaust valve of a four-stroke petrol engine.
(h) Coffee in a vacuum flask.
(i) A domestic boiler.

1.2 Wetsuits are usually made from 4 mm thick rubber ($k \approx 0.2$ W/m K). To improve the heat retention, several manufacturers offer a top-of-the-range model with titanium ($k = 20$ W/m K) flakes embedded in the rubber. Can you explain the purpose of these flakes?

1.3 The wall of a house 7 m wide and 6 m high is made from 0.3 m thick brick with $k = 0.6$ W/m K. The surface temperature on the inside of the wall is 16°C and that on the outside 6°C. Find the heat flux through the wall and the total heat loss through it.

[20 W/m^2, 840 W]

1.4 A 20 mm dia. copper pipe is used to carry heated water. The external surface of the pipe is subjected to a convective heat transfer coefficient of $h = 6$ W/m^{2}K. Find the heat loss by convection per metre length of pipe when the external surface temperature is 80°C and the surroundings are at 20°C. Assuming black body radiation, what is the heat loss by radiation?

[22.6 W/m, 29.1 W/m]

1.5 Assuming black body behaviour, estimate the heat transfer by radiation from the surface of a 60 mm dia. spherical lamp with a surface temperature of 80°C, when the surroundings are at 20°C.

[5.2 W]

1.6 A plate 0.3 m long and 0.1 m wide, with a thickness of 12 mm is made from stainless steel ($k = 16$ W/m K). The top surface is exposed to an airstream of temperature 20°C. In an experiment, the plate is heated by an electrical heater (also 0.3 m by 0.1 m) positioned on the underside of the plate, and the temperature of the plate adjacent to the heater is maintained at 100°C. A voltmeter and ammeter are connected to the heater and these read 200 V and 0.25 A, respectively. Assuming that the plate is perfectly insulated on all sides except the top surface, what is the convective heat transfer coefficient?

[12.7 W/m^{2}K]

1.7 An electronic component dissipates 0.38 W through a heat sink by convection and (black body) radiation into surroundings at 20°C. What is the surface temperature of the heat sink if the convective heat transfer coefficient is 6 W/m^{2}K and the heat sink has an effective area of 0.001 m^2?

[50°C]

Chapter two

Conduction – Analytical methods and applications

2.1 ● Introduction

Conduction is the mode of heat transfer that occurs as a result of the motion of molecules, electrons and vibrations in the atomic lattice (see Section 1.2). For solids that are poor conductors of heat, such as plastics, thermal energy is transported through vibrations of the lattice structure. In good conductors, such as metals, the process is greatly enhanced by the motion of free electrons.

Although frequent example will be made of conduction in solids, it is important to note that conduction also occurs in liquids and gases. However, it is rarely the most significant mode of heat transport in fluids. Any heating will give rise to a density gradient and result in *convection* from bulk motion of the fluid itself, as opposed to motion on the atomic scale.

There are many examples of physical laws, derived empirically from observation, which relate the flow of some quantity (for example fluid flow in a pipe or of electrical current in a wire) to a potential (for example pressure difference, voltage). The microscopic events which manifest themselves as the phenomena of heat being transferred by conduction are related on the macroscopic scale by Fourier's law: that the heat transfer per unit area, q, or heat flux in a given direction is proportional to the temperature gradient in that direction. This is written in general terms as

$$q_n \propto \frac{\partial T}{\partial n} \tag{2.1}$$

where n denotes the direction of the outward normal. To be specific, for the x direction we would write for the area, A, that

$$\frac{\dot{Q}_x}{A} = q_x = -k\frac{\partial T}{\partial x} \ (\text{W/m}^2) \tag{2.2}$$

where $\dot{Q}$ is the rate of heat transfer in watts. A leading negative sign in Equation (2.2) ensures that the flux is positive in the direction of increasing x, and the constant of proportionality, k, is known as the thermal conductivity. In general, thermal conductivity depends on temperature, and in certain cases (laminates, wood and carbon fibre composite being some notable examples) there is anisotropy, where the thermal conductivity parallel to the laminate, grain or fibres differs from that in the perpendicular direction.

For engineering purposes, a conduction analysis is likely to involve the calculation of temperatures within a solid region where the external conditions (surface temperature, heat transfer by convection and/or radiation) are known. Steady-state conduction is the name used when temperatures are invariant with time. Transient, or time-dependent conduction is used when either temperatures vary in time but arrive at some thermal steady state, or vary periodically such as the daily cycle of heating and cooling of a wall by solar radiation. An analysis may be concerned with steady-state or time-dependent conduction or both. Conduction can often be treated as one- or two-dimensional, even though in the general case it is three-dimensional. Analytical solutions to two-dimensional time-dependent and three-dimensional steady-state conduction are outside the scope of most engineering texts. However, such problems, and those involving composite regions, anisotropy and temperature-dependent properties are amenable to solution using numerical techniques such as those described in Chapter 3. Alternatively, or as a complement, the classic text by Carslaw and Jaeger (1980) or the more recent book by Mikhailov and Özisik (1994) both provide a thorough catalogue of analytical solutions to these more complex problems.

This chapter begins with the derivation of the general conduction equation and then proceeds to examine one-dimensional steady-state, two-dimensional steady-state and transient solutions with illustrative examples. In the author's teaching experience, the act of simplifying is often a source of problems, as it appears that the necessary assumptions can only be made after the results of a complete analysis are known. This is not necessarily the case and, where possible, assumptions are explored and further guidance is given in the text.

2.2 ● Derivation of the general conduction equation

Consider the volume element of a material shown in Figure 2.1, where the sides are of dimension Δx, Δy and Δz. For completeness, assume heat $d\dot{Q}_g(q'_g \Delta x \Delta y \Delta z)$ is also generated within the element, as would be produced from an embedded heater or from chemical reaction. Applying the principle of conservation of energy to the element gives

$$d\dot{Q}_x + d\dot{Q}_y + d\dot{Q}_z + d\dot{Q}_g - d\dot{Q}_{x+\Delta x} - d\dot{Q}_{y+\Delta y} - d\dot{Q}_{z+\Delta z}$$

$$= \text{Rate of change of internal energy} \tag{2.3}$$

The right-hand side can be written in terms of the temperature T, volume V and, assuming a constant specific heat capacity C, the density, ρ as

$$\frac{\partial E}{\partial t} = V\rho C \frac{\partial T}{\partial t} = (\Delta x \Delta y \Delta z)\rho C \frac{\partial T}{\partial t} \tag{2.4}$$

At this point it is appropriate to introduce a concept that will be used throughout this book when deriving the various governing equations. A Taylor series expansion defines the value of a continuous function at a point with respect to the value at some other point, and the first, second, third, etc., derivatives of the function. This is expressed mathematically, for the function T (as for example temperature) evaluated at $x + \Delta x$ as

$$T_{x+\Delta x} = T_x + \frac{\partial T}{\partial x}\Delta x + \frac{1}{2!}\frac{\partial^2 T}{\partial x^2}\Delta x^2 + \frac{1}{3!}\frac{\partial^3 T}{\partial x^3}\Delta x^3 + \cdots$$

For a control volume approach we assume that Δx is small enough that the variation of temperature between x and $x + \Delta x$ may be considered linear. Hence we can omit the terms with second and higher derivatives, so

$$T_{x+\Delta x} = T_x + \frac{\partial T}{\partial x}\Delta x$$

which amounts to the statement that the value at $x + \Delta x$ is equal to the value at x added to the product of the gradient and the distance Δx.

Applying a Taylor series expansion to the heat flows at $x + \Delta x$, $y + \Delta y$ and $z + \Delta z$ in Figure 2.1 gives

$$d\dot{Q}_{x+\Delta x} = d\dot{Q}_x + \frac{\partial}{\partial x}(d\dot{Q}_x)\Delta x$$

$$d\dot{Q}_{y+\Delta y} = d\dot{Q}_y + \frac{\partial}{\partial y}(d\dot{Q}_y)\Delta y \tag{2.5}$$

$$d\dot{Q}_{z+\Delta z} = d\dot{Q}_z + \frac{\partial}{\partial z}(d\dot{Q}_z)\Delta z$$

Since the heat flow $d\dot{Q}_x$ crosses the plane at x, with area $\Delta y \Delta z$

$$d\dot{Q}_x - d\dot{Q}_{x+\Delta x} = -\frac{\partial}{\partial x}(d\dot{Q}_x)\Delta x$$

$$= \frac{\partial}{\partial x}\left(k\frac{\partial T}{\partial x}\right)\Delta x \Delta y \Delta z \tag{2.6}$$

The other terms may be likewise expanded and substitution into Equation (2.3) gives the result

$$\frac{\partial}{\partial x}\left(k\frac{\partial T}{\partial x}\right) + \frac{\partial}{\partial y}\left(k\frac{\partial T}{\partial y}\right) + \frac{\partial}{\partial z}\left(k\frac{\partial T}{\partial z}\right) + \frac{q'_{\mathrm{g}}}{k} = \rho C\frac{\partial T}{\partial t} \tag{2.7}$$

Equation (2.7) is the general form of the conduction equation. A number of different variants in three dimensions are listed below.

Fig. 2.1 ●
Volume element
for conduction

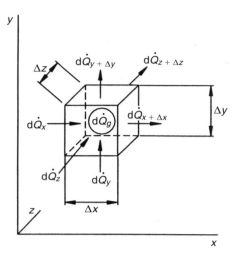

For a material where the thermal conductivity is constant (that is both independent of temperature and invariant through the conducting medium), Equation (2.7) becomes

$$\frac{\partial^2 T}{\partial x^2} + \frac{\partial^2 T}{\partial y^2} + \frac{\partial^2 T}{\partial z^2} + \frac{q'_g}{k} = \left(\frac{\rho C}{k}\right)\frac{\partial T}{\partial t} \tag{2.8}$$

Fourier's equation applies when there is no internal heat generation, which for constant thermal properties is

$$\frac{\partial^2 T}{\partial x^2} + \frac{\partial^2 T}{\partial y^2} + \frac{\partial^2 T}{\partial z^2} = \left(\frac{1}{\alpha}\right)\frac{\partial T}{\partial t} \tag{2.9}$$

where $\alpha = k/\rho C$ is known as the thermal diffusivity and has units of m^2/s.

Laplace's equation applies for steady-state conduction without internal heat generation, and for constant thermal conductivity is

$$\frac{\partial^2 T}{\partial x^2} + \frac{\partial^2 T}{\partial y^2} + \frac{\partial^2 T}{\partial z^2} = 0 \tag{2.10}$$

Poisson's equation applies for steady-state conduction with internal heat generation, for constant thermal conductivity:

$$\frac{\partial^2 T}{\partial x^2} + \frac{\partial^2 T}{\partial y^2} + \frac{\partial^2 T}{\partial z^2} + \frac{q'_g}{k} = 0 \tag{2.11}$$

The above equations were derived using a Cartesian coordinate system. In certain geometries, discs, shells, spheres and so on, it is advantageous to adopt a cylindrical coordinate system where $y = r\sin\theta$, $x = r\cos\theta$, which by substitution into, for example, Equation (2.9) or by derivation from first principles using an appropriate form of geometry for the control volume element, gives

$$\left(\frac{1}{r}\right)\frac{\partial T}{\partial r} + \frac{\partial^2 T}{\partial r^2} + \left(\frac{1}{r^2}\right)\frac{\partial^2 T}{\partial \theta^2} + \frac{\partial^2 T}{\partial z^2} = \left(\frac{1}{\alpha}\right)\frac{\partial T}{\partial t} \tag{2.12}$$

Of particular significance are the terms $1/r$ and $1/r^2$ which arise because the area of the element increases with the radial coordinate.

The equations cited above were derived with relative simplicity; however, their complete solution by analytical method is far more complicated. In many respects such a task is not the domain of the engineer, whose principal concern is the modelling of the physical processes in a particular design problem by appropriate equations, and pragmatisim rather than elegance or originality takes precedence in finding a solution. The requirements of design can be complex and it will probably be necessary to use numerical methods to predict temperature distributions in realistic geometries.

However, analytical solutions do have their place, and the text by Carslaw and Jaeger (1980) or more recently that of Mikhailov and Özisik (1994) should be consulted for comprehensive surveys of solutions to the conduction equation. Firstly, analytical solutions can be used to validate numerical codes; secondly, from examination of the terms, they can provide insight to a simpler and more expedient solution. Finally, some problems are unique and a solution may not be available in the published literature, but advanced texts like these will probably describe the techniques required to develop it.

2.3 ● Boundary conditions and initial conditions

General solutions to the conduction equations are of little practical use unless boundary conditions are specified and, for the case of time-dependent conduction, initial conditions as well.

As the name implies, initial conditions specify the temperature distribution throughout the entire region under consideration at time $t = 0$. For convenience they are usually abbreviated to this form, in the case of a 3D conduction problem:

$T(x, y, z, 0) =$ some known value(s) or some known function(s)

Boundary conditions can be formulated in a number of different ways, the most commonly used in heat conduction are referred to as boundary conditions of the first, second and third kind. The complexity of the mathematical statement increases from the first kind through the second kind to the third kind.

A boundary condition of the first kind defines the temperature on the boundary explicitly, for example $T(0, t) = 100°C$, which is shorthand for the temperature in a one-dimensional region at $x = 0$, and for time greater than $t = 0$, has the value of $100°C$.

A boundary condition of the second kind defines the temperature implicitly, through an explicit statement of the heat flux. For example, a heat flux of $1\,\text{kW/m}^2$ at the boundary $x = L$ could be written as $q(L, t) = 1000\,\text{W/m}^2$. Since the heat flux is related to the temperature gradient through Fourier's law, this feeds back information from the boundary to the temperature distribution inside the body, and the above statement would actually be expressed in terms of temperature gradient as

$$\left(\frac{\mathrm{d}T}{\mathrm{d}x}\right)_{x=L} = -\frac{1000}{k}$$

where k is the thermal conductivity of the material.

A boundary condition of the third kind specifies the temperature gradient at the surface in terms of an external heat transfer coefficient h, the (possibly unknown) surface temperature T_s and the fluid reference temperature T_{ref}. For example,

$$-k\left(\frac{\mathrm{d}T}{\mathrm{d}x}\right)_{x=0} = h(T_s - T_{\text{ref}})$$

informs us that at $x = 0$ the heat flux from conduction within the material escapes by convection into a surrounding fluid. Some examples of boundary conditions applied to engineering components are given in Figure 2.2.

It can be seen that the boundary conditions discussed above are linear relations of temperature and are therefore referred to as linear boundary conditions. Other boundary conditions used in conduction are, for example, where there is a radiative heat flux from or into the surface (written in terms of T^4), heat flux by free convection ($T^{5/4}$ or $T^{4/3}$) or where there is a change of phase such as melting where energy is absorbed but the temperature remains constant. For applications of these non-linear boundary conditions to conduction problems the reader is again referred to either Carslaw and Jaeger (1980) or Mikhailov and Özisik (1994).

Fig. 2.2
Boundary conditions for some engineering components:
(a) motorcycle cylinder; (b) printed circuit board;
(c) cooled gas turbine blade

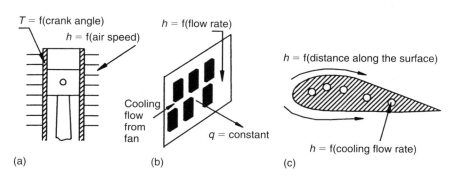

T = f(crank angle)
h = f(air speed)
h = f(flow rate)
h = f(distance along the surface)
Cooling flow from fan
q = constant
h = f(cooling flow rate)
(a)   (b)   (c)

2.4 ● **Dimensionless groups for conduction**

Dimensionless groups can be formed using Buckingham's π theorem or from examination of the governing equations and boundary conditions. The latter approach is used here, with a relatively simple example to extract the most important groups.

Figure 2.3 shows a one-dimensional material of thermal conductivity k, thermal diffusivity α and width L. At time $t = 0$, the temperature is T_0, the surface at $x = 0$ is insulated and the surface at $x = L$ experiences a heat flow by convection, where the convective heat transfer coefficient is h, into a surrounding fluid at temperature T_{ref}.

In dimensional form, the governing equation, initial condition and boundary conditions can be written as

$$\frac{\partial^2 T}{\partial x^2} = \left(\frac{1}{\alpha}\right) \frac{\partial T}{\partial t}$$

$$T(x, 0) = T_0 \tag{2.13}$$

$$\frac{\partial T}{\partial x} = 0 \qquad (x = 0, t > 0)$$

$$-k \left(\frac{\partial T}{\partial x}\right) = h(T_{\text{s}} - T_{\text{ref}}) \qquad (x = L, t > 0)$$

Fig. 2.3 ●
One-dimensional transient conduction:
(a) physical model;
(b) temperature profile

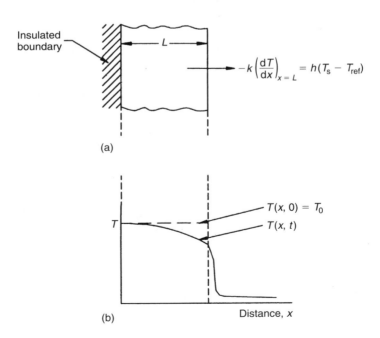

Choosing the following, and almost obvious, dimensionless groups:

$$X = \frac{x}{L}$$

$$\Theta = \frac{(T - T_{\text{ref}})}{(T_0 - T_{\text{ref}})}$$

(2.14)

and a dimensionless time,

$$\tau = \frac{\alpha t}{L^2}$$

which is not so obvious, but in view of the units of α, both t and L^2 become obvious choices, allows the governing equation, initial condition and boundary conditions (2.13) to be written as

$$\frac{\partial^2 \Theta}{\partial X^2} = \frac{\partial \Theta}{\partial \tau}$$

$$\Theta(X, 0) = 0$$

(2.15)

$$\frac{\partial \Theta}{\partial X} = 0 \qquad (X = 0, \tau > 0)$$

$$\left(\frac{\partial \Theta}{\partial X}\right) = -(hL/k)\Theta \qquad (X = 1, \tau > 0)$$

The advantage of recasting the equation in dimensionless form is that we have reduced the number of variables; the problem therefore becomes more general.

The physical interpretation of these groups is as follows:

X a dimensionless coordinate

τ the Fourier number, Fo = (Heat conducted across L)/(Heat stored across L)

$\dfrac{hL}{k}$ the Biot number, Bi, (Heat transferred by convection)/(Heat transferred by conduction)

A large Fourier number ($Fo \gg 1$), because of a long time period, a large value of thermal diffusivity or conduction across a small distance implies that a thermal steady state is being approached. A Fourier number of order unity implies transient effects are important. A small Fourier number ($Fo \ll 1$) implies that the bulk of the conducting material is unaware of the transient.

A large Biot number ($Bi \gg 1$) implies that there are large temperature gradients within the material. Conversely, a small Biot number ($Bi \ll 1$) implies a uniform temperature distribution.

The Fourier number and Biot number are of paramount importance in conduction analysis. An estimate of these quantities allows simplifying assumptions to be made with consequent reduction in the complexity of the analysis and should therefore be one of the first steps carried out.

2.5 ● One-dimensional steady-state conduction

Engineering components are three-dimensional and, generally speaking, so is the mechanism of conduction. But most conduction problems can be reduced to a two- or even one-dimensional analysis by consideration of the dimensions of the component, the boundary conditions and the degree of accuracy required. For example, conduction through the walls of a building (of thickness L_x, height L_y and width L_z) will occur in the direction normal to the wall (the x direction), across the wall towards the corners (the z direction) and along the wall in the direction from the roof to the floor (the y direction). The principal temperature difference that 'drives' the flow of heat by conduction is that between the inside and the outside air temperatures, ΔT. Since the length scales are of differing magnitudes ($L_x \ll L_y$ or L_z) and simplistically $q \approx \Delta T/L$, then as a first approximation most of the heat flow is in the x direction. Two-dimensional effects will occur, for example, if there is a significant variation of the heat transfer coefficient along or across the wall, or if the thickness is not constant, or more generally, if the Biot number varies. Three-dimensional effects will occur near the edges and corners where the walls meet and in the vicinity of openings for windows (often referred to as edge effects).

The question arises, Is it worthwhile to include all these effects? This should be answered with respect to the accuracy required and, as equally important, the accuracy of the information used. In particular, the actual heat transfer coefficient is unlikely to be represented by a correlation to an accuracy of better than 10%, and, depending on the application, introducing an extra level of complexity to gain an accuracy of say 2°C may not be worthwhile. There will also be tolerances on the value of thermal conductivity assumed and, in the example cited, variations would occur due to moisture content, differences in composition of the bricks, the influence of mortar and so on. A general answer is not therefore possible but relies on the application of experience to judge what level of complexity is necessary.

In the following sections we briefly examine various cases of one-dimensional steady-state conduction.

2.5.1 Simple plane walls

For conduction in one-dimension, Equation (2.10) becomes

$$\frac{d^2 T}{dx^2} = 0 \tag{2.16}$$

with $T_{(x=0)} = T_1$ and $-k(dT/dx)_{x=L} = h(T_2 - T_{\text{ref}})$. Upon integrating Equation (2.16),

$$\frac{dT}{dx} = \text{constant} = C_1$$

and

$$T = C_1 x + C_2$$

The constants C_1 and C_2 can be determined from the boundary conditions $T = T_1$ at $x = 0$ and $-k(dT/dx) = h(T_2 - T_{ref})$ at $x = L$. The solution for the temperature distribution in the wall then becomes

$$T = T_1 - \left\{ \frac{hx}{k}(T_2 - T_{ref}) \right\} \tag{2.17}$$

Comments 1. Equation (2.17) has the form of the equation of a straight line.
2. The heat flux, $q = -k(dT/dx) = h(T_2 - T_{ref})$, is constant through the wall, which is to be expected from the assumption of one-dimensional conduction.
3. Increasing the temperature difference $(T_2 - T_{ref})$, heat transfer coefficient h, or thermal conductivity k, leads to an increased heat flow through the wall. If the situation above models the heat flow through the walls of a house, then the corresponding physical mechanisms lead to an increase in the heat flow either by heating the room or for a cooler external temperature, by an increase in the speed of the wind blowing onto or along the outside wall, or by building the walls from a material with a higher thermal conductivity.

Example 2.1

The walls of the house shown in Figure 2.4 are 4 m high, 5 m wide and 0.3 m thick and made from brick with a thermal conductivity of $k = 0.6$ W/m K. The temperature of the air inside the house is 20°C and outside it is $-10°C$. There is a heat transfer coefficient of 10 W/m²K on the inside wall and 30 W/m²K on the outside wall. Calculate the inside and outside surface temperatures of the wall, the heat flux and total heat flow through the wall.

Assumptions 1. One-dimensional, steady-state conduction through the wall.
2. Constant thermal properties.

Solution Let the thickness of the wall be L, the heat transfer coefficients on the inside and outside h_i and h_o, respectively, and the corresponding values of inside and outside air temperatures T_i and T_o. Let the surface temperature on the inside surface of the wall be T_1 and that on the outside T_2. Recognising from the one-dimensional assumption that the heat flux, q, through the wall is constant, then it follows that

$$q = h_i(T_i - T_1)$$

$$\therefore (T_i - T_1) = \frac{q}{h_i} \tag{2.18a}$$

Fig. 2.4 ●
Steady-state
conduction
through a plane
wall (Example 2.1)

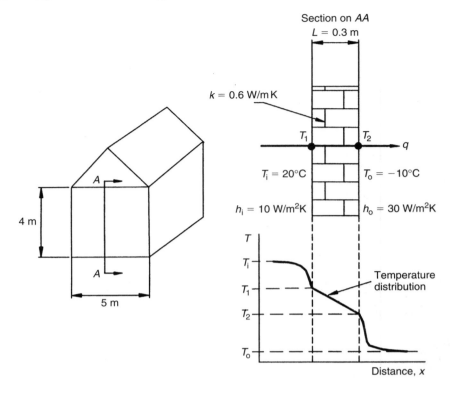

$$q = \frac{k(T_1 - T_2)}{L}$$

$$\therefore (T_1 - T_2) = \frac{qL}{k} \tag{2.18b}$$

$$q = h_o(T_2 - T_o)$$

$$\therefore (T_2 - T_o) = \frac{q}{h_o} \tag{2.18c}$$

By adding (2.18a), (2.18b) and (2.18c), we eliminate the unknown temperatures T_1 and T_2:

$$(T_i - T_o) = \frac{q}{h_i} + \frac{qL}{k} + \frac{q}{h_o} \tag{2.19}$$

$$= q\left(\frac{1}{h_i} + \frac{L}{k} + \frac{1}{h_o}\right)$$

hence

$$q = \frac{(T_i - T_o)}{\left(\dfrac{1}{h_i} + \dfrac{L}{k} + \dfrac{1}{h_o}\right)} \tag{2.20}$$

Substituting the appropriate quantities into Equation (2.20),

$$q = \frac{(20 - (-10))}{\left(\dfrac{1}{10} + \dfrac{0.3}{0.6} + \dfrac{1}{30}\right)}$$

$$= 47.4\,\text{W/m}^2$$

$$\dot{Q} = qA$$

$$= 47.4 \times 5 \times 4 = 947\,\text{W}$$

The wall surface temperatures can be obtained from (2.18a) and (2.18b):

$$T_1 = T_i - \frac{q}{h_i}$$

$$= 20 - \frac{47.4}{10} = 15.26°\text{C}$$

$$T_2 = T_1 - \frac{qL}{k}$$

$$= 15.26 - \left(\frac{47.4 \times 0.3}{0.6}\right) = -8.4°\text{C}$$

Comment The values cited for the heat transfer coefficients are representative of values for free convection on the inside wall and forced convection due to a wind blowing at 10 m/s parallel to the surface on the outside wall.

Recognising that Equation (2.19) has the form, potential = rate × resistance, the term $(1/h_i + L/k + 1/h_o)$ may be considered a thermal resistance, since this opposes the flow of heat against the imposed temperature difference. The reciprocal of this term, $(1/h + L/k + 1/h_o)^{-1}$, is referred to as the overall heat transfer coefficient, given the symbol U, and features in the design and specification of heat exchangers (see Chapter 9), building materials and insulation.

2.5.2 Composite plane walls

The foregoing may be easily generalised to the case of a composite wall, comprising a number of layers of differing widths and thermal conductivity. This is best illustrated by way of an example.

Example 2.2

The composite wall shown in Figure 2.5 is built from an inside layer and an outside layer of 0.15 m thick brick ($k = 0.6\,\text{W/m K}$). Between them is a

Fig. 2.5 ●
Steady-state
conduction
through a
composite wall
(Example 2.2)

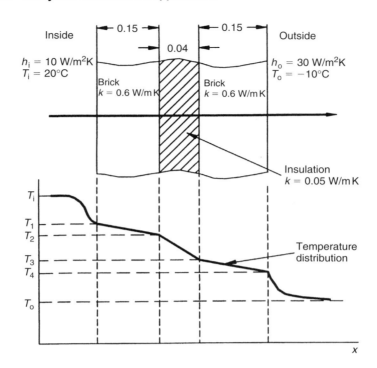

layer of insulation 0.04 m thick and $k = 0.05$ W/m K. The temperature of
the air inside the house is 20°C and outside -10°C; there is a heat transfer
coefficient of 10 W/m²K on the inside wall and 30 W/m²K on the outside
wall. Calculate the inside and outside surface temperatures of the wall, the
heat flux and total heat flow through the wall.

Assumptions **1.** One-dimensional, steady-state conduction through the wall.

2. Constant thermal properties.

3. Perfect thermal contact between adjacent layers.

Solution Since $q = $ constant,

$$q = h_i(T_i - T_1)$$

$$\therefore (T_i - T_1) = \frac{q}{h_i} \tag{2.21a}$$

$$q = \frac{k_a(T_1 - T_2)}{L_a}$$

$$\therefore (T_1 - T_2) = \frac{qL_a}{k_a} \tag{2.21b}$$

$$q = \frac{k_b(T_2 - T_3)}{L_b}$$

$$\therefore (T_2 - T_3) = \frac{qL_b}{k_b} \tag{2.21c}$$

$$q = \frac{k_c(T_3 - T_4)}{L_c}$$

$$\therefore (T_3 - T_4) = \frac{qL_c}{k_c} \tag{2.21d}$$

$$q = h_o(T_4 - T_o)$$

$$\therefore (T_4 - T_o) = \frac{q}{h_o} \tag{2.21e}$$

As before, adding (2.21a) to (2.21e):

$$(T_i - T_o) = \frac{q}{h_i} + \frac{qL_a}{k_a} + \frac{qL_b}{k_b} + \frac{qL_c}{k_c} + \frac{q}{h_o} \tag{2.22}$$

$$= q\left(\frac{1}{h_i} + \frac{L_a}{k_a} + \frac{L_b}{k_b} + \frac{L_c}{k_c} + \frac{1}{h_o}\right) \tag{2.23}$$

hence

$$q = \frac{(T_i - T_o)}{\left(\dfrac{1}{h_i} + \dfrac{L_a}{k_a} + \dfrac{L_b}{k_b} + \dfrac{L_c}{k_c} + \dfrac{1}{h_o}\right)}$$

$$q = \frac{(20 - (-10))}{\left(\dfrac{1}{10} + \dfrac{0.15}{0.6} + \dfrac{0.04}{0.05} + \dfrac{0.15}{0.6} + \dfrac{1}{30}\right)}$$

$$= 20.9\,\text{W/m}^2$$

$$\dot{Q} = qA = 20.9 \times 5 \times 4$$

$$= 419\,\text{W}$$

The wall surface temperatures can be obtained from (2.21a) and (2.21e):

$$T_1 = T_i - \frac{q}{h_i}$$

$$= 20 - \frac{20.9}{10}$$

$$= 17.91°\text{C}$$

$$T_4 = T_o + \frac{q}{h_o}$$

$$= -10 + \frac{20.9}{30}$$

$$= -9.3°\text{C}$$

Comment Comparison with Example 2.1 shows that there is a significant reduction (56%) in the heat loss due to a relatively thin layer of insulation. The inside wall temperature is higher and the outside lower, compared with the uninsulated wall.

2.5.3 One-dimensional radial conduction

Pipes, cables, annular fins, reaction and pressure vessels are examples of engineering systems where conduction occurs radially. Simple examination of radial (or cylindrical) and Cartesian geometries highlights the important fact that in a radial system the area normal to the heat flux increases with the coordinate, whereas in a Cartesian system it is constant. Consequently (unless the wall thickness is much less than the radius), Equations (2.7) to (2.11) cannot be used; the appropriate equation with all the terms retained is Equation (2.12). A cross-section through a thick-walled pipe is shown in Figure 2.6. Making the assumptions of one-dimensional, steady-state conduction with constant thermal conductivity (note ordinary derivatives replace the partial derivatives in Equation (2.12) since T is solely a function of radius, r):

$$\frac{1}{r}\frac{\mathrm{d}T}{\mathrm{d}r} + \frac{\mathrm{d}^2 T}{\mathrm{d}r^2} = 0 \tag{2.24}$$

which is equivalent to

$$\frac{\mathrm{d}}{\mathrm{d}r}\left(r\frac{\mathrm{d}T}{\mathrm{d}r}\right) = 0 \tag{2.25}$$

This has the general solution for the temperature distribution

$$T = C_1 \ln r + C_2 \tag{2.26}$$

where C_1 and C_2 are constants which are found from the boundary conditions. For example, if, as shown in Figure 2.6, the surface temperature at $r = r_1$ is T_1 and that at $r = r_2$ is T_2, then

$$C_1 = \frac{(T_2 - T_1)}{\ln(r_2/r_1)} \tag{2.27a}$$

Fig. 2.6 ●
Radial conduction
through a pipe
wall

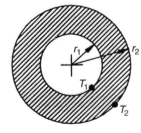
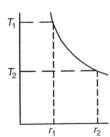

$$C_2 = T_1 - C_1 \ln(r_1) \tag{2.27b}$$

The temperature distribution through the walls can now be obtained from substituting Equations (2.27a) and (2.27b) into Equation (2.26), from which

$$\frac{(T - T_1)}{(T_2 - T_1)} = \frac{\ln(r/r_1)}{\ln(r_2/r_1)} \tag{2.28}$$

The heat flow, $\dot{Q}$ (watts), through the cylindrical wall is obtained from differentiating Equation (2.28):

$$\dot{Q} = -kA\frac{dT}{dr}$$

where $A = 2\pi rL$ is the surface area over a length L:

$$\dot{Q} = \frac{-2\pi Lk(T_2 - T_1)}{\ln(r_2/r_1)} \tag{2.29}$$

Example 2.3

A cable of radius r_1 and resistance $R(\Omega/\text{m})$ and carrying a current $I(\text{A})$ is surrounded by an insulator of radius r_2 and thermal conductivity k. The external heat transfer coefficient and air temperature are h_o and T_{air}, respectively. Derive an expression for the temperature distribution in the insulator.

A 1 mm dia. copper wire of resistance $0.02\,\Omega/\text{m}$ is surrounded by a 2.3 mm dia. plastic coating of $k = 0.2\,\text{W/m K}$. The outside surface of the coating is cooled by air, where the convective heat transfer coefficient is 16 W/m K. Determine the maximum current if the surface-to-air temperature difference is to be limited to 35°C. What is the temperature of the copper wire?

Solution From the general Equation (2.26):

$$T = C_1 \ln r + C_2$$
$$\text{At } r = r_1, \; I^2R = -2\pi r_1 k\left(\frac{dT}{dr}\right) \tag{2.30}$$

Differentiating Equation (2.30):

$$\left(\frac{dT}{dr}\right)_{r=r_1} = \frac{C_1}{r_1}$$

$$\therefore C_1 = -\frac{I^2R}{2\pi k}$$

At $r = r_2$, $T = T_2$, so from Equation (2.30):

$$T_2 = C_1 \ln r_2 + C_2$$

$$C_2 = T_2 + \left(\frac{I^2 R}{2\pi k}\right) \ln r_2$$

So, substituting for C_1 and C_2 in Equation (2.30), gives

$$T = \left(\frac{I^2 R}{2\pi k}\right)\{\ln(r_2/r)\} + T_2 \tag{2.31}$$

At the surface $r = r_2$, heat is convected to the surrounds hence

$$I^2 R = 2\pi r_2 h_o (T_2 - T_{air})$$

from which

$$T_2 = T_{air} + \left(\frac{I^2 R}{2\pi r_2 h_o}\right) \tag{2.32}$$

Substituting for T_2, from Equation (2.32) into Equation (2.31) gives the general equation for the temperature distribution in the plastic coating surrounding the wire as

$$T = \left(\frac{I^2 R}{2\pi}\right)\left\{\frac{\ln(r_2/r)}{k} + \frac{1}{r_2 h_o}\right\} + T_{air} \tag{2.33}$$

From Equation (2.33) and with $r = r_2$:

$$I^2 = \frac{2\pi(T_2 - T_{air})}{R\{\ln(r_2/r_2)/k + 1/r_2 h_o\}}$$

$$= \frac{2\pi \times 35}{0.02\left\{\dfrac{\ln(1.15/1.15)}{0.2} + \dfrac{1}{0.00115 \times 16}\right\}}$$

$$= \frac{2\pi \times 35}{\{0.02 \times 54.35\}}$$

$$I = 14.2\,A$$

To determine the temperature of the wire, put $I = 14.2$ A, $T = T_1$ and $r = r_1$ in Equation (2.33):

$$T_1 = \left(\frac{I^2 R}{2\pi}\right)\left\{\frac{\ln(r_2/r_1)}{k} + \frac{1}{r_2 h_o}\right\} + T_{air}$$

$$= \left(\frac{(14.2)^2 \times 0.02}{2\pi}\right)\left\{\frac{\ln(1.15/0.5)}{0.2} + \frac{1}{0.00115 \times 16}\right\} + 35$$

$$= 72.6°C$$

Comment   The above values are representative of wire used for a domestic lighting circuit, where the supply is unlikely to exceed 5 A. Note the square law relation between temperature difference and current. The external heat transfer coefficient is evaluated from Equation (6.30), free convection from a horizontal cylinder in air.

2.5.4 One-dimensional radial conduction in composite systems

A similar method to that applied to plane walls can be used for radial systems except, as indicated by Equation (2.29), it is the heat flow rate $\dot{Q}$, not the heat flux q, that is constant.

Example 2.4

A typical domestic central heating installation utilises approximately 50 m of 15 mm outside diameter 1 mm wall thickness copper pipe to convey water at 70°C. A cross-section through such a pipe is depicted in Figure 2.7. Calculate the heat loss from this length of pipe, with a 15 mm radial thickness of insulation and compare this to the value without insulation. Take the ambient air temperature as 15°C and the respective values of internal and external heat transfer coefficients as 100 W/m²K and 8 W/m²K. The thermal conductivity of copper is 400 W/m K and the thermal conductivity of the insulation is 0.05 W/m K.

Assumptions
1. One-dimensional steady-state radial conduction with constant thermal properties.
2. Neglect heat transfer by radiation.
3. The external heat transfer coefficient is independent of the pipe outside diameter.
4. Perfect thermal contact between adjacent layers.

Fig. 2.7 ●
Heat loss from an insulated heating pipe (Example 2.4)

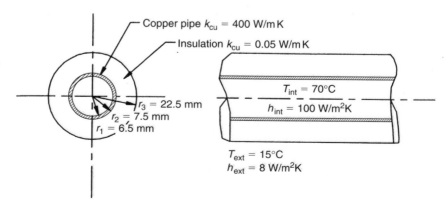

Copper pipe k_{cu} = 400 W/m K
Insulation k_{cu} = 0.05 W/m K

r_3 = 22.5 mm
r_2 = 7.5 mm
r_1 = 6.5 mm

T_{int} = 70°C
h_{int} = 100 W/m²K

T_{ext} = 15°C
h_{ext} = 8 W/m²K

Solution   Let T_1 and T_2 be the surface temperatures on the copper at $r = r_1$ and $r = r_2$, respectively, and T_3 be the temperature on the outside of the insulation at $r = r_3$. Let the ambient air temperature be T_{ext} and the water temperature be T_{int}. The thermal conductivity of the copper is k_{cu}, that of the insulation k_{ins}, and the internal and external heat transfer coefficients, h_{int}, h_{ext}.

With insulation, since the heat flow, $\dot{Q}$, is constant

$$\dot{Q} = \frac{2\pi L k_{cu}(T_1 - T_2)}{\ln(r_2/r_1)} \tag{2.34a}$$

$$= \frac{2\pi L k_{ins}(T_2 - T_3)}{\ln(r_3/r_2)} \tag{2.34b}$$

$$= 2\pi L h_{ext} r_3 (T_3 - T_{ext}) \tag{2.34c}$$

$$= 2\pi L h_{int} r_1 (T_{int} - T_1) \tag{2.34d}$$

After eliminating the unknown surface temperatures by adding (2.34a), (2.34b), (2.34c) and (2.34d), this can be rearranged to the form

$$q = \frac{\dot{Q}}{A} = U\Delta T$$

where

$$\frac{1}{U} = \left\{ \left(\frac{r_3}{h_{int} r_1} \right) + \left(\frac{r_3 \ln(r_2/r_1)}{k_{cu}} \right) + \left(\frac{r_3 \ln(r_3/r_2)}{k_{ins}} \right) + \left(\frac{1}{h_{ext}} \right) \right\}$$

$$\Delta T = (T_{int} - T_{ext})$$

$$A = 2\pi r_3 L \tag{2.35}$$

Substituting values for the thermal conductivity, heat transfer coefficient and radii gives

$$\frac{1}{U} = \left(\frac{0.0225}{0.0075 \times 100} \right) + \left(\frac{0.0225 \ln(0.0075/0.0065)}{400} \right)$$

$$+ \left(\frac{0.0225 \ln(0.0225/0.0075)}{0.05} \right) + \frac{1}{8}$$

$$= 0.03 + (8 \times 10^{-6}) + 0.494 + 0.125 = 0.649 \, \text{m}^2\text{K/W}$$

$$U = 1.54 \, \text{W/m}^2\text{K}$$

$$\Delta T = 70 - 15 = 55°\text{C}$$

$$A = 2\pi \times 0.0225 \times 50 = 7.068 \, \text{m}^2$$

Hence, with insulation the heat loss $\dot{Q}$ is

$$Q = 1.54 \times 7.068 \times 55$$

$$= 600 \, \text{W}$$

For a bare pipe, the term $r_3 \ln(r_3/r_2)/k_{\text{ins}}$ is absent from the overall heat transfer coefficient. Since the most significant of the remaining terms is $1/h_{\text{ext}}$, the overall heat transfer coefficient can be said to be dominated, or governed by the external heat transfer coefficient and $U \approx h_{\text{ext}}$, from which

$$\dot{Q} = 8 \times (2\pi \times 0.0075 \times 50) \times 55 = 1037 \, \text{W}$$

Comment   The addition of insulation to the pipe gives a 42% reduction in the heat loss. The thermal resistances due to conduction through the copper pipe and from convection on the inside are negligible. The value of the heat transfer coefficient acting on the external surface of the pipe is assumed to remain independent of the pipe radius, which is reasonable given the form of Equation (6.30). Although thermal radiation from the pipe to the surrounds has been ignored, this could be accounted for using the 'radiative heat transfer coefficient', as defined in Equation (1.6), and added to the convective heat transfer coefficient. However, from Equation (2.34c) the surface temperature of the insulation, T_3, is approximately 25°C, whereas for the bare pipe it is approximately 70°C. Including the effect of thermal radiation in the analysis is therefore likely (depending on the surface finish and hence emissivity of the copper pipe) to increase the above value of reduction in the total heat loss.

From Example 2.4 it may be thought that adding more insulation always results in a decrease in the heat flow through the pipe walls. This, however, is not the case, since a thicker layer of insulation also increases the external surface area exposed to the ambient surrounds (and for simplicity we have assumed that, when insulated, h is unaffected by both an increase in the outside diameter of the pipe and a decrease in the surface-to-air temperature difference). Rewriting Equation (2.35) in full:

$$\dot{Q} = \frac{2\pi L \Delta T}{\left(\dfrac{1}{h_{\text{int}}r_1}\right) + \left(\dfrac{\ln(r_2/r_1)}{k_{\text{cu}}}\right) + \left(\dfrac{\ln(r_3/r_2)}{k_{\text{ins}}}\right) + \left(\dfrac{1}{h_{\text{ext}}r_3}\right)}$$

As the outer radius, r_3, is increased by the addition of more insulation, the minimum thermal resistance (or maximum heat flow) is found from the condition $d\dot{Q}/dr_3 = 0$, which may be verified to occur when

$$r_{3,\text{crit}} = \frac{k_{\text{ins}}}{h_{\text{ext}}} \tag{2.36}$$

The radius determined from Equation (2.36) is referred to as the critical insulation radius. Using values of k_{ins} and h_{ext} appropriate to expanded foam insulation and free convection in air (as in Example 2.4), $r_3 = 0.05/8 = 6.25 \, \text{mm}$. The critical insulation radius is therefore only likely to be a consideration in small diameter pipes and current-carrying wires.

2.6 ● Thermal contact resistance

In Examples 2.2 and 2.4 the assumption of perfect thermal contact between adjacent conducting layers was made. In reality there will not be intimate contact owing to imperfections in the surfaces, leaving microscopic gaps filled with fluid (most probably air). These effects depend upon the surface roughness, contact pressure and temperature and separating medium, and the reader is referred to Krieth and Bohn (1993) and Fried (1969) for further information. It is, however, necessary to recognise that imperfect thermal contact between mating surfaces can be handled by way of a thermal contact resistance (units m^2K/W). Table 2.1 gives some representative values.

Table 2.1 ●

Typical values of thermal contact resistance

Source: Adapted from Fried (1969)

Material	Thermal contact resistance (m^2K/W) Contact pressure	
	10^5 Pa	10^7 Pa
Aluminium	0.00015–0.0005	0.00002–0.00004
Copper	0.0001–0.001	0.00001–0.00005
Magnesium	0.00015–0.00035	0.00002–0.00004
Stainless steel	0.0006–0.0025	0.00007–0.0004

2.7 ● Temperature-dependent thermal conductivity

Another assumption made in the previous examples is that the thermal conductivity remains constant. Inspection of tabulated values of thermal conductivity will show that this is not usually the case. For example, the thermal conductivity of 0.5% carbon steel is 55 W/m K at 0°C, 52 W/m K at 100°C, 48 W/m K at 200°C, 42 W/m K at 400°C, 35 W/m K at 600°C and 29 W/m K at 1000°C. Although not linear across the range 0°C to 1000°C, it is, however, reasonable to assume a linear variation across a limited range of interest of the form

$$k = k_0(1 + C_0 T) \tag{2.37}$$

where the values of C_0 and k_0 can be found from published data of the variation of thermal conductivity with temperature. In the example above, for the range 100°C to 200°C, $k_0 = 56$ W/m K and $C_0 = -0.0007143/°C$.

Example 2.5

Assuming one-dimensional, steady-state conduction, obtain an expression for the heat flux through a material of thickness L, with a temperature-dependent thermal conductivity given by $k = k_0(1 + C_0 T)$ where C_0 and k_0

are known constants. The surface temperature at $x = 0$ is T_1 and the surface temperature at $x = L$ is T_2.

Calculate the heat flux through a 2 mm thick, 0.5% carbon steel plate when one surface is at 100°C and the other at 102°C, given that $k_0 = 56$ W/m K and $C_0 = -0.0007143/°C$; compare this with the value obtained assuming a constant value of $k = 56$ W/m K.

Solution   For non-constant thermal conductivity, steady-state conduction

$$\frac{d}{dx}\left(\frac{k\,dT}{dx}\right) = 0$$

which, after substituting the expression for k, gives

$$k_0\left\{\frac{d}{dx}(1 + C_0T)\frac{dT}{dx}\right\} = 0$$

Integrating gives

$$k_0(1 + C_0T)\frac{dT}{dx} = \text{constant} = -q$$

Integrating again with $T = T_1$ at $x = 0$ and $T = T_2$ at $x = L$, gives

$$-k_0\left\{(T_2 - T_1) + \frac{C_0}{2}(T_2^2 - T_1^2)\right\} = qL$$

which can be rearranged to give

$$q = -\frac{k_0(T_2 - T_1)}{L}\left\{1 + \left(\frac{C_0}{2}\right)(T_2 + T_1)\right\} \tag{2.38}$$

From Equation (2.38),

$$q = -\frac{56(102 - 100)}{0.002}\left\{1 + \left(\frac{-0.000\,7143}{2}\right)(102 + 100)\right\}$$

$$= -51.96\,\text{kW/m}^2$$

For constant k,

$$q = -\frac{56(102 - 100)}{0.002}$$

$$= -56\,\text{kW/m}^2$$

Comment   The above is of more than just academic interest as it is relatively easy to verify that the same result is obtained from using a value of thermal conductivity evaluated at the mean temperature, $\frac{1}{2}(T_2 + T_1)$. The error in the heat flux due to ignoring the temperature dependence of thermal conductivity is roughly 8%, which could be significant, depending on the circumstances.

2.8 ● **Fins and extended surfaces**

Sometimes it is necessary to enhance the overall surface heat transfer rate. In general terms this may be achieved by lowering the convective resistance (by increasing the convective heat transfer coefficient) or from lowering the conductive resistance (by increasing the surface area). The latter can be achieved by extending the surface area through the addition of fins or spines – some examples are illustrated in Figure 2.8. Fins may be classified as longitudinal (Figure 2.8a and b), radial (Figure 2.8e), annular (Figure 2.8f), or spines (Figure 2.8c and d), often referred to as pin fins. Fins and extended surfaces are used in transformer shields, integrated circuit heat sinks, heat exchanger tubes, engine, electric motor and gearbox casings. An increase in surface area, however, does not always guarantee an increase in the overall heat transfer, as the fin material must have a relatively high thermal conductivity so that the increase in the conduction path length does not cause an insulation effect. Projections from buildings and also where internal walls join onto an external wall, are less obvious examples of fins, where the increase in heat transfer represents an undesirable or parasitic loss.

2.8.1 **Analysis of steady-state heat transfer from extended surfaces**

A schematic diagram depicting the heat flows by conduction and convection through a general extended surface is shown in Figure 2.9 The simple

Fig. 2.8 ●
Fins and spines:
(a) rectangular fin;
(b) triangular fin;
(c) circular spine;
(d) conical spine;
(e) radial fins;
(f) annular fins

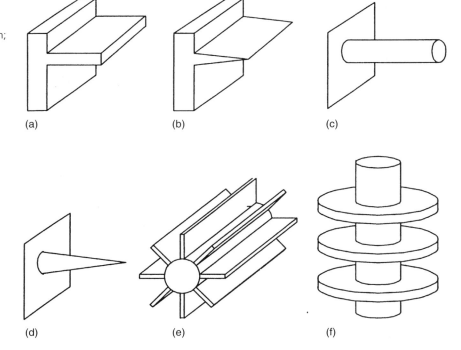

(a)   (b)   (c)

(d)   (e)   (f)

Fig. 2.9 ●

Heat flow through a general extended surface

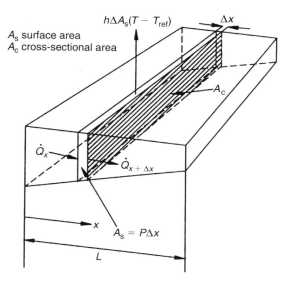

A_s surface area
A_c cross-sectional area

$h\Delta A_s(T - T_{ref})$

Δx

A_c

$\dot{Q}_x$

$\dot{Q}_{x+\Delta x}$

x

$A_s = P\Delta x$

L

one-dimensional heat conduction equation (2.16) is not valid in these circumstances because heat is convected from the lateral surface, i.e. the fin's external surface area.

An appropriate equation for steady-state heat conduction can be derived by making the following assumptions:

1. The temperature of the surface is a function of one dimension only (from the comments following Equation (2.15), this implies $Bi \ll 1$).

2. The fin is homogeneous and heat transfer by radiation is either negligible or accounted for by a radiation heat transfer coefficient.

3. The heat transfer coefficient is constant along the length of the fin.

4. The fluid surface temperature is also constant along the length of the fin.

Making a heat balance on an element of thickness Δx of the fin,

Heat conducted into the element	=	Heat conducted out of the element	+	Loss by convection to surrounding fluid
$\dot{Q}_x$	=	$\dot{Q}_{x+\Delta x}$	+	$h\Delta A_s(T - T_{ref})$

$$(2.39)$$

where the surface area,

$$\Delta A_s = (\text{perimeter}) \times (\Delta x) = P\Delta x \tag{2.40}$$

Using a Taylor series expansion and Fourier's law,

$$\dot{Q}_x - \dot{Q}_{x+\Delta x} = k\frac{d}{dx}\left(A_c\frac{dT}{dx}\right)\Delta x$$

where A_c is the cross-sectional area. To maintain generality, A_c varies along the length of the fin. For example, the rectangular fin, circular spine and radial fin (Figure 2.8a, c and e) have constant cross-sectional areas. For the triangular fin and conical spine (Figure 2.8b and d), A_c decreases with increasing distance from the base, x. For the annular fin, A_c increases with radius.

Hence collecting together the terms and rearranging gives

$$k\frac{\mathrm{d}}{\mathrm{d}x}\left(A_c\frac{\mathrm{d}T}{\mathrm{d}x}\right)\Delta x - h\Delta A_s(T - T_{ref}) = 0 \tag{2.41}$$

After expanding out the term on the left-hand side (using the rule for differentiating a product), and as Δx tends to zero, the general fin equation can be expressed as

$$\frac{\mathrm{d}^2 T}{\mathrm{d}x^2} + \frac{1}{A_c}\frac{\mathrm{d}A_c}{\mathrm{d}x}\frac{\mathrm{d}T}{\mathrm{d}x} - \frac{h}{A_c k}\frac{\mathrm{d}A_s}{\mathrm{d}x}(T - T_{ref}) = 0 \tag{2.42}$$

Solutions to Equation (2.42) require further consideration of the geometry, in particular the behaviour of cross-sectional area and surface area with coordinate x. The reader is referred to Mikhailov and Özisik (1994) for analytical solutions to many different fin geometries, and to Bejan (1993) for applications of optimising a fin geometry for known (functional) relations of the heat transfer coefficient.

The simplest geometry to consider is where the surface and cross-sectional areas are uniform, as with short radial fins found on the external casings of electric motors and gearboxes (Figure 2.8e), and with rectangular fins (Figure 2.8a) used in electronic heat sinks. Here

$$A_s = Px$$

$$\therefore \frac{\mathrm{d}A_s}{\mathrm{d}x} = P$$

$$A_c = \text{constant}$$

$$\therefore \frac{\mathrm{d}A_c}{\mathrm{d}x} = 0$$

so, Equation (2.42) becomes

$$\frac{\mathrm{d}^2 T}{\mathrm{d}x^2} - \frac{hP}{A_c k}(T - T_{ref}) = 0 \tag{2.43}$$

For convenience, define $\Theta = (T - T_{ref})$ and $m^2 = (hP/A_c k)$, then Equation (2.43) becomes

$$\frac{\mathrm{d}^2\Theta}{\mathrm{d}x^2} - m^2\Theta = 0 \tag{2.44}$$

for which a general solution is

$$\Theta = C_1 e^{mx} + C_2 e^{-mx} \tag{2.45}$$

where C_1 and C_2 are constants which depend upon the particular boundary conditions imposed. (An alternative solution would be to use trigonometric functions. However, a solution containing exponential functions is favoured owing to the boundary condition for the infinite fin, where $C_1 = 0$.) Since there are two constants, two boundary conditions are required. For the first, at the base of the fin ($x = 0$) the base temperature is usually known. For the second it is customary to make use of conditions at the tip of the fin; the fin may convect into the surrounds, it may be insulated or it may have a known temperature. A further simplification results under certain conditions, when the fin may be considered to be of infinite length. The first of these conditions will be derived below; results for the others are quoted, it being a relatively simple matter to derive them using the method illustrated.

Additional heat transfer by the contribution of convection from the fin tip

$$x = 0, \quad \Theta = \Theta_b$$

$$x = L, \quad -k\left(\frac{d\Theta}{dx}\right) = h_{tip}\Theta$$

Substitution of these boundary conditions into the fin equation (2.45) leads to the following relations:

$$\Theta_b = C_1 + C_2$$

$$C_1 e^{mL} - C_2 e^{-mL} = \frac{h_{tip}}{km}(C_1 e^{mL} + C_2 e^{-mL})$$

After eliminating C_1 and C_2, then rearranging, we obtain Equation (2.46). Remember that $\cosh x = \frac{1}{2}(e^x + e^{-x})$ and $\sinh x = \frac{1}{2}(e^x - e^{-x})$.

$$\frac{\Theta}{\Theta_b} = \frac{\cosh m(L - x) + \left(\dfrac{h_{tip}}{mk}\right)\sinh m(L - x)}{\cosh mL + \left(\dfrac{h_{tip}}{mk}\right)\sinh mL} \tag{2.46}$$

Of particular interest is the rate of heat flow, $\dot{Q}_b$, through the base of the fin, which from Fourier's law is $\dot{Q}_b = -kA_c(dT/dx)_{x=0}$

$$\dot{Q}_b = (hPkA_c)^{1/2}\Theta_b \frac{\sinh mL + \left(\dfrac{h_{tip}}{mk}\right)\cosh mL}{\cosh mL + \left(\dfrac{h_{tip}}{mk}\right)\sinh mL} \tag{2.47}$$

Adiabatic tip

$(dT/dx)_{x=L} = 0$ is a reasonable approximation if the fin is thin in relation to its length, or the tip is effectively insulated by a support to prevent vibration.

$$\frac{\Theta}{\Theta_b} = \frac{\cosh m(L - x)}{\cosh mL} \tag{2.48}$$

$$\dot{Q}_b = (hPkA_c)^{1/2}\Theta_b \tanh mL \tag{2.49}$$

Known tip temperature
$\Theta_{x=L} = \Theta_L$, as for example in an evaporating or condensing fluid.

$$\frac{\Theta}{\Theta_b} = \frac{\left(\dfrac{\Theta_L}{\Theta_b}\right)\sinh mx + \sinh m(L-x)}{\sinh mL} \tag{2.50}$$

$$\dot{Q}_b = \frac{(hPkA_c)^{1/2}\Theta_b\left\{\cosh mL - \dfrac{\Theta_L}{\Theta_b}\right\}}{\sinh mL} \tag{2.51}$$

Infinite fin
If the fin is sufficiently long such that $T \to T_\infty$ as $L \to \infty$, then C_1 in Equation (2.45) will equal zero and

$$\frac{\Theta}{\Theta_b} = e^{-mx} \tag{2.52}$$

$$\dot{Q}_b = (hPkA_c)^{1/2}\Theta_b \tag{2.53}$$

2.8.2 Fin performance

Where augmentation of heat transfer is required, fins can be added to increase the surface area. However, since the additional surface area represents an increase in the thermal resistance there is no guarantee that the addition of fins will lead to any improvement in the thermal performance and the increase in surface area may be outweighed by the reduction in mean surface temperature.

Two parameters that quantify the usefulness of fins, which are used extensively in design calculations are:

1. the fin effectiveness, ε_{fin},
2. the fin efficiency, η_{fin}.

The fin effectiveness, ε_{fin}, is defined as:

$$\frac{\text{Fin heat transfer rate}}{\text{Heat transfer rate that would occur in the absence of the fin}} \tag{2.54}$$

Hence,

$$\varepsilon_{\text{fin}} = \frac{\dot{Q}_b}{hA_c\Theta_b}$$

which for the case of the *infinite fin* (2.53) becomes

$$\varepsilon_{\text{fin}} = \left(\frac{Pk}{hA_c}\right)^{1/2} \tag{2.55}$$

Examination of Equation (2.55) informs us that the effectiveness (which should be as large as possible and certainly never less than about 2) increases with the thermal conductivity of the fin material, hence copper and aluminium are obvious choices. A fin geometry is also more effective if the ratio of perimeter to cross-sectional area is large, hence thin and closely spaced fins tend to be used. Equally, small values of heat transfer coefficient (for example free convection) act to increase the fin effectiveness. The effectiveness is also inversely proportional to the square root of the Biot number. An effective fin will therefore have a small Biot number which justifies assumption 1 at the beginning of Section 2.8.1.

The fin efficiency, η_{fin}, is defined as

$$\frac{\text{Actual heat transfer through the fin}}{\substack{\text{Heat that would be transferred if the entire fin} \\ \text{area were at the base temperature}}} \tag{2.56}$$

Hence

$$\eta_{\text{fin}} = \frac{\dot{Q}_b}{hA_s\Theta_b}$$

which for the case of the *infinite fin* becomes

$$\eta_{\text{fin}} = \left(\frac{PkA_c}{hA_s^2}\right)^{1/2} \tag{2.57}$$

where A_s is the external surface area of the fin bathed by the surrounding fluid.

Equations (2.55) and (2.57) for fin effectiveness and efficiency apply to the case of an infinite fin with constant surface and cross-sectional area. For this geometry, it is a simple matter to derive similar relations corresponding to the remaining three boundary conditions. For other geometries, such as annular fins, concave and convex surface profiles, triangular and conical fins, it is more convenient to use a fin efficiency chart. The pioneering work in this field was done by Gardner (1945), and the larger heat transfer texts generally include one or two of his figures; Mikhailov and Özisik (1994) present a number of such charts where the efficiency is correlated against a similar parameter to the right-hand side of Equation (2.57).

Example 2.6

An aluminium heat sink for electronics components has a base of width $B = 50\,\text{mm}$ and length $W = 70\,\text{mm}$. Attached to the base are eight ($N = 8$) aluminium ($k = 180\,\text{W/m K}$) fins of length $L = 12\,\text{mm}$ and thickness $t = 3\,\text{mm}$. The fins are cooled by air at 25°C with a convective heat transfer coefficient of $h = 10\,\text{W/m}^2\,\text{K}$ (Figure 2.10). Assuming that the

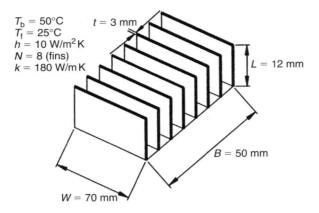

Fig. 2.10 ●
Electronics heat
sink (Example 2.6)

$T_b = 50°C$ $t = 3$ mm
$T_f = 25°C$
$h = 10$ W/m^2 K
$N = 8$ (fins)
$k = 180$ W/m K

$L = 12$ mm

$B = 50$ mm

$W = 70$ mm

same value of heat transfer coefficient acts on the tip of the fin as along the rest of the external surface, determine:

(a) The heat flow through the heat sink for a base temperature of 50°C.

(b) The fin effectiveness.

(c) The fin efficiency.

(d) The length of the fin so the heat flow is 95% of the heat flow for an infinite fin.

Solution   Total heat flow = Heat flow through unfinned portion of the base + Heat flow through the fins

$$\dot{Q} = \dot{Q}_u + \dot{Q}_f$$

$$\dot{Q}_u = hA_u\Theta_b$$

$$= hW(B - Nt)\Theta_b$$

$$= 10 \times 0.07\{0.05 - (8 \times 0.003)\} \times (50 - 25)$$

$$= 0.455\,\text{W}$$

From Equation (2.47)

$$\dot{Q}_f = N(hPkA_c)^{1/2}\,\frac{\Theta_b\left\{\sinh mL + \left(\dfrac{h}{mk}\right)\cosh mL\right\}}{\cosh mL + \left(\dfrac{h}{mk}\right)\sinh mL}$$

$$P = 2(W + t) = 0.146\,\text{m}$$

$$A_c = Wt = 2.1 \times 10^{-4}\,\text{m}^2$$

$$m = \left(\frac{hP}{kA_c}\right)^{1/2} = 6.2148\,\text{m}^{-1}$$

$$mL = 0.074\,578$$

$$\cosh mL = 1.002\,78$$

$$\sinh mL = 0.074\,65$$

$$(hPkA_c)^{1/2} = 0.235\,\text{W/K}$$

$$\frac{h}{mk} = 8.939 \times 10^{-3}$$

$$\dot{Q}_f = 8 \times 0.235 \times 25 \times \frac{\{0.074\,65 + (8.939 \times 10^{-3} \times 1.002\,78)\}}{1.002\,78 + (8.939 \times 10^{-3} \times 0.074\,65)}$$

$$= 3.916\,\text{W}$$

$$\dot{Q} = 0.455 + 3.916 = 4.4\,\text{W}$$

$$\text{Effectiveness} = \frac{\dot{Q}_b}{hA_c\Theta_b}$$

$$= \frac{3.916/8}{10 \times 0.003 \times 0.07 \times 25} = 9.32$$

$$\text{Efficiency} = \frac{\dot{Q}_b}{hA_s\Theta_b}$$

$$A_s = 2L(W + t) + Wt = 1.962 \times 10^{-3}\,\text{m}^2$$

$$\eta_{\text{fin}} = \frac{3.916/8}{10 \times 1.962 \times 10^{-3} \times 25} = 0.998$$

For the heat flow to be within 95% of the infinite fin solution

$$\frac{\sinh mL + \left\{\left(\dfrac{h}{mk}\right)\cosh mL\right\}}{\cosh mL + \left\{\left(\dfrac{h}{mk}\right)\sinh mL\right\}} \geqslant 0.95$$

Since $h/mk \ll 1$, the above inequality may be simplified to $\tanh mL \geqslant 0.95$, which occurs when $mL \geqslant 1.83$ or $L \geqslant 295\,\text{mm}$.

Comment Without fins and assuming the same value of heat transfer coefficient, the aluminium plate would dissipate

$$\dot{Q} = h(WB)\Theta_b = 10 \times 0.05 \times 0.07 \times 25 = 0.875\,\text{W}$$

A fivefold increase in the net heat transfer is obtained by the addition of fins. Since $L \ll 295\,\text{mm}$ the fin clearly cannot be treated as an infinite fin, and Equation (2.53) gives

$$\dot{Q}_f = N\dot{Q}_b \approx 47\,\text{W}$$

which is over ten times that predicted by Equation (2.47) for a finite length fin.

2.9 ● Two-dimensional steady-state conduction

The previous sections have treated conduction as a one-dimensional phenomenon, with temperature gradients in only one spatial coordinate. Consider now the rectangular slab shown in Figure 2.11, of length L in the x direction and W in the y direction. Three of the sides are maintained at the same temperature, suppose it is 50°C. The side $y = W$ of the slab is maintained at a temperature above the other sides, which to be general we will assume is a function of x, so $T(x,W) = f(x)$. If this temperature increased sinusiodally from $T(0,W) = 50°\text{C}$ to $T(L,W) = 50°\text{C}$ with a maximum value of 100°C at $x = L/2$, then

$$f(x) = 50 + 50\sin(\pi x/L)$$

In the physical world these constant temperature (or isothermal) boundary conditions could be used to model a heat sink or two-dimensional fin, where the three sides at constant temperature are immersed in a fluid with a very high heat transfer coefficient.

The equation for steady-state conduction with constant thermal conductivity in the region depicted by Figure 2.11 is given by the two-dimensional form of Laplace's equation (Equation 2.10):

$$\frac{\partial^2 T}{\partial x^2} + \frac{\partial^2 T}{\partial y^2} = 0 \tag{2.58}$$

Fig. 2.11 ●
Two-dimensional steady-state conduction with temperature specified at the boundaries (2.71)

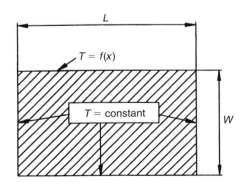

The standard way to solve this *partial differential equation* is to use the method of separation of variables, which reduces the order of complexity to that of solving two *ordinary differential equations*. First it is necessary to examine the boundary conditions as these provide clues to the solution. By inspection we see symmetry in the boundary conditions at $x = 0$ and $x = L$, which implies that the solution in this direction will require periodic functions (sine and cosine). Next, we define a new variable as the excess temperature, in this case $\Theta = T - 50$. The transformed equation then becomes

$$\frac{\partial^2 \Theta}{\partial x^2} + \frac{\partial^2 \Theta}{\partial y^2} = 0 \tag{2.59}$$

subject to the following boundary conditions:

$$\Theta(0,y) = \Theta(L,y) = \Theta(x,0) = 0 \tag{2.60a}$$

$$\Theta(x,W) = f(x) - 50 = g(x) \tag{2.60b}$$

The technique of separation of variables begins by separating the solution into two component parts: $X(x)$ which depends solely on x, and $Y(y)$ which depends solely on y. We can then write

$$\Theta(x,y) = X(x)\,Y(y) \tag{2.61a}$$

Differentiating Equation (2.61a), we get

$$\frac{\partial^2 \Theta}{\partial x^2} = Y\frac{d^2 X}{dx^2} \tag{2.61b}$$

$$\frac{\partial^2 \Theta}{\partial y^2} = X\frac{d^2 Y}{dy^2} \tag{2.61c}$$

Substituting Equations (2.61b) and (2.61c) back into Equation (2.59) and dividing by the product XY gives

$$\frac{1}{X}\frac{d^2 X}{dx^2} = -\frac{1}{Y}\frac{d^2 Y}{dy^2} \tag{2.62}$$

Inspection of Equation (2.62) does indeed show that it has been separated: the left side is a function only of x, the right only of y (hence the use of ordinary derivatives). Each side must then be a constant, which we will denote as $-p^2$. The sign of the constant is crucial since, as noted above, the solution in the x direction must involve periodic functions, and this will be satisfied with a negative value of the constant. Hence it follows that

$$\frac{d^2 X}{dx^2} + p^2 X = 0 \tag{2.63a}$$

$$\frac{d^2 Y}{dy^2} - p^2 Y = 0 \tag{2.63b}$$

Recalling from Equation (2.61a) that $\Theta(x,y) = X(x)Y(y)$, the general solution to Equations (2.63a) and (2.63b) is the product of the individual solutions:

$$\Theta = (C_1 \cos px + C_2 \sin px)(C_3 e^{-py} + C_4 e^{py}) \tag{2.64}$$

where C_1, C_2, C_3 and C_4 are constants; their values will now be determined from the boundary conditions.

From applying the condition that at $x = 0$, $\Theta(0,y) = 0$, it is apparent that $C_1 = 0$. From the condition that at $y = 0$, $\Theta(x,0) = 0$, we get

$$0 = C_2 \sin px(C_3 + C_4)$$

This can occur only if either $C_2 = 0$ (which would remove any dependence of the temperature in the x direction) or $C_3 = -C_4$. Recognising that $\sinh z = \frac{1}{2}(e^{-z} - e^z)$, Equation (2.64) then simplifies to

$$\Theta = C_0 \sin px \sinh py \qquad \text{where } C_0 = \frac{1}{2}C_2 C_3 \tag{2.65}$$

From the condition at $x = L$, $\Theta(L,y) = 0$, we obtain

$$0 = C_0 \sin pL \sinh py \tag{2.66}$$

Equation (2.66) can only be satisfied for all values of y if

$$\sin(pL) = 0 \tag{2.67a}$$

This implies that p must be an integer multiple of π/L, that is,

$$p = \frac{n\pi}{L} \qquad (n = 1,2,3,\ldots) \tag{2.67b}$$

The implication of Equation (2.67b) is that there is a separate solution for each value of integer n. The complete solution is then expressed as a series or summation of the separate solutions, hence

$$\Theta = \sum_{n=1}^{\infty} C_n \sin\left(\frac{n\pi x}{L}\right) \sinh\left(\frac{n\pi y}{L}\right) \tag{2.68}$$

The remaining constant is obtained from the boundary condition along $y = W$ where $T = f(x)$ or in the transformed variable $\Theta = g(x)$. Applying this condition yields the result

$$g(x) = \sum_{n=1}^{\infty} C_n \sin\left(\frac{n\pi x}{L}\right) \sinh\left(\frac{n\pi W}{L}\right) \tag{2.69}$$

The last term in Equation (2.69) is a constant, so the above expression may be identified as a Fourier series expansion of $g(x)$. The constant C_n and the sinh term may be grouped together as a Fourier coefficient (conventionally mathematics texts usually refer to this as b_n), from which

$$C_n = \frac{2}{L \sinh(n\pi W/L)} \int_0^L g(x) \sin\left(\frac{n\pi x}{L}\right) dx \tag{2.70}$$

Substituting Equations (2.69) and (2.70) into (2.68), the complete solution becomes

$$\Theta = \frac{2}{L} \sum_{n=1}^{\infty} \frac{\sin\left(\frac{n\pi x}{L}\right) \sinh\left(\frac{n\pi y}{L}\right)}{\sinh\left(\frac{n\pi W}{L}\right)} \int_0^L g(x) \sin\left(\frac{n\pi x}{L}\right) dx \qquad (2.71)$$

The reader should recognise that since we have referred to the temperature distribution along $y = W$ as a general function $g(x)$, the integral in Equation (2.71) needs to be evaluated for the specific temperature distribution. For the case of $g(x) = \text{constant} = \Theta_0$, say (and noting that $-\cos(n\pi) = (-1)^{n+1}$)

$$\int g(x) \sin\left(\frac{n\pi x}{L}\right) dx = \frac{\Theta_0 L}{n\pi} \{(-1)^{n+1} - 1\}$$

in which case the temperature field is given by

$$\Theta = \frac{2\Theta_0}{\pi} \sum_{n=1}^{\infty} \frac{\{(-1)^{n+1} - 1\} \sin\left(\frac{n\pi x}{L}\right) \sinh\left(\frac{n\pi y}{L}\right)}{n \sinh(n\pi W/L)} \qquad (2.72)$$

The boundary condition imposed on the three sides with a specified temperature is, however, of limited practical use. A more useful boundary condition is when one of the sides exchanges heat by convection to a surrounding fluid, at $T = 0$, two of the other sides are insulated, the fourth side has a known temperature distrubution. This is illustrated in Figure 2.12 and formally, these boundary conditions are written as

$$\Theta(x,0) = g(x)$$

$$\left(\frac{\partial \Theta}{\partial y}\right) = 0 \qquad (0 < x < L, y = W)$$

$$\left(\frac{\partial \Theta}{\partial x}\right) = 0 \qquad (0 < y < W, x = 0)$$

$$\left(\frac{\partial \Theta}{\partial y}\right) = -\frac{h\Theta}{k} \qquad (0 < y < W, x = L)$$

where the heat transfer coefficient is h, and the thermal conductivity k. The solution to Equation (2.58) with the above boundary conditions is

$$\Theta = 2 \sum_{n=1}^{\infty} \frac{\left\{\left(\frac{h}{k}\right)^2 + p_n^2\right\} \cos p_n x \cosh p_n(W - y)}{\left[\left\{\left(\frac{h}{k}\right)^2 + p_n^2\right\} L + \left(\frac{h}{k}\right)\right] \cosh p_n W} \int_0^L g(x) \cos p_n x \, dx \qquad (2.73)$$

where p_n are the positive roots of

$$p_n \tan p_n L = \frac{h}{k} \qquad (2.74)$$

Fig. 2.12 ●
Two-dimensional
steady-state
conduction with
a convective
b. c. (2.73)

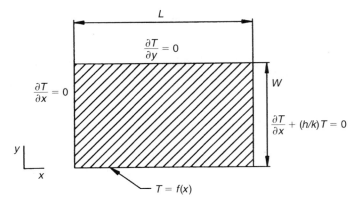

If $g(x)$ is a constant, say Θ_0, then

$$\Theta = \frac{2\Theta_0 h}{k} \sum_{n=1}^{\infty} \frac{\cos p_n x \cosh p_n(W - y)}{\left[\left\{\left(\frac{h}{k}\right)^2 + p_n^2\right\} L + \left(\frac{h}{k}\right)\right] \cos p_n L \cosh p_n W} \tag{2.75}$$

Example 2.7

Figure 2.13 shows a schematic diagram of the extrusion of plastic ($k = 0.3$ W/m K) components 20 mm high and 30 mm wide. The upper and lower surfaces are maintained at a constant temperature of 120°C. The sides experience an effective heat transfer coefficient of $h = 10$ W/m²K to air at 20°C. Calculate the temperature inside the component at $x = 10$ mm, $y = 7.5$ mm.

Solution First we recognise the symmetry in the problem, which reduces the region for analysis to that shown by the shaded area in Figure 2.13. The boundary conditions are indicated in Figure 2.13 and comparison with this example shows that Equation (2.75) is appropriate.

For the purpose of this calculation it is preferable to rewrite Equation (2.75) as

$$\Theta = \frac{2\Theta_0 h}{k} \sum_{n=1}^{\infty} \frac{A_n}{B_n}$$

where the coefficients A_n and B_n are

$$A_n = \cos p_n x \cosh p_n(W - y)$$

$$B_n = \left[\left\{\left(\frac{h}{k}\right)^2 + p_n^2\right\} L + \left(\frac{h}{k}\right)\right] \cos p_n L \cosh p_n W \qquad (\text{m}^{-1})$$

Fig. 2.13 ●
Extrusion of plastic
(Example 2.7)

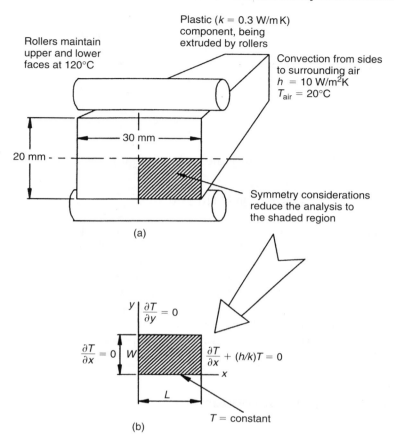

Plastic ($k = 0.3$ W/m K) component, being extruded by rollers

Rollers maintain upper and lower faces at 120°C

Convection from sides to surrounding air
$h = 10$ W/m²K
$T_{air} = 20°C$

Symmetry considerations reduce the analysis to the shaded region

(a)

y | $\frac{\partial T}{\partial y} = 0$

$\frac{\partial T}{\partial x} = 0$ W

$\frac{\partial T}{\partial x} + (h/k)T = 0$

x

L

$T = $ constant

(b)

Solving Equation (2.74) for $n = 1$, $h = 10\,\text{W/m}^2\text{K}$, $k = 0.3\,\text{W/m K}$ and $L = 0.015\,\text{m}$, gives $p_1 = 43.551\,44\,\text{m}^{-1}$. Solving Equation (2.75) for $x = 0.01\,\text{m}$, $y = 0.0075\,\text{m}$ and $W = 0.01\,\text{m}$ gives $A_1 = 0.912\,0324$, $B_1 = 68.299\,93\,\text{m}^{-1}$ and $A_1/B_1 = 1.335\,334 \times 10^{-2}\,\text{m}$. It is convenient to set out these results in a table (Table 2.2).

It can be seen from the right-hand column of Table 2.2 that the series converges rapidly and five terms will give adequate accuracy. Summing the elements of the right-hand column gives

$$\Theta = \frac{2\Theta_o h}{k} \sum_{n=1}^{5} \frac{A_n}{B_n}$$

$$= 2 \times (120 - 20) \times \left(\frac{10}{0.3}\right) \times (0.013\,538\,32)$$

$$= 90.25°C$$

Table 2.2 ●

n	p_n (m^{-1})	A_n	B_n (m^{-1})	$\dfrac{A_n}{B_n}$ (m)
1	43.551 44	0.912 0324	68.299 93	$1.335\,334 \times 10^{-2}$
2	219.4873	−0.674 549	−3471.84	$1.942\,915 \times 10^{-4}$
3	424.108	−0.734 092	95 196.14	$-7.711\,37 \times 10^{-6}$
4	631.8324	2.527 93	−1 672 196	$-1.511\,743 \times 10^{-6}$
5	840.4009	2.168 383	23 743 912	$-9.132\,368 \times 10^{-8}$

Hence the temperature at $x = 0.01$ m, $y = 0.075$ m is

$$T = \Theta + T_{\text{air}}$$
$$= 90.25 + 20 = 110.25°\text{C}$$

Comment The solution to this two-dimensional steady-state conduction problem is quite involved. For the point considered here ($x = 0.01$ m, $y = 0.0075$ m), it was necessary to consider only 5 terms. However, near the boundary where the temperature is constant ($0 \leqslant x \leqslant L$, $y = 0$), it is necessary to consider at least 50 terms. Equation (2.74) has no direct solution and must be solved iteratively (a Newton–Raphson method was used in this example). A listing of a program written in BASIC to solve Equations (2.74) and (2.75) is available in Appendix C, named SS2D(NR).bas.

2.10 ● Transient conduction

The steady-state condition may be all that is required of a particular analysis (to estimate the heating requirement of a building, to predict fin performance, calculate insulation requirements, and so on) and it may not be worth the added complexity to address transient, or time-dependent, conduction phenomena.

In some applications, transient conduction can be critically important. For example, in gas turbine engines, it is necessary to maintain fine clearances between the rotating blades and stationary casing. These clearances depend on thermal expansion, hence they also depend on temperature. And since the temperatures vary with time, this is a transient problem. The differential growth between the blades and casing may also lead to the possibility of contact between rotating and stationary components, clearly a situation to be avoided. By now the reader will no doubt have grasped the complexity of the situation. The turbine example is cited not to discourage, but to illustrate a complex, yet achievable, thermal modelling.

In three dimensions, and for a material with constant thermal properties, the transient heat conduction, or Fourier's, equation is

$$\frac{\partial^2 T}{\partial x^2} + \frac{\partial^2 T}{\partial y^2} + \frac{\partial^2 T}{\partial z^2} = \frac{1}{\alpha} \frac{\partial T}{\partial t} \tag{2.76}$$

where $\alpha = k/\rho C$ is the thermal diffusivity. Analytical solutions with relatively simple boundary conditions, or even the 2D approximation, pose a formidable mathematical task and are beyond the scope of this book. Much of the work on analytical solutions took place long before the arrival of computers. Today, it is not only the specialist research departments that are equipped with massive computational resources, and numerical techniques available in software packages can be applied to obtain solutions in three dimensions. Nonetheless, analytical solutions to transient conduction problems do have their place. In this section we focus on various approximations as they not only provide insight into the phenomena of transient conduction but can also provide adequate accuracy for the purpose of design. They are also useful in the experimental measurement of heat transfer (see Chapter 10).

2.10.1 One-dimensional approximation

Consider the common case of a plane wall or plate, with constant thermal properties, thickness $2L$, and in effect infinite in the other directions. At the two surfaces ($x = -L$ and $x = L$) the wall exchanges heat by convection with a surrounding fluid where the fluid temperature is T_f and the convective heat transfer coefficient is h. Consideration of symmetry reveals that we need consider only the region $0 \leqslant x \leqslant L$. (For $-L \leqslant x \leqslant 0$, the transient response is identical.) The appropriate form of Fourier's equation is

$$\frac{\partial^2 T}{\partial x^2} = \frac{1}{\alpha} \frac{\partial T}{\partial t} \tag{2.77}$$

Recalling Equation (2.14):

$$X = x/L$$

$$\Theta = \frac{(T - T_f)}{(T_0 - T_f)}$$

$$Fo = \alpha t/L^2$$

it is convenient to transform this to the dimensionless form. A certain economy is achieved as the solution will then be independent of the actual numerical values of temperature, length and relevant thermal properties.

$$\frac{\partial^2 \Theta}{\partial X^2} = \frac{\partial \Theta}{\partial Fo} \tag{2.78}$$

with the initial condition

$$\Theta(X, 0) = 1 \tag{2.79a}$$

and the boundary conditions

$$\frac{\partial \Theta}{\partial X} = 0 \qquad (X = 0, Fo > 0) \tag{2.79b}$$

$$\frac{\partial \Theta}{\partial X} = -Bi\,\Theta \qquad (X = 1,\ Fo > 0) \tag{2.79c}$$

The solution to Equation (2.78) can be obtained using the method of separation of variables, where the complete solution is expressed as the product of two separate solutions, i.e. $\Theta(X,Fo) = F(X)\,G(Fo)$, where $F(X)$ is a function solely of X and $G(Fo)$ is a function solely of Fo.

The mechanics of the solution procedure can be found in more comprehensive books than this, such as Bejan (1993), and Incropera and DeWitt (1996). The solution takes the form of an infinite series

$$\Theta = \sum_{n=1}^{\infty} A_n \exp(-p_n^2 Fo) \cos(p_n X) \tag{2.80a}$$

where the coefficients

$$A_n = \frac{4 \sin p_n}{2p_n + \sin 2p_n} \tag{2.80b}$$

And the values of p_n (or eigenvalues) can be obtained from the transcendental equation

$$p_n \tan p_n = Bi \tag{2.80c}$$

The first six roots of Equation (2.80c) are given in Table 2.3, and a listing of a computer program written in BASIC that calculates these roots is presented in Appendix C, named Pn(nr).bas.

Of further interest may be the instantaneous surface heat transfer rate at time t and the total heat transfer from the surface up to time t. The former is obtained from

$$q = -k \left(\frac{\partial T}{\partial x} \right)_{x=L}$$

from which

$$\frac{q}{T_f - T_0} = \frac{k}{L} \sum_{n=1}^{\infty} A_n p_n \sin p_n \exp(-p_n^2 Fo) \tag{2.81}$$

Table 2.3 ● The first six roots of $p_n \tan p_n = Bi$

Bi	p_1	p_2	p_3	p_4	p_5	p_6
0	0	π	2π	3π	4π	5π
0.01	0.0998	3.1448	6.2848	9.4258	12.5672	15.7086
0.1	0.3111	3.1731	6.2991	9.4354	12.5743	15.7143
1	0.8603	3.4256	6.4373	9.5293	12.6453	15.7713
3	1.1925	3.8088	6.7040	9.7240	12.7966	15.8945
10	1.4289	4.3058	7.2281	10.2003	13.2142	16.2594
30	1.5202	4.5615	7.6057	10.6543	13.7085	16.7691
100	1.5552	4.6658	7.7764	10.8871	13.9981	17.1093
∞	$\pi/2$	$3\pi/2$	$5\pi/2$	$7\pi/2$	$9\pi/2$	$11\pi/2$

and the latter from

$$\int_0^L \rho C(T - T_f)\,dx$$

and noting that, initially, the heat contained per unit surface area, $\dot{Q}_0$, is simply

$$\dot{Q}_0 = \rho L C(T_0 - T_f)$$

$$\frac{\dot{Q}}{\dot{Q}_0} = \sum_{n=1}^{\infty} \frac{1}{p_n} \frac{\sin^2 p_n \exp(-p_n^2 Fo)}{p_n + \sin p_n \cos p_n} \tag{2.82}$$

A listing of a computer program which solves Equations (2.80), (2.81) and (2.82) is given in Appendix C, named onedxt.bas.

The strategy for obtaining solutions to other geometries such as the sphere and cylinder is similar, and the reader is referred to Carslaw and Jaeger (1980) and Wong (1977). Solutions are also available in graphical form, called Heisler charts (temperature against Fourier number for various values of inverse Biot number) and Gröber charts (temperature against Biot number for various values of Fourier number). Although the common use of powerful desktop computers has replaced these charts, one for the plane wall is included in Figure 2.14 for reference.

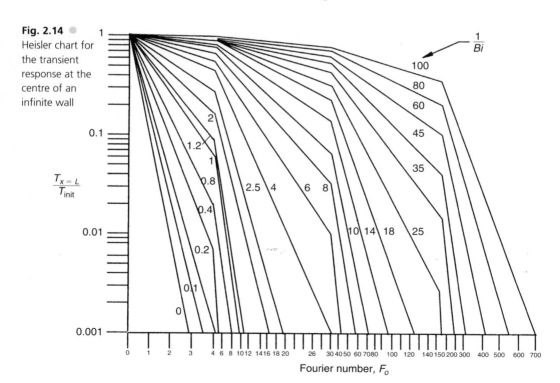

Fig. 2.14 Heisler chart for the transient response at the centre of an infinite wall

Example 2.8

In the heat treatment of a steel alloy, a plate 30 mm thick and initially at 600°C is immersed in a liquid at 20°C where the heat transfer coefficient is 3200 W/m²K. Calculate the temperature at the centre of the plate after 50 s (for the steel take $\rho = 7850 \text{ kg/m}^3$, $C = 445 \text{ J/kgK}$ and $k = 16 \text{ W/m K}$).

Solution The temperature distribution with time is given by evaluating the series expression

$$\Theta = \sum_{n=1}^{\infty} A_n \exp(-p_n^2 Fo) \cos(p_n X)$$

Evaluating the relevant quantities:

$$\alpha = \frac{k}{\rho C} = \frac{16}{7850 \times 445} = 4.6 \times 10^{-6} \text{ m}^2/\text{s}$$

$$L = \frac{0.03}{2} = 0.015 \text{ m}$$

$$Fo = \frac{\alpha t}{L^2} = \frac{4.6 \times 10^{-6} \times 50}{2.25 \times 10^{-4}} = 1.0222$$

$$Bi = \frac{hL}{k} = \frac{3200 \times 0.015}{16} = 3$$

From Table 2.3, for $Bi = 3$, Equation (2.80b) for A_n, and since $x = 0$, the cosine term is unity at the centreline:

$p_1 = 1.1925$	$A_1 = 1.2102$	$\exp(-p_1^2 Fo) = 0.233\,7213$
$p_2 = 3.8088$	$A_2 = -0.28815$	$\exp(-p_2^2 Fo) = 4 \times 10^{-7}$
$p_3 = 6.7040$	$A_3 = 0.11545$	$\exp(-p_3^2 Fo) = 1.1 \times 10^{-20}$
$p_4 = 9.7240$	$A_4 = -0.01473$	$\exp(-p_4^2 Fo) = 1.1 \times 10^{-42}$
$p_5 = 12.7966$	$A_5 = 0.03506$	$\exp(-p_5^2 Fo) = 2.1 \times 10^{-73}$
$p_6 = 15.8945$	$A_6 = -0.023107$	$\exp(-p_6^2 Fo) = 0$

Since after $n = 2$ the exponential term is in effect zero, summation of the product of the last two columns is then equivalent to

$$\sum_{n=1}^{2} A_n \exp(-p_n^2 Fo)$$

giving the non-dimensional temperature of

$$(1.2102 \times 0.233\,7213) + (-0.288\,15 \times 4 \times 10^{-7}) = 0.282\,85$$

Since

$$\Theta = \frac{T - T_f}{T_0 - T_f}$$

we have

$$T = 0.282\,85\,(T_0 - T_f) + T_f$$
$$= 0.282\,85(600 - 20) + 20$$
$$= 184.05°C$$

Comment This answer can be verified from the Heisler chart (Figure 2.14), where for $1/Bi = 0.333$ and $Fo = 1.022$, $\Theta \approx 0.3$. Note also how rapidly the series converges: in this example (Fo is of order unity) it is necessary to include only two terms; for $Fo \ll 1$ it may be necessary to include more than this.

2.10.2 Low Biot number or lumped mass approximation

Consider a body having much lower *internal* thermal resistance than at the surface due to the convective heat transfer coefficient, so that $L/k \ll 1/h$. If we examine the temperature distribution within the body, we would expect that at any instant it will be relatively uniform, since the low resistance requires very little potential or temperature difference to convey the flow of heat. The above inequality can be rewritten as $Bi \ll 1$, and under these conditions the temperature distribution within the body can be considered as uniform at any instant in time. The body thus behaves as a lumped mass (sometimes also referred to as lumped capacitance, derived from the capacity to store heat) and the temperature–time history of the body is governed by the external conditions. Strictly we should include radiation as well as convection at the surface. Since the radiative flux depends upon the fourth power of the absolute temperature, this leads to an expression which, although simple to derive, is complicated to solve by analytical methods. If radiation is judged to be significant in relation to convection, then it is quite simple to solve the equation numerically. The low Biot number approximation also provides a simple technique to measure heat transfer rates (from the viewpoint of measurements, radiation can be minimised by using a suitable surface finish). This is discussed further in Chapter 10.

Using the above arguments it is possible to apply a simple equation for conservation of energy for a solid of area A, mass m and specific heat C, exchanging heat by convective heating or cooling to surrounds at temperature T_∞, such as

$$\dot{Q} = mC\frac{\mathrm{d}}{\mathrm{d}t}(T_s - T_\infty) = -hA(T_s - T_\infty) \tag{2.83}$$

For the case of forced convection, where the heat transfer coefficient does not vary with $(T_s - T_\infty)$, Equation (2.83) can be solved for the variation of the temperature ratio θ:

$$\ln \theta = -\lambda h t \quad \text{or} \quad \theta = \exp(-\lambda h t)$$

where

$$\theta = \frac{T_s - T_\infty}{T_{s,i} - T_{\infty,i}} \quad \text{and} \quad \lambda = \frac{A}{mC} \tag{2.84a}$$

and the extra subscript 'i' is used to denote conditions at time $t = 0$. An identical result is obtained by recognising that $Bi = hL/k$; $\alpha = k/\rho C$; $L = $ volume/area and $Fo = \alpha t/L^2$; from which

$$\ln \theta = -Bi \cdot Fo \quad \text{or} \quad \theta = \exp(-Bi \cdot Fo) \tag{2.84b}$$

For the case of free convection (where the heat transfer coefficient does indeed depend on $T_s - T_\infty$), it can be shown that

$$\theta^{-n} = 1 + (nh_i\lambda)t \tag{2.85}$$

where h_i is the value of the heat transfer coefficient at time $t = 0$ and the value of the exponent n is either $\frac{1}{4}$ for laminar free convection or $\frac{1}{3}$ for turbulent free convection (see Chapter 6).

Although derived from consideration of a surface being cooled by the surrounding fluid, Equations (2.84) and (2.85) also apply if the fluid heats the body.

Example 2.9

A titanium alloy blade from an axial compressor for which $k = 25\,\text{W/m K}$, $\rho = 4500\,\text{kg/m}^3$ and $C = 520\,\text{J/kg K}$, is initially at $60°\text{C}$. The effective thickness, from pressure to suction side, is $10\,\text{mm}$, and in a gas stream at $600°\text{C}$ the blade experiences a heat transfer coefficient of $500\,\text{W/m}^2\text{K}$. Use the low Biot number approximation to estimate the temperature of the blade after $t = 1$, 5, 20 and $100\,\text{s}$. Compare this result with the full analy-tical solution, using the program onedxt.bas, listed in Appendix C, if necessary.

Solution   The first step is to check that $Bi \ll 1$:

$$Bi = \frac{hL}{k} = \frac{500 \times 0.005}{25} = 0.1$$

Hence it is possible to use the low Biot number approximation

$$\theta = \exp(-Bi \cdot Fo)$$

Evaluating the relevant quantities:

$$\alpha = \frac{k}{\rho C} = \frac{25}{4500 \times 520} = 10.68 \times 10^{-6}\,\text{m}^2/\text{s}$$

$$L = \frac{0.01}{2} = 0.005\,\text{m}$$

$$Fo = \frac{\alpha t}{L^2} = \frac{10.68 \times 10^{-6} \times t}{25 \times 10^{-6}} = 0.427t$$

Table 2.4
Comparison between low Biot number and analytical solutions

Time (s)	Fo	Temperature (°C)		Error[a] (%)
		Analytical solution	Low Biot number	
1	0.427	73.7	82.6	65
5	2.135	153.8	163.9	11
20	8.54	360.1	370.3	3.4
100	42.7	591.2	592.5	0.24

[a] The error is defined as $\dfrac{T_{low\ Biot} - T_{analytical}}{T_{analytical} - T_\infty}$.

At $t = 1\,\text{s}$, $\theta = \exp(-0.0427) = 0.9582$. Since

$$\theta = \frac{T_s - T_\infty}{T_{s,i} - T_\infty}$$

we have

$$T_s = \theta(T_{s,i} - T_\infty) - T_\infty$$
$$= 0.9582 \times (600 - 60) + 600$$
$$= 82.6°\text{C}$$

The results for $t = 1$, 5, 20 and 100 s are set out in Table 2.4, together with those from the analytical solution; these are obtained from solving (2.80a), (2.80b) and (2.80c), but the series can be truncated after two terms.

Comment   It can be seen that the error associated with the low Biot number approximation improves as the Fourier number increases. To be more specific, for $Fo < 1$ the temperature at $x = 0$ or, in the case of this example, the midplane of the compressor blade, remains unaffected by the propagation of the isotherms from the surface. Clearly there is a significant temperature variation within the blade material for these early times and the lumped mass technique is invalid. We should therefore extend the criterion for the validity of this technique to $Fo > 1$ and $Bi \ll 1$. When $t = 100$ s, or $Fo = 42.7$, the blade is approaching its steady-state temperature.

2.10.3 The semi-infinite solid

The comment at the end of Example 2.9 raises the possibility of the limiting case of a material that is semi-infinite in the direction of the heat flow. That is, the boundary physically at the centreline is considered to be far enough away from the surface, so for the time period being considered, the surface conditions do not penetrate, and the boundary is maintained at the initial temperature T_i, that is $T(\infty,t) = T_i$. This leads to a number of closed-form solutions to Equation (2.77) depending upon which boundary condition ($T = $ constant, $q = $ constant or $h = $ constant) is represented at the surface.

These solutions are listed below and involve two new functions, the Gaussian error function of x written as erf(x), and the complementary error function, erfc(x) defined as

$$\text{erfc}(x) = 1 - \text{erf}(x)$$

where

$$\text{erf}(x) = \frac{2}{\sqrt{\pi}} \int_{u=0}^{x} \exp(-u^2)\, du = \frac{2}{\sqrt{\pi}} \left\{ x - \frac{x^3}{3} + \frac{1}{2!}\frac{x^5}{5} - \frac{1}{3!}\frac{x^7}{7} + \cdots \right\}$$

(2.86)

A simple BASIC program that calculates error functions is listed in Appendix C, named erf.bas.

Constant surface temperature

$$T(0,t) = T_s$$

$$q(0,t) = -k\left(\frac{\partial T}{\partial x}\right)_{x=0} = \frac{k(T_s - T_i)}{\sqrt{(\pi\alpha t)}}$$

$$\frac{T(x,t) - T_s}{T_i - T_s} = \text{erf}\left\{ \frac{x}{(4\alpha t)^{1/2}} \right\}$$

(2.87)

Constant surface heat flux

$$q(0,t) = q_s$$

$$T(x,t) - T_i = \left[\left\{ \frac{q_s}{k}\left(\frac{4\alpha t}{\pi}\right)^{1/2} \right\} \exp\left(\frac{-x^2}{4\alpha t}\right) \right] - \left[\left(\frac{q_s x}{k}\right) \text{erfc}\left\{ \frac{x}{(4\alpha t)^{1/2}} \right\} \right]$$

(2.88)

Constant surface heat transfer coefficient

$$-k\left(\frac{\partial T}{\partial x}\right)_{x=0} = h(T_\infty - T_s)$$

$$\frac{T(x,t) - T_i}{T_\infty - T_i}$$

$$= \text{erf c}\left\{ \frac{x}{(4\alpha t)^{1/2}} \right\} - \left[\exp\left(\frac{hx}{k} + \frac{h^2\alpha t}{k^2}\right) \text{erfc}\left\{ \frac{x}{(4\alpha t)^{1/2}} + \frac{h}{k}(\alpha t)^{1/2} \right\} \right]$$

(2.89)

Example 2.10

Use the semi-infinite solid approximation to estimate the midplane temperature of the compressor blade in Example 2.9.

Solution   From Equation (2.89),

$$\frac{T(x,t) - T_i}{T_\infty - T_i}$$

$$= \text{erfc}\left\{\frac{x}{(4\alpha t)^{1/2}}\right\} - \left[\exp\left(\frac{hx}{k} + \frac{h^2\alpha t}{k^2}\right)\text{erfc}\left\{\frac{x}{(4\alpha t)^{1/2}} + \frac{h}{k}(\alpha t)^{1/2}\right\}\right]$$

After one second and for $x = 5\,\text{mm}$

$$\frac{T(x,t) - T_i}{T_\infty - T_i} = \text{erfc}\left\{\frac{0.005}{(4 \times 10.68 \times 10^{-6})^{1/2}}\right\}$$

$$- \left[\exp\left(\frac{500 \times 0.005}{25} + \frac{500^2 \times 10.68 \times 10^{-6}}{25^2}\right)\right]$$

$$\times \left[\text{erfc}\left\{\frac{0.005}{(4 \times 10.68 \times 10^{-6})^{1/2}} + \frac{500(10.68 \times 10^{-6})^{1/2}}{25}\right\}\right]$$

$$= \text{erfc}\{0.765\} - [\exp(0.1 + 0.004\,27)][\text{erfc}(0.765 + 0.06536)]$$

$$= \text{erfc}\{0.765\} - [\exp(0.104\,27)][\text{erfc}(0.830\,36)]$$

From tables, or using the program erf.bas listed in Appendix C, $\text{erfc}(0.765) = 0.2794$ and $\text{erfc}(0.830\,36) = 0.2404$, hence

$$\frac{T(x,t) - T_i}{T_\infty - T_i} = 0.012\,63$$

From which

$$T = T_i + (T_\infty - T_i)0.012\,63$$

$$= 60 + (600 - 60)0.012\,63$$

$$= 66.8°\text{C}$$

The midplane temperatures evaluated at $t = 5$, 20 and 100 s are

$$T_{t=5s} = 100.7°\text{C}$$

$$T_{t=20s} = 163.8°\text{C}$$

$$T_{t=100s} = 277.4°\text{C}$$

Comment   Comparison with the results given in Table 2.4 shows that the semi-infinite solution is reasonably accurate for $Fo < 1$, but when $Fo > 1$ the boundary condition at the midplane becomes violated and therefore the semi-infinite solution is not valid for $Fo > 1$.

2.11 ● Closing comments

This chapter has introduced the mechanism of heat transfer we refer to as conduction. In the context of engineering applications, this is more likely to be of concern in solids than in fluids (where convection will dominate). Conduction may be treated as either steady-state or transient. A very large value of Fourier number (of the order 100) will imply that a thermal steady state has been attained. Careful consideration of the geometry and boundary conditions is required in order to reduce the complexity of a physical situation to an appropriate conduction model.

Simple analytical solutions are easily derived for one-dimensional steady-state conduction in both Cartesian (plates and walls) and cylindrical (pipes) coordinates. Two-dimensional steady-state analytical solutions are considerably more complex to derive, although a number of texts are available that have catalogued many of these. However, given the availability and cost of modern personal computers, a numerical method will almost certainly be easier.

Fins and extended surface are an important engineering application of a one-dimensional conduction analysis. The design engineer will be concerned with calculating the heat conducted through a fin, the fin effectiveness and efficiency. Some fairly simple relations have been presented for fins where the surface and cross-sectional area are constant.

In this chapter, transient conduction has been treated as one-dimensional. In reality this may not be the case, and although analytical solutions are available for two-dimensional transient conduction, in practice a numerical method is preferred as being both easier and quicker. The analytical solution to one-dimensional transient conduction, can be further simplified depending on the value of the Fourier and Biot numbers. For $Fo > 1$ and $Bi < 1$, the simple lumped mass technique can be used. For $Fo < 1$ the semi-infinite model can be used. However, for $Fo > 1$ and $Bi > 1$, the full analytical solution needs to be used.

A serious limitation on analytical methods is imposed by the difficulties in implementing realistic boundary conditions (for example the heat transfer coefficient may vary along a surface), representing the geometry of a complex component and accounting for variable thermophysical properties. These aspects are considerably easier to handle using numerical methods, which are introduced in the following chapter. Nonetheless, analytical solutions do have their place, and results from a numerical solution should always be compared with those from an appropriate analytical solution – to detect errors in the computer code and as a check on the overall accuracy.

2.12 ● References

Bejan, A. (1993). *Heat Transfer*. New York: Wiley
Carslaw H.S. and Jaeger J.C. (1980). *Conduction of Heat in Solids*. Oxford: Oxford University Press

Fried, E. (1969). Thermal conduction contribution to heat transfer at contacts. In *Thermal Conductivity* (Tyre, R.P., ed.) Vol. 2. London: Academic Press

Gardner, K. (1945). Efficiency of extended surfaces. *Trans. ASME*, **67**

Incropera, F.P. and DeWitt, D.P. (1996). *Fundamentals of Heat and Mass Transfer* 4th edn. New York: Wiley

Kreith, F. and Bohn, M.S. (1993). *Principles of Heat Transfer* 5th edn. St Paul, MN: West Publishing

Mikhailov, M.D. and Özisik, M.N. (1994). *Unified Analysis and Solutions of Heat and Mass Diffusion*. New York: Dover Publications

Wong, H.Y. (1977) *Handbook of Essential Formulae and Data on Heat Transfer for Engineers*. London: Longman

2.13 ● End of chapter questions

2.1 Using an appropriate control volume, show that the time-dependent conduction equation in cylindrical coordinates for a material with constant thermal conductivity, density and specific heat is given by

$$\frac{\partial^2 T}{\partial r^2} + \frac{1}{r}\frac{\partial T}{\partial r} + \frac{\partial^2 T}{\partial z^2} = \frac{1}{\alpha}\frac{\partial T}{\partial t}$$

2.2 An industrial freezer is designed to operate with an internal air temperature of $-20°C$ when the external air temperature is $25°C$; the internal and external heat transfer coefficients are $12\,W/m^2K$ and $8\,W/m^2K$, respectively. The walls of the freezer are of composite construction, comprising an inner layer of plastic ($k = 1\,W/m\,K$ and thickness 3 mm) and an outer layer of stainless steel ($k = 16\,W/m\,K$ and thickness 1 mm). Sandwiched between these two layers is a layer of insulation material with $k = 0.07\,W/m\,K$. Find the width of insulation that is required to reduce the convective heat loss to $15\,W/m^2$.

[195 mm]

2.3 A pipe of 60 mm dia. carries oil at $230°C$ where the heat transfer coefficient is $250\,W/m^2K$. The pipe is to be insulated using a material of $k = 0.06\,W/m\,K$. On the outer surface of the insulation is a plastic coating of 1 mm radial thickness and with $k = 1.6\,W/m\,K$. Calculate the radial thickness of insulation which will reduce the outside temperature of the coating to $50°C$, when the ambient air temperature is $20°C$ and the convective and radiative heat transfer coefficients on the outside are 6 and $8\,W/m^2K$, respectively (ignore the thermal resistance of the pipe). Calculate the percentage reduction in the heat loss with this thickness of insulation.

[20 mm, 75%]

2.4 A 3 mm thick, $k = 2\,W/m\,K$ glass cover for an automobile headlamp has a thin electrical heater embedded 1 mm from the front surface. It is required to maintain a front surface temperature of $5°C$ when the outside air temperature is $-20°C$, the air temperature inside the headlamp shell is $20°C$ and the outside and inside convective heat transfer coefficients are $100\,W/m^2K$ and $10\,W/m^2K$, respectively. Calculate the electrical power required by the heater for a surface area of $0.06\,m^2$.

[158 W]

2.5 By considering an air temperature probe as a solid cylindrical fin (or spine) of thermal conductivity k, diameter D and length L, show that the temperature difference between the tip and surrounding fluid will be 0.5% or less than the temperature difference between the base of the probe and the surrounding fluid when

$$\left(\frac{hL^2}{kD}\right)^{1/2} \geqslant 2.65$$

where h is the surrounding convective heat transfer coefficient. Calculate the temperature recorded by a probe of length $L = 20\,mm$, $k = 19\,W/m\,K$, $D = 3\,mm$, when there is an external heat transfer coefficient of $h = 50\,W/m^2K$, an actual air temperature of $50°C$ and the surface temperature at the base of the probe is $60°C$.

[55.39°C]

2.6 An electric motor casing has a diameter of 0.36 m and length of 0.4 m. The casing is made from cast steel ($k = 60$ W/m K) and a number of fins are required which will dissipate 400 W of heat into surrounding air where the heat transfer coefficient is 10 W/m^{2}K. Each fin is to run the entire length of the motor casing, be 8 mm thick and have a radial length of 10 mm. Determine the number of fins required to maintain a temperature difference between the casing surface and the surrounding air of 30°C. Comment on the proposed design.

[108]

2.7 A plate is bounded by the lines $x = 0$, $y = 0$, $x = a$ and $y = b$. The three sides $x = 0$, $0 \leqslant y \leqslant b$; $y = 0$, $0 \leqslant x \leqslant a$ and $x = a$; $0 \leqslant y \leqslant b$ are maintained at a constant temperature of T_o. The temperature distribution along the side $y = b$, $0 \leqslant x \leqslant a$ is given by the expression

$$T = T_o \sin\left(\frac{\pi x}{a}\right)$$

Show that the two-dimensional steady-state temperature distribution in the plate is given by

$$T = T_o + \left\{ \sin\left(\frac{\pi x}{a}\right) \sinh\left(\frac{\pi y}{a}\right) \bigg/ \sinh\left(\frac{\pi b}{a}\right) \right\}$$

2.8 The expression for the critical insulation radius (2.36) was derived assuming the external heat transfer coefficient is independent of the insulation radius. Derive an equivalent expression for the general case of $h = Cr^n$, where C and n are constants and r is the insulation radius. Continue this to show that the ratio of the critical insulation radius for variable and constant heat transfer coefficients is given by

$$\frac{r_{crit,var}}{r_{crit,const}} = \left(\frac{h_o r_1}{k_{ins}}\right)^{n/(n+1)} (n+1)^{1/(n+1)}$$

where h_o is the external heat transfer coefficient for the uninsulated pipe of radius r_1 and k_{ins} is the thermal conductivity of the insulation.

2.9 In free convection the heat transfer coefficient varies with the surface-to-fluid temperature difference $(T_s - T_f)$. Using the low Biot number approximation and, assuming this variation to be of the form $h = C(T_s - T_f)^n$, where C and n are constants, show that the variation of the

dimensionless temperature ratio with time will be given by

$$\theta^{-n} = 1 + (nh_{init}\lambda)t$$

where

$$\theta = \frac{T - T_f}{T_{init} - T_f}$$

$$\lambda = \frac{\text{Area}}{(\text{Mass} \times \text{Specific heat capacity})}$$

h_{init} = the heat transfer coefficient at $t = 0$

Use the above equation to determine the time taken for an aluminium ($\rho = 2750$ kg/m^3 $C = 870$ J/kg K) motorcycle fin of effective area 0.04 m^2 and thickness 2 mm to cool from 120°C to 40°C in surrounding air at 20°C when the initial external heat transfer coefficient due to laminar free convection is 16 W/m^{2}K. Compare this with the time estimated from Equation (2.84b), which assumes a constant value of heat transfer coefficient.

[590 s and 479 s]

2.10 In the manufacture of a composite plastic component, a 3 mm thick layer of glass reinforced plastic (GRP) is to be bonded to a much thicker layer of structural foam ($k = 0.04$ W/m K) which, for the purpose of this question, may be considered a perfect insulator. To achieve an effective bond, the surface of the GRP is heated using an air jet, where the temperature of the jet is 100°C and the convective heat transfer coefficient is 1000 W/m^{2}K. Taking the following properties of GRP: $\rho = 1050$ kg/m^3, $C = 2000$ J/kg K and $k = 0.4$ W/m K, and assuming an initial temperature of 20°C, use the full analytical solution (2.80) to calculate the surface temperature and the temperature at the joint between GRP and the foam after 50 s. Compare this answer with the low Biot number approximation and give reasons for the difference.

[97.62°C and 86.92°C (analytical solution), 99.97°C (low Biot number)]

2.11 The composite component in Question 2.10 is allowed to cool by laminar free convection in air at 20°C, assuming an initial uniform temperature of 92°C and an initial value of the heat transfer coefficient $h_{init} = 10$ W/m^{2}K. Use the relation developed in Question 2.9 to estimate the time to cool to 30°C. Compare this with the answer obtained assuming a constant value of heat transfer coefficient of 10 W/m²K.

[1608 s and 1244 s]

2.12 A design of an apartment block at a ski resort requires a balcony projecting from each of the 350 separate apartments. The walls of the building are 0.3 m wide and made from a material with $k = 1$ W/m K. Use the fin approximation to examine the implications on heat transfer for two proposed designs and for the option of doing nothing. In each of the first two cases, the balcony projects 2 m from the building and has a length (parallel to the wall) of 4 m. Assume an inside air temperature of 20°C and an outside air temperature of −5°C; the overall (convective + radiative) heat transfer coefficient on the inside of the building is 8 W/m²K and that on the outside of the building 20 W/m²K.

(a) A balcony constructed from solid concrete and built into the wall, 0.2 m thick, $k = 2$ W/m K

(b) A balcony suspended from 3 steel beams, $k = 40$ W/m K, built into the wall and projecting out by 2 m, each of effective cross-sectional area $A_c = 0.01$ m², perimeter $P = 0.6$ m (the actual floor of the balcony in this case may be considered to be insulated from the wall).

(c) No balcony.

[(a) $q = 77.2$ W/m², $\dot{Q} = 6891$ W;
(b) $q = 182$ W/m², $\dot{Q} = 1915$ W;
(c) $q = 52.6$ W/m²]

Chapter three

Conduction – Numerical methods and applications

3.1 ● Introduction

The analytical method introduced in the previous chapter yields solutions that are exact at any point in the conducting material or domain. But as seen, even for quite simple two-dimensional conduction, with idealised boundary conditions, the analytical method leads to a solution which itself is by no means simple to evaluate. For engineers, the value of the analytical method is threefold:

1. It informs us of the important governing parameters (for example, the Biot number Bi, and the Fourier number Fo).

2. It provides an overall view of the behaviour of a thermal model (for example, the exponential variation of temperature with time).

3. Since the answers are exact, they can be used as part of the validation process of a computer code based on numerical methods.

In the context of engineering, the components we design are often not simple idealised shapes such as plane walls and infinite cylinders, and there is usually no suitable one- or two-dimensional representation. The actual boundary conditions tend also to be complicated. For example, the heat transfer coefficient may vary along the surface, as may the surface temperature and local air temperature; the initial conditions may vary throughout the material; there may be a periodic variation in the heat flux, and so on.

Numerical techniques are an expedient to this. A numerical solution gives approximate answers at 'grid points'. The starting point is from the appropriate governing differential equation. We then invoke a particular discretisation process (a Taylor series expansion, conservation within a control volume or minimising a function) to transform this into a set of *algebraic* equations. It is the solution of these equations which provides the temperature field at specific points on the imaginary grid within the material.

Numerical techniques were pioneered in the 1940s, and prior to the introduction of digital computers, large hall-like offices were staffed by hundreds of employees, calculating solutions to these algebraic equations by hand (earning the equivalent of 50 pence a grid point). Since then, mechanical and electronic calculators have eased the task somewhat, and in 1999 a desktop personal computer has the processing power to carry out a three-dimensional time-dependent conduction analysis in a matter of minutes.

Numerical methods, then, are ideal for use on a modern computer. It is not at all difficult to write one's own code in any of the common languages. Most personal computers are supplied with the programming language BASIC, whereas FORTRAN is usually available on workstations and larger computers. In fact, writing one's own numerical program can be an easier task than writing a program to compute the analytical solution; and as we have seen, this also involves some degree of approximation, for example in where we decide to truncate the infinite series or in the level of accuracy required in calculating the roots of a transcendental equation. Sections C.5 and C.6 list two BASIC programs associated with this chapter.

It is also possible to purchase software to carry out conduction modelling (often called diffusion problems). This will frequently have a 'front-end' to set up the geometry, material specifications, grid and boundary conditions, and a graphical post-processor for viewing the output results as contour plots and time animations. The component of software that performs the calculations may be either finite element or finite difference based. The essential principles of the numerical method are far easier to understand using the finite difference method, and these are used for the discussion in this chapter. Finite element methods (which grew from applications in structural and stress analysis) lend themselves to irregular geometries. The book by Owen and Hinton (1980) is an excellent introductory text to finite element methods.

The finite difference approximation to a conducting region or domain may be derived using either a Taylor series expansion of the derivatives $\partial^2 T / \partial x^2$, $\partial T / \partial t$, and so on, or a control volume formulation. For the same thermal model with the same grid they yield identical equations, but the control volume approach is easier to apply where the domain is a composite of different material properties. In order to illustrate the principles, this chapter concentrates solely on the approach that uses the Taylor series expansion; the reader is referred to Patankar (1980) for a thorough treatment of the control volume approach.

Figure 3.1 shows a graph of hypothetical temperature distribution in a material represented by the function $T = T_{(x)}$. Using a Taylor series expansion of the value of T at the point $x + \delta x$, written as $T_{(x+\delta x)}$:

$$T_{(x+\delta x)} = T_{(x)} + \frac{\mathrm{d}T}{\mathrm{d}x}\delta x + \frac{1}{2!}\frac{\mathrm{d}^2 T}{\mathrm{d}x^2}\delta x^2 + \frac{1}{3!}\frac{\mathrm{d}^3 T}{\mathrm{d}x^3}\delta x^3 + \frac{1}{4!}\frac{\mathrm{d}^4 T}{\mathrm{d}x^4}\delta x^4 + \dots$$

$$(3.1)$$

Fig. 3.1 ●
The continuous
function $T = T_{(x)}$

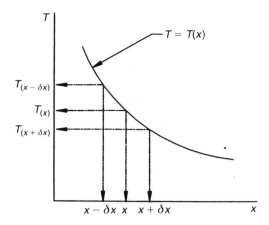

Similarly, at the point $x - \delta x$,

$$T_{(x-\delta x)} = T_{(x)} - \frac{\mathrm{d}T}{\mathrm{d}x}\delta x + \frac{1}{2!}\frac{\mathrm{d}^2 T}{\mathrm{d}x^2}\delta x^2 - \frac{1}{3!}\frac{\mathrm{d}^3 T}{\mathrm{d}x^3}\delta x^3 + \frac{1}{4!}\frac{\mathrm{d}^4 T}{\mathrm{d}x^4}\delta x^4 + \cdots$$

(3.2)

By adding Equations (3.1) and (3.2)

$$T_{(x+\delta x)} + T_{(x-\delta x)} = 2T_{(x)} + \frac{\mathrm{d}^2 T}{\mathrm{d}x^2}\delta x^2 + \text{(terms involving } \delta x^4 \text{ and}$$
$$\text{higher-order powers of } \delta x)$$

(3.3)

For most problems in conduction, it is sufficient to ignore terms containing powers of δx^4 and higher-order terms, and by rearranging Equation (3.3), we obtain the *finite difference approximation to the second derivative*:

$$\frac{\mathrm{d}^2 T}{\mathrm{d}x^2} \approx \frac{T_{(x+\delta x)} - 2T_{(x)} + T_{(x-\delta x)}}{\delta x^2}$$

(3.4)

By subtracting Equation (3.2) from equation (3.1) we get what is known as the central difference approximation to the first derivative:

$$\frac{\mathrm{d}T}{\mathrm{d}x} \approx \frac{T_{(x+\delta x)} - T_{(x-\delta x)}}{2\delta x}$$

(3.5)

Since Equations (3.4) and (3.5) retain powers of δx^2 (even though in Equation (3.5) terms involving δx^2 are absent because they cancel when Equation (3.2) is subtracted from (3.1)), they are said to be of second-order accuracy. Reducing the grid size (i.e. δx) by half will bring about a fourfold decrease in the *discretisation* error, which is inherent in transforming the continuous differential equation into algebraic expressions which apply only at discrete points.

3.2 ● Finite difference equations for two-dimensional steady-state conduction

Consider the two-dimensional region shown in Figure 3.2. An imaginary grid is laid over the region with m grid lines in the horizontal, x, direction and n in the vertical, y, direction. Correspondingly, there are a total of nm nodes. Let us use the indices, i and j ($i = 1, 2, 3, \ldots, m$ and $j = 1, 2, 3, \ldots, n$) to locate our position on the mesh.

In this two-dimensional representation:

1. The general interior point, i, j, has neighbours $(i + 1, j), (i - 1, j)$, $(i, j + 1)$ and $(i, j - 1)$.

2. The four boundaries are located at $(i = 1, j = 2, 3, \ldots, n - 1)$, $(i = m, j = 2, 3, \ldots, n - 1)$, $(i = 2, 3, \ldots, m - 1, j = 1)$ and $(i = 2, 3, \ldots, m - 1, j = n)$.

3. The corner points are at $(i = 1, j = 1)$, $(i = m, j = 1), (i = 1, j = n)$ and $(i = m, j = n)$.

To derive the appropriate set of discretised equations, we must consider separately the interior points, the boundaries and the corner points.

Fig. 3.2 ●
Finite difference grid for a two-dimensional steady-state conduction analysis with m grid points in the x direction, n grid points in the y direction

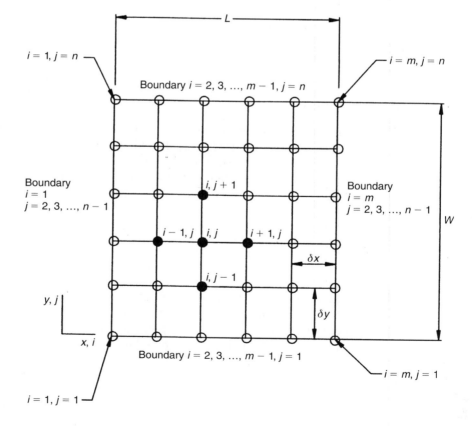

3.2.1 Interior points

For the general interior point at $x = (i\,\delta x)$ and $y = (j\,\delta y)$, in this two-dimensional region, Laplace's equation (2.58) applies, hence

$$\frac{\partial^2 T}{\partial x^2} + \frac{\partial^2 T}{\partial y^2} = 0 \tag{3.6}$$

Using Equation (3.4) for both the second derivatives, this may be *discretised* as

$$\frac{T_{(x+\delta x, y)} - 2T_{(x,y)} + T_{(x-\delta x, y)}}{\delta x^2}$$

$$+ \frac{T_{(x, y+\delta y)} - 2T_{(x,y)} + T_{(x, y-\delta y)}}{\delta y^2} = 0 \tag{3.7}$$

Or, in the i, j notation of the discrete grid,

$$\frac{T_{(i+1, j)} - 2T_{(i, j)} + T_{(i-1, j)}}{\delta x^2}$$

$$+ \frac{T_{(i, j+1)} - 2T_{(i, j)} + T_{(i, j-1)}}{\delta y^2} = 0 \tag{3.8}$$

Collecting together the coefficients for each of the five adjacent points (i, j), $(i+1, j)$,$(i-1, j)$, $(i, j+1)$ and $(i, j-1)$ gives

$$T_{(i-1, j)} + \frac{\delta x^2}{\delta y^2} T_{(i, j-1)} - 2\left(1 + \frac{\delta x^2}{\delta y^2}\right) T_{(i, j)} + \left(\frac{\delta x^2}{\delta y^2}\right) T_{(i, j+1)}$$

$$+ T_{(i+1, j)} = 0 \tag{3.9}$$

This can also be written as $A_w T_w + A_s T_s + A_p T_p + A_n T_n + A_e T_e = 0$, where the coefficients A_w, A_s, A_n and A_e take their names from the points of the compass (west, south, north and east) and denote their relative position to the grid point p. For a two-dimensional Cartesian grid:

$$A_w = A_e = 1$$

$$A_s = A_n = \frac{\delta x^2}{\delta y^2}$$

$$A_p = -2\left(1 + \frac{\delta x^2}{\delta y^2}\right)$$

Equation (3.9) will apply to all the interior mesh points, but not the boundaries and corners where the problem-specific boundary conditions need to be applied and incorporated into the solution. An equivalent equation can be derived for a two-dimensional region in cylindrical polar (r–z) coordinates, and this is worked through in Example 3.1, later in this chapter. It is useful to note the symmetry in the above equation: the central point has the largest coefficient and the coefficients for the vertical grid line are identical, as are those for the horizontal grid line.

3.2.2 Boundary conditions

The classification of boundary conditions is explained in Section 2.3. Boundary conditions of the first kind, where the boundary temperature is specified are readily written down as $T(i, j) = $ (a known value). For boundary conditions of the second and third kinds, the temperature is implicitly defined in terms of a gradient:

$$-k\frac{\partial T}{\partial x}\left(\text{or,} \ -k\frac{\partial T}{\partial y}\right) = (\text{A known value of heat flux})$$

$$-k\frac{\partial T}{\partial x}\left(\text{or,} \ -k\frac{\partial T}{\partial y}\right) = (\text{A known heat transfer coefficient}) \\ \times (\text{Surface-to-fluid temperature difference})$$

Since the algebraic equations used for the interior points are of second-order accuracy, to maintain this accuracy at the boundaries we require a numerical formulation of the gradients $\partial T/\partial x$ and $\partial T/\partial y$ which is also a second-order approximation.

The central difference approximation (3.5) gives the required degree of accuracy. Applying this to the convecting boundary node $(x = 0, \ y = j\delta y)$ in Figure 3.3a, where the boundary condition is

$$-k\frac{\partial T}{\partial x} = h(T_{(i, j)} - T_{\text{amb}})$$

gives the result by evaluating the gradient in the direction of the outward normal.

$$\frac{k}{2\,\delta x}\left\{T_{(i+1, j)} - T_{(i-1, j)}\right\} = h\left\{T_{(i, j)} - T_{\text{amb}}\right\} \tag{3.10}$$

The point $(i - 1, j)$ is referred to as a *fictitious point* since, as it lies outside the boundary of the conducting region, it clearly does not exist. However, we may use this concept to eliminate the fictitious point by rearranging Equation (3.10) as

$$T_{(i-1, j)} = T_{(i+1, j)} - \left[\frac{2\,\delta x\,h}{k}\left\{T_{(i, j)} - T_{\text{amb}}\right\}\right] \tag{3.11}$$

Substituting Equation (3.11) into the general equation (3.9), we obtain the required formulation of this boundary condition in terms of the grid points that physically exist:

$$\frac{\delta x^2}{\delta y^2}T_{(i, j-1)} - 2\left(1 + \frac{\delta x^2}{\delta y^2} + \frac{\delta x h}{k}\right)T_{(i, j)} + \frac{\delta x^2}{\delta y^2}T_{(i, j+1)} + 2T_{(i+1, j)}$$

$$= \frac{-2\,\delta x\,hT_{\text{amb}}}{k} \tag{3.12}$$

A great advantage over analytical techniques is that values of *local heat transfer coefficient* and *local fluid temperature* can be handled with relative simplicity, whereas analytical solutions would require sophisticated mathematical treatment. A boundary condition modelling an insulated, or

Fig. 3.3 ●
Second-order
finite difference
approximations
of boundary
conditions: (a)
heat transfer
coefficient;
(b) heat flux

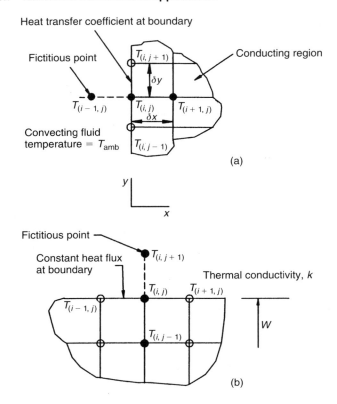

adiabatic surface, $(\partial T/\partial x)_{x=0} = 0$, is obtained by putting $h = 0$ in Equation (3.12).

The foregoing technique is easily applied to a boundary condition of the second kind (constant heat flux), for example to the boundary nodes $y = W$, $x = i\,\delta x$ in Figure 3.3b, where

$$-k\frac{\partial T}{\partial y} = q$$

$$= \frac{-k}{2\,\delta y}\left\{T_{(i,j+1)} - T_{(i,j-1)}\right\}$$

which, after substitution for $T_{(i,j+1)}$ into Equation (3.9), gives the result

$$T_{(i-1,j)} + 2\frac{\delta x^2}{\delta y^2}T_{(i,j-1)} - 2\left(1 + \frac{\delta x^2}{\delta y^2}\right)T_{(i,j)} + T_{(i+1,j)} = \frac{2q\,\delta x^2}{k\,\delta y}$$

(3.13)

The appropriate equation for the corner points is derived in a similar manner except that there will be two fictitious grid points. This is illustrated in Example 3.1.

3.3 ● Application to two-dimensional steady-state conduction problems

The first stage in setting up a numerical solution to a particular conduction problem is the specification of a grid. Although this is quite simple, the reader is encouraged when writing a computer program to ensure that the geometry for the grid is described in terms of the variables of the problem itself. That is, for a region of length L in the x direction, and W in the y direction, and with m and n grid points of grid size δx and δy in the respective x and y directions

$$\delta x = \frac{L}{(m-1)} \quad \text{and} \quad \delta y = \frac{W}{(n-1)}$$

This makes provision for important tests to be carried out on the effect of the grid size on the numerical results, so-called grid dependency tests.

The finite difference equations derived in Section 3.2 for the interior, boundary and corner points are coded into a program; program logic and structure are used to select the correct equation for the grid point under consideration. This can be best illustrated by the following example, which may be solved using the program FDRZ.bas in Appendix C.

Example 3.1

A gas turbine compressor disc is made of an alloy steel with $k = 20\,\text{W/m K}$. The disc is 30 mm thick in the axial direction, has an outer radius of $r_{\text{outer}} = 300\,\text{mm}$ and a bore radius of $r_{\text{inner}} = 150\,\text{mm}$. In a thermal steady state, the faces of the disc (extending from the bore to the outer radius) are cooled by convection into surrounding air at $T_{\text{air}} = 450°\text{C}$, for which at radius r (in metres) the *local* heat transfer coefficient is $h = 133r^{0.6}\,\text{W/m}^2\text{K}$. The outer radius is at a constant temperature of $T_{\text{rim}} = 600°\text{C}$ and the bore can be considered adiabatic. Using a finite difference mesh with four points in the axial direction and four in the radial direction:

(a) Derive appropriate finite difference equations for the interior points.

(b) Derive the finite difference equations for the boundaries and corner points.

(c) Write down these equations in matrix form.

Solution A schematic representation of the physical model and the finite difference grid is illustrated in Figure 3.4. The boundary conditions are symmetrical about the midplane of the disc at $z = 15\,\text{mm}$. Consequently we need consider only half of the disc: $0 \leqslant z \leqslant 15\,\text{mm}$.

(a) The appropriate governing equation is (from Equation 2.12):

$$\frac{\partial^2 T}{\partial r^2} + \frac{1}{r}\frac{\partial T}{\partial r} + \frac{\partial^2 T}{\partial z^2} = 0 \tag{3.14}$$

Fig. 3.4 ●
Thermal model for
Example 3.1:
(a) schematic
diagram of the
compressor disc;
(b) finite difference
grid (4 axial and
4 radial nodes)

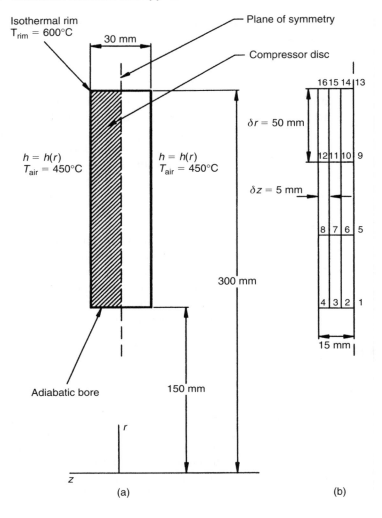

(a)

(b)

Considering a point inside the disc, denoted in general by the indices i, j where the local coordinates are $r = r_{inner} + (j - 1)\,\delta r$ and $z = (i - 1)\,\delta z$, then

$$\frac{\partial^2 T}{\partial z^2} \approx \frac{T_{(i+1,j)} - 2T_{(i,j)} + T_{(i-1,j)}}{\delta z^2} \tag{3.15a}$$

$$\frac{\partial^2 T}{\partial r^2} \approx \frac{T_{(i,j+1)} - 2T_{(i,j)} + T_{(i,j-1)}}{\delta r^2} \tag{3.15b}$$

$$\frac{1}{r}\frac{\partial T}{\partial r} \approx \frac{T_{(i,j+1)} - T_{(i,j-1)}}{2r(i,j)\,\delta r} \tag{3.15c}$$

From Equation (3.14) it is clear that the sum of (3.15a), (3.15b) and (3.15c) is zero. Gathering together the coefficients $T_{(i,j)}$, $T_{(i+1,j)}$, etc.,

Fig. 3.5 ●
Grid point and
finite difference
nomenclature

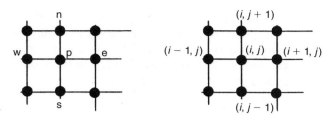

we obtain the general finite difference equation for the interior points:

$$T_{(i,j+1)}\left\{\frac{1}{\delta r^2}+\frac{1}{2r_{(i,j)}\,\delta r}\right\}+T_{(i,j-1)}\left\{\frac{1}{\delta r^2}-\frac{1}{2r_{(i,j)}\,\delta r}\right\}$$

$$+\,T_{(i,j)}\left\{\frac{-2}{\delta r^2}-\frac{2}{\delta z^2}\right\}+T_{(i-1,j)}\left\{\frac{1}{\delta z^2}\right\}+T_{(i+1,j)}\left\{\frac{1}{\delta z^2}\right\}=0$$

(3.16)

Note the use of braces { } to indicate the terms known as *temperature coefficients*. For ease of programming it is convenient to change our notation. Considering Figure 3.5 where the specific point, p, is flanked by point n (for north) above, point s (south) below, e (east) to the right and w (west) on the left side, Equation (3.16) may then be rewritten as

$$A_n T_n + A_s T_s + A_p T_p + A_e T_e + A_w T_w = 0 \qquad (3.17)$$

where
$$A_n = \left\{\frac{1}{\delta r^2}+\frac{1}{2r_{(i,j)}\,\delta r}\right\}$$

$$A_s = \left\{\frac{1}{\delta r^2}-\frac{1}{2r_{(i,j)}\,\delta r}\right\}$$

$$A_p = \left\{\frac{-2}{\delta r^2}-\frac{2}{\delta z^2}\right\}$$

$$A_e = \left\{\frac{1}{\delta z^2}\right\}$$

$$A_w = \left\{\frac{1}{\delta z^2}\right\}$$

(b) Equations for boundary and corner points

(i) *Points 5 and 9 (plane of symmetry)*
The point e in this case is fictitious, but can be eliminated from Equation (3.17) by applying the boundary condition which applies here. This is $-k\partial T/\partial z = 0$, or in finite difference form $(T_w - T_e) = 0$; hence $T_w = T_e$. Substituting this into Equation (3.17), we get

$$A_n T_n + A_s T_s + A_p T_p + (A_w + A_e)T_w = 0 \qquad (3.18)$$

(ii) *Points 8 and 12 (convection to surrounding fluid at T_{air})*

The point w is now the fictitious point which needs to be eliminated from the general equation by applying the boundary condition $-k\partial T/\partial z = h(T - T_{air})$. In finite difference notation this statement becomes

$$\frac{k}{2\,\delta z}(T_e - T_w) = h(T_p - T_{air})$$

$$T_w = T_e - \frac{2\,\delta z\,h}{k}(T_p - T_{air})$$

which, on substituting into Equation (3.17), gives the result:

$$A_n T_n + A_s T_s + \left(A_p - \frac{2A_w\,\delta z\,h}{k}\right)T_p + (A_e + A_w)T_e$$

$$= -\frac{2A_w\,\delta z\,h}{k}\,T_{air} \tag{3.19}$$

(iii) *Points 2 and 3 (insulated or adiabatic boundary condition)*

$$\frac{1}{r}\frac{\partial T}{\partial r} = 0,$$

from which

$$\frac{(T_n - T_s)}{2\,\delta r\,r} = 0 \quad \text{and} \quad T_n = T_s$$

hence the equation at this boundary is

$$(A_n + A_s)T_n + A_p T_p + A_e T_e + A_w T_w = 0 \tag{3.20}$$

(iv) *Points 13, 14, 15 and 16 (fixed value of temperature)*

$$T_p = T_{rim} \tag{3.21}$$

(v) *Corner point, node 1*

At this corner, two boundary conditions apply:

$$\frac{\partial T}{\partial z} = 0 \quad \text{and} \quad \frac{\partial T}{\partial r} = 0$$

making the appropriate substitutions gives the result

$$(A_n + A_s)T_n + A_p T_p + (A_e + A_w)T_w = 0 \tag{3.22}$$

(vi) *Corner point, node 4*

Again we must form an equation that takes account of the two neighbouring boundary conditions $\partial T/\partial r = 0$ and $-k\partial T/\partial z = h(T - T_{air})$. Expressing these in finite difference form and then substituting into Equation (3.17), we obtain

$$(A_n + A_s)T_n + \left(A_p - \frac{2A_w\,\delta z\,h}{k}\right)T_p + (A_e + A_w)T_e$$

$$= -\frac{2A_w\,\delta z\,h}{k}\,T_{air} \tag{3.23}$$

(c) Matrix form of the equations

Recognise that when applied to each interior grid point, together with the boundary and corner points, Equations (3.17) to (3.23) will generate a matrix equation of the form $\mathbf{AT} = \mathbf{B}$. The coefficient matrix $\mathbf{A}$ is made up from the individual values of A_n, A_s, A_p, A_e and A_w applied to each point, and $\mathbf{B}$ is a matrix containing the left side is the equations, which contains temperature-independent boundary condition information. To obtain the numerical values of these matrix elements, we consider each point of the 16 finite difference nodes in turn and write the appropriate equation. Taking for example point 10 in Figure 3.4, an interior node, Equation (3.17)

$$A_n T_n + A_s T_s + A_p T_p + A_e T_e + A_w T_w = 0$$

applies, from which (since the north point is point 14, the south, point 6, the east point 9 and the west, point 11)

$$A_{14} T_{14} + A_6 T_6 + A_{10} T_{10} + A_9 T_9 + A_{11} T_{11} = B_{10}$$

Evaluating these coefficients in accordance with Equation (3.17), with $r = 250 \, \text{mm}$, $\delta r = 50 \, \text{mm}$ and $\delta z = 5 \, \text{mm}$, gives

$$440 T_{14} + 360 T_6 - 80\,800 T_{10} + 40\,000 T_9 + 40\,000 T_{11} = 0$$

The entire system of equations is written out as follows:

$$-80\,800 T_1 + 80\,000 T_2 + 800 T_5 = 0$$
$$(A_3, A_4, A_6, \ldots, A_{16} = 0)$$
$$40\,000 T_1 - 80\,800 T_2 + 40\,000 T_3 + 800 T_6 = 0$$
$$(A_4, A_5, A_7, \ldots, A_{16} = 0)$$
$$40\,000 T_2 - 80\,800 T_3 + 40\,000 T_4 + 800 T_7 = 0$$
$$(A_1, A_5, A_6, A_8, \ldots, A_{16} = 0)$$
$$80\,000 T_3 - 81\,652.19 T_4 + 800 T_8 = -383\,485.4$$
$$(A_1, A_2, A_5, A_6, A_7, A_9, \ldots, A_{16} = 0)$$
$$350 T_1 - 80\,800 T_5 + 80\,000 T_6 + 450 T_9 = 0 \quad \text{(etc.)}$$
$$350 T_2 + 40\,000 T_5 - 80\,800 T_6 + 40\,000 T_7 + 450 T_{10} = 0$$
$$350 T_3 + 40\,000 T_6 - 80\,800 T_7 + 40\,000 T_8 + 450 T_{11} = 0$$
$$350 T_4 + 80\,000 T_7 - 81\,812.74 T_8 + 450 T_{12} = -455\,734.7$$
$$360 T_5 - 80\,800 T_9 + 80\,000 T_{10} + 440 T_{13} = 0$$
$$360 T_6 + 40\,000 T_9 - 80\,800 T_{10} + 40\,000 T_{11} + 440 T_{14} = 0$$
$$360 T_7 + 40\,000 T_{10} - 80\,800 T_{11} + 40\,000 T_{12} + 440 T_{15} = 0$$
$$360 T_8 + 80\,000 T_{11} - 81\,957.84 T_{12} + 440 T_{16} = -521\,024.5$$
$$T_{13} = 600$$
$$T_{14} = 600$$
$$T_{15} = 600$$
$$T_{16} = 600$$

$$(3.24)$$

Comment   A grid of $4 \times 4 = 16$ mesh points was chosen here to illustrate the method of applying finite difference approximations to a partial differential equation. For practical applications (since a second-order method was used where the accuracy could be improved by a factor of 4 by halving δz and δr) there would probably be too few mesh points to yield accurate enough answers, and the discretisation error will be significant. Although an analytical solution would provide the 'true' solution, for these boundary conditions this would not be a simple matter to implement. Consequently, the discretisation error can be examined by carrying out numerical tests on successively finer grids. The temperature obtained at some fixed location is monitored and a graph drawn of this value against grid size, or number of grid points. As δr and δz approach zero, an error will still be present and this is termed the *truncation error*, which is inherent in the Taylor series expansion. In a second-order method the series is truncated after second-order terms; if the truncation error is unacceptable then more grid points are required. If using more grid points places severe restrictions on computing time, then a higher-order discretisation scheme should be used. However, when δr and δz are made small enough, the overall error can actually grow, owing to what is known as *round-off*, which is caused by the representation of numbers on a computer and the consequence of repeated operations on them.

The above 16 simultaneous *algebraic* equations are equivalent to the matrix equation $\mathbf{AT} = \mathbf{B}$. Most of the elements of $\mathbf{A}$ are zero and for clarity these have been omitted. But it is easy to recognise, for example, that $A_{4,3} = 80\,000$, $A_{4,4} = -81\,652.19$, $A_{4,8} = 800$ and $B_4 = -383\,485.4$; the first subscript denotes the row and hence specific mesh point number, the second (not present in the vector $\mathbf{B}$) the column number and hence identity of the neighbouring node. The notation is similar to that of influence coefficients used in structural analysis: $A_{4,8}$, for example, represents the influence of the temperature at node 8 on that at node 4. The solution to this system of equations will provide the values of the unknown steady-state temperature field within the compressor disc (Figure 3.7), and the program FDRZ.bas in Appendix C may be used to solve this problem. This topic is discussed in Section 3.4.

3.4 ● Solution methods for problems in numerical conduction

Practical application of a finite difference solution to acceptable accuracy usually requires a large number of mesh points. Thermal models that simulate conduction within internal combustion engines, for example, will have tens of thousands of points. Such a detailed analysis is made possible only by the relatively inexpensive, very fast and widespread use of computing power. Although it is possible to use manual calculation techniques, this is not recommended today unless, as is often the case, it is necessary for the development of a computer program.

There are two different approaches to solving the system $\mathbf{AT} = \mathbf{B}$: direct methods and indirect or iterative methods. A direct method solves without any iterative sequence – the most common (although not recommended for large matrices) is matrix inversion, where $\mathbf{T} = \mathbf{BA}^{-1}$ and $\mathbf{A}^{-1}$, is the inverse of $\mathbf{A}$ (derived by a sequence of algebraic operations). This facility is commonly available on spreadsheet programs. In terms of computer storage and processor time, a more efficient direct solution of this system of equations is Gaussian elimination. Here an equivalent system of equations is formulated from a series of operations on $\mathbf{A}$, and produces the modified matrix $\mathbf{A}^{\text{mod}}$. The resulting system of modified simultaneous equations now contain a first equation with one unknown, a second with two, and so on. Simple back substitution is used to obtain the solution to the unknown values. This technique is relatively simple to program (Scratton (1984) is a very useful text for this), or software can be purchased, or obtained from newsgroup sites on the Internet.

3.4.1 Iterative techniques

Iterative techniques are very simple to program. The simplest iterative scheme, but the most inefficient in terms of computational time, is to make an initial guess for all the values of temperature and recalculate updated values by cycling through the equations with this initial guess. This sequence is repeated until there is sufficiently small change in the values (or residuals) from successive iterations, and the solution is said to be converged. The numerical value of the small change, or residual, at which the computation is halted is referred to as a *convergence criterion*.

With respect to the set of equations given in Equation (3.24), where for row 1

$$A_{1,1} T_1 + A_{1,2} T_2 + A_{1,3} T_3 = B_1$$

we could write

$$T_1 = \frac{(B_1 - A_{1,2} T_2 - A_{1,3} T_3)}{A_{1,1}} \tag{3.25a}$$

Similarly for mesh points (or rows) 2, 3 and 4:

$$T_2 = \frac{(B_2 - A_{2,1} T_1 - A_{2,3} T_3 - A_{2,6} T_6)}{A_{2,2}} \tag{3.25b}$$

$$T_3 = \frac{(B_3 - A_{3,2} T_2 - A_{3,4} T_4 - A_{3,7} T_7)}{A_{3,3}} \tag{3.25c}$$

$$T_4 = \frac{(B_4 - A_{4,3} T_3 - A_{4,8} T_8)}{A_{4,4}} \tag{3.25d}$$

The extension to the remaining equations should, by now, be obvious.

There are two possible approaches to executing this algorithm, formally known as Jacobi and Gauss–Seidel. Successive overrelaxation (SOR) is a refinement of Gauss–Seidel which dramatically improves the rate of convergence.

Jacobi

We may carry out one whole iteration inserting the values of T obtained from the previous cycle. This method is simple to program:

1. Isolate each variable in turn.
2. New value of each variable $=$ respective function evaluated with old values of all variables.
3. Continue until the difference between new and old values is acceptable.

The drawback is that convergence takes a large number of iterations. In general terms we can express the iterative cycle for the $(n+1)$th estimate of T_i as

$$T_i^{(n+1)} = \frac{B_i - \sum_{j=1}^{n} A_{i,j} \, T_j^{(n)}}{A_{i,i}} \qquad (j \neq i) \tag{3.26a}$$

Gauss–Seidel

If the latest values are used in each equation (i.e. use the result for T_1 from (3.25a) in (3.25b), that for T_2 from (3.25b) in (3.25c), and so on), convergence rates will be improved:

$$T_i^{(n+1)} = \frac{B_i - \sum_{j=1}^{i-1} A_{i,j} \, T_j^{(n+1)} - \sum_{j=i+1}^{n} A_{i,j} \, T_j^{(n)}}{A_{i,i}} \tag{3.26b}$$

By adding and subtracting $T_i^{(n)}$ to the right-hand side of Equation (3.26b) and rearranging, we are able to rewrite the Gauss–Seidel sequence as

$$T_i^{(n+1)} = T_i^{(n)} + R_i \tag{3.26c}$$

where R_i is known as the *residual* or the change in T_i over one iteration.

Successive overrelaxation

Successive overrelaxation can be used to significantly improve the rate of convergence achieved by either Jacobi or Gauss–Seidel. This is done by weighting the influence of the residual to the $(n+1)$th estimate. In practice, this involves the use of an acceleration parameter, Ω, in the form

$$T_i^{(n+1)} = T_i^{(n)} + \Omega R_i$$

The equation for the iterative sequence then becomes

$$T_i^{(n+1)} = (1 - \Omega) T_i^{(n)} + \frac{\Omega}{A_{i,i}} \left\{ B_i - \sum_{j=1}^{i-1} A_{i,j} \, T_j^{(n+1)} - \sum_{j=i+1}^{n} A_{i,j} \, T_j^{(n)} \right\} \tag{3.27}$$

In linear equations such as the conduction equation, the best value for Ω is somewhere in the range $1 < \Omega < 2$.

The path to convergence, iteration by iteration, for node 7 ($r = 200$ mm, $z = 10$ mm), in Example 3.1 is illustrated in Figure 3.6. The initial guess for the entire unknown temperature field T_i, $i = 1, 2, \ldots, 12$, was set at $\frac{1}{2}(T_{rim} + T_{air}) = 525°C$, although it is quite easy to devise more enlightened initial estimates which, for example, could assume a radial variation of temperature. The Jacobi iteration is shown for reference, and even after 200 iterations the results are not converged. The Gauss–Seidel method approaches a converged solution more rapidly but still requires in excess of 200 iterations. The results for $\Omega = 1.7$, 1.8 and 1.9 clearly demonstrate the effectiveness of using this acceleration parameter. Although analytical techniques exist to calculate an optimum value of Ω for a particular problem (Bayley *et al.* (1972) present a formula for calculating the optimum value of Ω), when more complex geometries and boundary conditions are modelled, it is more practical to optimise Ω by trial and examination as in Figure 3.6. And in this case a value of $\Omega \approx 1.8$ appears to achieve optimum convergence.

The successive overrelaxation iterative sequence described above has been applied to the compressor disc example (Example 3.1) and the

Fig. 3.6 ●
Comparison of convergence history for the temperature at node 7 of the compressor disc in Example 3.1

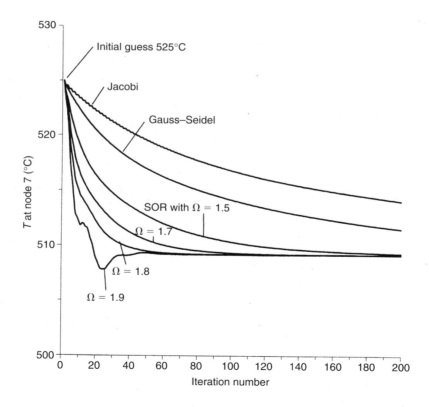

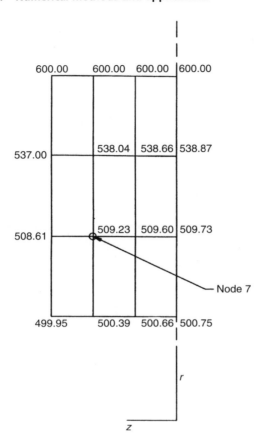

Fig. 3.7 ●
Steady-state
temperature field
(°C) for the disc
in Example 3.1,
converged to
2 d.p.

converged values of steady-state temperature field (in °C) are shown in Figure 3.7.

3.5 ● Numerical methods for transient conduction

For problems in transient conduction, we start with a known set of initial conditions and 'march' forward in time. In addition to discretisation in space, it is now necessary to discretise time, using a 'time step'. Since we use discrete values of the time step, we will obtain solutions of the temperature field at each time step, as the solution proceeds in time from one time step to the next.

The governing partial differential equations for transient conduction were introduced in Chapter 2. For transient conduction, in one spatial dimension these are

$$\frac{\partial^2 T}{\partial x^2} = \frac{1}{\alpha} \frac{\partial T}{\partial t} \qquad \text{(Cartesian coordinates)} \qquad (3.28a)$$

$$\frac{\partial^2 T}{\partial r^2} + \frac{1}{r}\frac{\partial T}{\partial r} = \frac{1}{\alpha}\frac{\partial T}{\partial t} \qquad \text{(cylindrical coordinates)} \qquad (3.28b)$$

Let us denote values of temperature obtained at the current time step, that is at time t, by the superscript (n), and values to be obtained at the next time step (at time $t + \delta t$) by the superscript $(n + 1)$. The temporal derivatives (on the right-hand side) can be expressed using a forward difference approximation as

$$\frac{1}{\alpha}\frac{\partial T}{\partial t} \approx \frac{T_i^{(n+1)} - T_i^{(n)}}{\alpha\,\delta t} \qquad (3.29)$$

The spatial derivatives (on the left-hand side of each equation) can be expressed using a second-order finite difference approximation (3.4) in an identical form to the steady-state equation. For example, in Equation (3.28a), we have

$$\frac{\partial^2 T}{\partial x^2} \approx \frac{T_{(i+1)} - 2T_{(i)} + T_{(i-1)}}{\delta x^2} \qquad (3.30)$$

As was the case for steady-state problems, boundary conditions are required at each physical boundary: a one-dimensional problem requires two boundary conditions; two-dimesional problems, four, and so on. These are dealt with in an identical way to those for the steady state, by invoking the concept of fictitious points (see Section 3.2.2).

Recognise that it is possible to express the derivatives in Equation (3.30) in terms of the temperatures at time t (i.e. 'old' and therefore known values), or $t + \delta t$ (i.e. 'new' and therefore unknown values) or a mixture of both. As will be discussed below, this ultimately depends on available computing resources, the degree of accuracy required and problems with numerical stability of the results.

3.5.1 Explicit solutions

An explicit formulation of the one-dimensional heat conduction equation is obtained from Equations (3.29) and (3.30) with all the temperatures in the spatial derivatives evaluated at (the current) time, t

$$\frac{T_i^{(n+1)} - T_i^{(n)}}{\alpha\,\delta t} = \frac{T_{(i+1)}^{(n)} - 2T_{(i)}^{(n)} + T_{(i-1)}^{(n)}}{\delta x^2} \qquad (3.31)$$

Equation (3.31) can be conveniently rearranged to give the one unknown temperature at the next time step in terms of the known values of temperature at the current time step:

$$T_i^{(n+1)} = \delta Fo \cdot T_{(i+1)}^{(n)} + (1 - 2\,\delta Fo)T_{(i)}^{(n)} + \delta Fo \cdot T_{(i-1)}^{(n)} \qquad (3.32)$$

where δFo ($\delta Fo = \alpha\,\delta t/\delta x^2$) is a Fourier number based on the grid size δx, and time step length δt, or Fourier step length.

Since *all* the temperatures at time step n are known, then the unknown temperature at the grid point i at time $t + \delta t$, can be calculated immediately and without solving a set of simultaneous equations. Equation (3.32) is therefore known as an *explicit scheme*. The advantage of an explicit scheme is its simplicity, leading to easy coding of a computer program and smaller storage requirements. The disadvantage is apparent from inspection of Equation (3.32): if $\delta Fo \geqslant 0.5$ then the central coefficient will be zero or negative. A zero or negative value of this coefficient will introduce instabilities in the solution leading to physically impossible solutions, hence the value of δFo is constrained by stability requirements. A convective boundary condition will require a more stringent condition for stability. The central coefficient is negative when $\delta Fo(1 + \delta Bi) \geqslant 0.5$, where δBi ($\delta Bi = h\, \delta x/k$) is a Biot number based on the grid size δx. In practice we usually require a small enough value of δx to minimise the discretisation error. By implication from the above two stability criteria, this requires a small value of time step. Consequently, for an explicit solution, a large number of time steps will be required to advance to a particular time. In many heat transfer textbooks written ten or more years ago, the reader was often (and at the time quite rightly so) led to believe that the small value of time step required by an explicit method would impose such a severe penalty on computational time as to make this method unworthwhile. To the author's knowledge, this does not appear to be the case today, where desktop computers have many times the computational speed of the mainframes of a decade ago. These factors of simple programming and the fast execution speed of contemporary desktop computers combine to make explicit methods an attractive proposition.

3.5.2 Implicit and semi-implicit solutions

If the spatial derivatives are evaluated using the temperatures at both the current and the subsequent time steps, then there is no longer a simple equation with only one unknown temperature. Instead there will be several unknown temperatures, and applied to each grid point this will lead to a set of simultaneous equations, that can be solved using either an iterative method or matrix algebra (Section 3.4). A general implicit formulation of the transient heat conduction equation is obtained by evaluating some fraction, θ, of the spatial derivatives at the subsequent $(n + 1)$th time step and the remaining $(1 - \theta)$ at the current (nth) time step. The finite difference approximation to Equation (3.28a) then becomes

$$\frac{(T_i^{(n+1)} - T_i^{(n)})}{\alpha\, \delta t} = \theta\, \frac{\left[T_{(i+1)}^{(n+1)} - 2T_{(i)}^{(n+1)} + T_{(i-1)}^{(n+1)} \right]}{\delta x^2}$$

$$+ (1 - \theta)\, \frac{\left[T_{(i+1)}^{(n)} - 2T_{(i)}^{(n)} + T_{(i-1)}^{(n)} \right]}{\delta x^2} \qquad (3.33)$$

After rearranging so that the 'new' (unknown) temperatures are on the left and the 'old' (known) temperatures on the right, we have

$$(\theta\,\delta Fo)T_{(i-1)}^{(n+1)} - (2\theta\,\delta Fo + 1)T_{(i)}^{(n+1)} + (\theta\,\delta Fo)T_{(i+1)}^{(n+1)}$$

$$= (\theta - 1)\,\delta Fo \cdot T_{(i-1)}^{(n)} + \{2(1-\theta)\,\delta Fo - 1\}T_{(i)}^{(n)} + (\theta-1)\,\delta Fo \cdot T_{(i+1)}^{(n)}$$

$$(3.34)$$

Unlike the explicit method, the above implicit finite difference approximation is unconditionally stable for $\theta \geqslant 0.5$. For $\theta = 1$ the solution is known as fully implicit, and although stable, it can give poor accuracy. It is easy to verify that *for $\theta = 0$, the explicit* solution is obtained. A popular scheme known as the *Crank–Nicolson* method is obtained by setting $\theta = 0.5$. This combines the stability of the fully implicit method with the accuracy of the explicit scheme.

Example 3.2

In the oil industry, thermal radiation from gas flames can create a major safety hazard both to personnel and to equipment. As an illustrative example, consider a building which houses staff and control equipment. The walls of the building are built from brick 0.2 m thick, with $\rho = 1920\,\text{kg/m}^3$, $C = 835\,\text{J/kg K}$ and $k = 0.72\,\text{W/m K}$. During periods of routine testing of the well head, the outside surface of one wall will receive a net heat flux of $6\,\text{kW/m}^2$. During these periods the convective heat transfer coefficient on the inside surface will be $10\,\text{W/m}^2\text{K}$ and the air temperature inside the building will be 20°C. For safety reasons it is necessary to limit the inside wall temperature to 65°C. How long will it take for the inside surface temperature to reach this temperature, and is the outside surface temperature at an acceptable level?

The BASIC program FDEXPXT.bas in Appendix C is used in the solving of this example.

Solution Figure 3.8 shows the finite difference grid for five grid points, $\delta x = 0.05\,\text{m}$. At boundary $x = L$ there is a constant heat flux of $6\,\text{kW/m}^2$; at boundary $x = 0$ there is a constant heat transfer coefficient of $10\,\text{W/m}^2\text{K}$. Prior to testing the well head, the wall remains unheated, so a constant initial temperature in the wall of 20°C (the 'initial condition') is assumed as this is equal to the ambient air temperature. To illustrate the advantages and limitations of the explicit and implicit methods, a computation will be carried out for each.

Firstly, we need to derive the appropriate finite difference equations for the interior points $i = 2, 3, 4$, and for the boundaries $i = 1$ and $i = 5$.

Interior points
The general finite difference equation (3.34) is equivalent to

$$A_{\text{w}}T_{\text{w}}^{(n+1)} + A_{\text{p}}T_{\text{p}}^{(n+1)} + A_{\text{e}}T_{\text{e}}^{(n+1)} = C_{\text{w}}T_{\text{w}}^{(n)} + C_{\text{p}}T_{\text{p}}^{(n)} + C_{\text{e}}T_{\text{e}}^{(n)} \qquad (3.35)$$

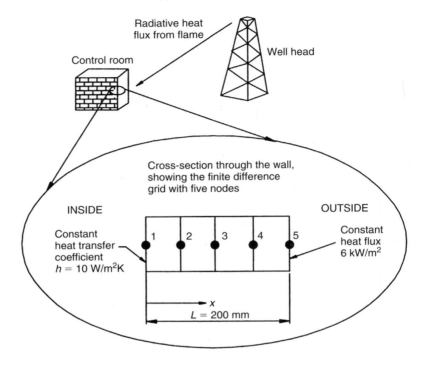

Fig. 3.8 ●
Transient one-
dimensional
conduction
(Example 3.2)

where $\quad A_e = A_w = (\theta\ \delta Fo)$
$A_p = -(2\theta\ \delta Fo + 1)$
$C_e = C_w = (\theta - 1)\delta Fo$
$C_p = \{2(1 - \theta)\ \delta Fo - 1\}$

Boundary x = L
The boundary condition is $q = -k\ (\partial T/\partial x)_{x=L}$, which using a central difference approximation (3.5) gives

$$q \approx \frac{-k}{2\,\delta x}\left(T_{(i-1)} - T_{(i+1)}\right)$$

the point $T_{(i+1)}$ does not physically exist, so rearranging gives

$$T_{(i+1)} = \left(T_{(i-1)} + \frac{2\,\delta x\,q}{k}\right)$$

or

$$T_e = \left(T_w + \frac{2\,\delta x\,q}{k}\right) \tag{3.36}$$

The boundary condition is implemented by substituting Equation (3.36) back into Equation (3.35) for both $T_e^{(n+1)}$ and $T_e^{(n)}$. From which at $x = L$

$$A_p T_p^{(n+1)} + (A_w + A_e) T_w^{(n+1)} = C_p T_p^{(n)} + (C_w + C_e) T_w^{(n)} + \frac{2\,\delta x\, q}{k}(C_e - A_e)$$

(3.37)

where the coefficients A_p, C_p, etc., have the standard one-dimensional Cartesian values given by Equation (3.35).

Boundary x = 0
For the boundary $x = 0$, the boundary condition statement is

$$h(T_p - T_{amb}) = -k\left(\frac{\partial T}{\partial x}\right)_{x=0}$$

which, after rearranging to separate the fictitious temperature T_w, is equivalent to

$$T_w = T_e - \frac{2\,\delta x\, h}{k}\,(T_p - T_{amb})$$

(3.38)

From substituting Equation (3.38) into the general finite difference equation (3.34), the equation to apply at the boundary $x = 0$, is

$$\left(A_p - \frac{2 A_w\, \delta x\, h}{k}\right) T_p^{(n+1)} + (A_w + A_e) T_e^{(n+1)}$$

$$= \left(C_p - \frac{2 C_w\, \delta x\, h}{k}\right) T_p^{(n)} + (C_w + C_e) T_e^{(n)} + \frac{2\,\delta x\, h}{k} T_{amb}(C_w - A_w)$$

(3.39)

Explicit solution
For an explicit solution, $\theta = 0$, and for stability reasons we require $\delta Fo(1 + \delta Bi) < 0.5$. For a grid of five nodes, $\delta x = 0.05\,\text{m}$, along with a convective heat transfer coefficient of $10\,\text{W/m}^2\text{K}$ and thermal conductivity of $k = 0.72\,\text{W/mK}$, this implies that

$$\delta t < \frac{0.5\, \delta x^2}{\alpha(1 + \delta x\, h/k)} = \frac{0.5 \times 0.05 \times 0.05}{4.49 \times 10^{-7} \times 1.695} = 1642\,\text{s}$$

A value of $\delta t = 500\,\text{s}$, should therefore not cause problems with stability.

Using Equations (3.35), (3.37) and (3.39), the five algebraic expressions are

$$T_1^{(n+1)} = C_{1,1} T_1^{(n)} + C_{1,2} T_2^{(n)} + B_1$$
$$T_2^{(n+1)} = C_{2,1} T_1^{(n)} + C_{2,2} T_2^{(n)} + C_{2,3} T_3^{(n)} + B_2$$
$$T_3^{(n+1)} = C_{3,2} T_2^{(n)} + C_{3,3} T_3^{(n)} + C_{3,4} T_4^{(n)} + B_3$$
$$T_4^{(n+1)} = C_{4,3} T_3^{(n)} + C_{4,4} T_4^{(n)} + C_{4,5} T_5^{(n)} + B_4$$
$$T_5^{(n+1)} = C_{5,4} T_4^{(n)} + C_{5,5} T_5^{(n)} + B_5$$

(3.40)

where, for $\delta x = 0.05\,\text{m}$, $h = 10\,\text{W/m}^2\text{K}$, $k = 0.72\,\text{W/m K}$ and $\delta t = 500\,\text{s}$,

$C_{1,1} = 0.695\,6088$

$C_{1,2} = 0.179\,6407$

$C_{2,1} = 8.982\,036E - 02$

$C_{2,2} = 0.820\,3593$

$C_{2,3} = 8.982\,036E - 02$

$C_{3,2} = 8.982\,036E - 02$

$C_{3,3} = 0.820\,3593$

$C_{3,4} = 8.982\,036E - 02$

$C_{4,3} = 8.982\,036E - 02$

$C_{4,4} = 0.820\,3593$

$C_{4,5} = 8.982\,036E - 02$

$C_{5,4} = 0.179\,6407$

$C_{5,5} = 0.820\,3593$

$B_1 = 2.495\,01$

$B_2 = 0$

$B_3 = 0$

$B_4 = 0$

$B_5 = 74.8503$

Using the program FDEXPXT.bas in Appendix C, equation (3.40) is started with $T_1^{(0)} = T_2^{(0)} = T_3^{(0)} = T_4^{(0)} = T_5^{(0)} = 20°\text{C}$, from which the direct solution for the temperatures after one time step of $500\,\text{s}$, gives $T_1^{(1)} = T_2^{(1)} = T_3^{(1)} = T_4^{(1)} = 20°\text{C}$ and $T_5^{(1)} = 94.8503°\text{C}$.

The temperatures after $15\,000$ and $15\,500\,\text{s}$ (after 30 and 31 time steps) span the required limit for the inside surface temperature ($T_1 = 65°\text{C}$).

Solution for time step 30, time $15\,000\,\text{s}$:

at $x = 0$ $T_1 = 62.146\,75°\text{C}$

at $x = 0.05\,\text{m}$ $T_2 = 109.4201°\text{C}$

at $x = 0.1\,\text{m}$ $T_3 = 221.1074°\text{C}$

at $x = 0.15\,\text{m}$ $T_4 = 432.8741°\text{C}$

at $x = 0.2\,\text{m}$ $T_5 = 776.8492°\text{C}$

Solution for time step 31, time $15\,500\,\text{s}$.

$T_1 = 65.381\,13°\text{C}$

$T_2 = 115.2057°\text{C}$

$T_3 = 230.0966°\text{C}$

$T_4 = 444.7491°\text{C}$

$T_5 = 789.9075°\text{C}$

By linear interpolation, the time to reach 65°C is 15 441 s (4 hours 17.5 minutes).

Comment   The outside surface temperature after this time is in excess of 780°C, and this is more likely to be a safety consideration than the inside temperature.

The model described incorporates the effects of thermal radiation emitted from the outside surface of the wall and also the presence of convection on the outside wall, by the collective term 'net' heat flux. In practice this would be a crude approximation, notably because the radiative flux emitted will depend on the surface temperature of the wall. This is an example of a more complicated boundary condition, where the flux at the boundary will actually change from one time step to the next. It is not too difficult to include such detail in a numerical model but this has been omitted here for the sake of focusing on the method itself.

Implicit $\theta = 0.5$, Crank–Nicolson solution
For $\theta = 0.5$, and with δt taken as 3000 s:

$$A_{1,1} = -1.913\,174 \quad C_{1,1} = -8.682\,641E - 02$$
$$A_{1,2} = 0.538\,922 \quad\quad C_{1,2} = -0.538\,9221$$

$$A_{2,1} = 0.269\,461 \quad\quad C_{2,1} = -0.269\,4611$$
$$A_{2,2} = -1.538\,922 \quad C_{2,2} = -0.461\,0779$$
$$A_{2,3} = 0.269\,4611 \quad\; C_{2,3} = -0.269\,4611$$

$$A_{3,2} = 0.269\,4611 \quad\; C_{3,2} = -0.269\,4611$$
$$A_{3,3} = -1.538\,922 \quad C_{3,3} = -0.461\,0779$$
$$A_{3,4} = 0.269\,4611 \quad\; C_{3,4} = -0.269\,4611$$

$$A_{4,3} = 0.269\,4611 \quad\; C_{4,3} = -0.269\,4611$$
$$A_{4,4} = -1.538\,922 \quad C_{4,4} = -0.461\,0779$$
$$A_{4,5} = 0.269\,4611 \quad\; C_{4,5} = -0.269\,4611$$

$$A_{5,4} = 0.538\,9221 \quad\; C_{5,4} = -0.538\,9221$$
$$A_{5,5} = -1.538\,922 \quad C_{5,5} = -0.461\,0779$$

$$B_1 = -14.970\,06$$
$$B_2 = 0$$
$$B_3 = 0$$
$$B_4 = 0$$
$$B_5 = -449.1018$$

This obviously requires an iterative solution, and Gauss–Seidel was used for the results shown here. The calculated temperatures were found to be converged (a change of less than one part in 10^6 between successive iterations of the surface-to-air temperature difference) after 10 iterations at each time step.

Solution for time step 5, time 15 000 s:

at $x = 0$ $T_1 = 62.494\,08°C$

at $x = 0.05\,m$ $T_2 = 109.6583°C$

at $x = 0.1\,m$ $T_3 = 220.3963°C$

at $x = 0.15\,m$ $T_4 = 431.0393°C$

at $x = 0.2\,m$ $T_5 = 774.5294°C$

Solution for time step 6, time 18 000 s:

$T_1 = 82.313\,82°C$

$T_2 = 144.7095°C$

$T_3 = 273.01°C$

$T_4 = 500.0872°C$

$T_5 = 849.9617°C$

By interpolation, the time taken for the inside surface temperature to reach 65°C is 15 279 s (4 hours 16.3 minutes).

Comment   There is a small difference (about 1%) between the times calculated using the explicit and the implicit Crank–Nicolson methods. Often during the development of a finite difference code, it is advisable to check the results. One way to do this is to recalculate the boundary conditions. The gradient at the surface $(\partial T/\partial x)_{x=0}$ can be evaluated using a second-order backward difference formula:

$$\frac{\partial T}{\partial x} \approx \frac{3T_i - 4T_{i-1} + T_{i-2}}{2\,\delta x} \tag{3.41}$$

Applying this to the results from the explicit $(\theta = 0)$ solution after $t = 15\,000\,s$ gives

$$q_{(x=0)} = -k\left(\frac{\partial T}{\partial x}\right)_{x=0}$$

$$= -0.72 \times \left\{ \frac{(3 \times 62.146\,75) - (4 \times 109.4201) + (221.1074)}{(2 \times 0.05)} \right\}$$

$$= 216.96\,W/m^2$$

from which the convective heat transfer coefficient,

$$h_{(x=0)} = \frac{q_{(x=0)}}{T_1 - T_{amb}}$$

$$= \frac{216.96}{62.146\,75 - 20}$$

$$= 5.15\,W/m^2K$$

and for the constant heat flux boundary condition at $x = L$,

$$q_{(x=L)} = -k\left(\frac{\partial T}{\partial x}\right)_{x=L}$$

$$= -0.72 \times \left\{\frac{(3 \times 776.8492) - (4 \times 432.8741) + (221.1074)}{(2 \times 0.05)}\right\}$$

$$= 5905 \text{ W/m}^2$$

These values, in particular the heat transfer coefficient, are sufficiently different from those used as the boundary conditions to indicate a problem with insufficient grid size. An explicit solution with 21 grid points ($\delta x = 0.01$ m) gives acceptable resolution. However, because we have reduced δx by a factor of 5, in order to maintain stability we must also decrease the time step. A value of $\delta t = 60$ s is within the stability criterion $\delta Fo(1 + \delta Bi) < 0.5$. The results after 259 time steps, $t = 15\,540$ s, are listed in Table 3.1; and from Equation (3.41) we obtain $h_{(x=0)} = 9.83$ W/m²K and $q_{(x=L)} = 5999$ W/m². These values are significantly closer to the imposed boundary condition values of $h = 10$ W/m²K and $q = 6000$ W/m². Reducing the grid size from $\delta x = 0.05$ m to $\delta x = 0.01$ m has therefore brought about an increase in the accuracy of the solution.

Table 3.1 ●

Grid point	x (m)	Temperature (°C)
1	0	64.984 35
2	0.01	71.964 07
3	0.02	80.616 07
4	0.03	91.160 54
5	0.04	103.8298
6	0.05	118.8682
7	0.06	136.5312
8	0.07	157.084
9	0.08	180.7991
10	0.09	207.9535
11	0.10	238.825
12	0.11	273.6888
13	0.12	312.8126
14	0.13	356.4526
15	0.14	404.8484
16	0.15	458.2188
17	0.16	516.7573
18	0.17	580.6282
19	0.18	649.9633
20	0.19	724.8586
21	0.20	805.3727

3.6 ● Closing comments

This chapter has introduced the finite difference method for dealing with two-dimensional steady-state and one-dimensional transient conduction. The governing partial differential equations and boundary conditions are transformed into a system of algebraic equations. This can be conveniently represented in matrix form and solved using either matrix algebra or iterative methods.

Numerical methods are particularly suitable for engineering applications given the widespread use of modern personal computers and the relatively simple programming task in implementing the algorithms.

The overall accuracy of any numerical method depends on a number of factors:

1. Grid size

2. Truncation error

3. Convergence

4. Numerical stability

Confidence in the results will follow from addressing each of these.

The discussion in this chapter has been intentionally restricted to simple cases in numerical heat conduction, the objective being to illustrate the method without the reader getting lost in the detail inherent in more complex problems involving, for example, non-linear boundary conditions (e.g. radiative heat transfer and free convection boundary conditions), composite regions (where the thermal properties vary), multidimensional transient conduction and change of phase. Derivation of the appropriate finite difference equations, although tedious, is not especially difficult, it being an extension of the principles discussed in this chapter to either more spatial dimensions, a more general form of the governing partial differential equation (i.e. $\partial/\partial x(-k\partial T/\partial x)$ rather than $-k\partial^2 T/\partial x^2$) and so on. The book by Özışık (1994) is recommended for further information. Although finite difference techniques can handle curvilinear boundaries, the derivation of the equations becomes complex and beyond the scope of this text. In such circumstances it is probably worthwhile to purchase software.

3.7 ● References

Bayley, F.J., Owen, J.M. and Turner, A.B. (1972). *Heat Transfer*. London: Thomas Nelson

Owen, D.R.J. and Hinton, E. (1980). *A Simple Guide to Finite Elements*. London: Pineridge Press

Özışık, M.N. (1994). *Finite Difference Methods in Heat Transfer*. Boca Raton, FL: CRC Press

Patankar, S. V. (1980). *Numerical Heat Transfer and Fluid Flow*. Washington, DC: Hemisphere

Scratton, R.E. (1984). *Basic Numerical Methods*. London: Edward Arnold

3.8 ● End of chapter questions

3.1 The lumped mass approximation (Section 2.10.2) can be used to predict the temperature change with time of a body when $Bi \ll 1$. An alloy plate of mass $m = 1$ kg, surface area $A = 0.1 \text{m}^2$ and specific heat capacity $C = 1000$ J/kg K is cooled by convection from an initial temperature of 100°C by air at 0°C where the convective heat transfer coefficient is $h = 100$ W/m²K. Discretise the lumped mass approximation $(dT/dt = -(hA/mC)T)$ and, for $\delta t = 1$ s, compare the results over $1 \leqslant t \leqslant 1000$ s with the analytical solution. Comment on limiting values of δt for (a) accuracy and (b) numerical stability.

Time (s)	T°C (analytical)	T°C (numerical)
1	99.00	99.00
2	98.02	98.01
5	95.13	95.10
10	90.48	90.44
20	81.87	81.79
50	60.65	60.50
100	36.79	36.60
200	13.53	13.39
500	0.673	0.657
1000	0.0045	0.0043

$$\begin{bmatrix} \text{(a)} & 10\text{ s approximate limit for accuracy (1\%} \\ & \text{difference at } t = 20\text{ s)} \\ \text{(b)} & \text{For numerical stability } hA\,\delta t/mC \ll 1 \end{bmatrix}$$

3.2 Spines or cylindrical fins (Section 2.8) are commonly used to augment convective heat transfer from surfaces. The governing equation for the steady-state temperature distribution in a spine of constant cross-sectional area is

$$\frac{d^2T}{dx^2} - m^2(T - T_f) = 0 \qquad \text{where} \qquad m^2 = \frac{4h}{kD}$$

In a particular application, a number of cylindrical steel spines are to be attached to a plate which has a surface temperature of $T_o = 100$°C. Each spine has a diameter of $D = 10$ mm, thermal conductivity $k = 54$ W/m K and length $L = 200$ mm. The surrounding fluid has a temperature of $T_f = 0$°C, the convective heat transfer coefficient is $h = 10$ W/m²K and the tip of the spine may be considered to be insulated.

(a) By discretising the above equation, show that

$$T_i = \frac{(T_{i-1} + T_{i+1} - m^2\delta x^2 T_f)}{(2 + m^2\delta x^2)}$$

(b) Using a grid of $\delta x = 50$ mm show that the general finite difference equation for the interior points $(i = 2, 3, 4)$ is given by

$$T_i = 0.4576(T_{i-1} + T_{i+1})$$

(c) Using a second-order approximation for the boundary condition at the tip $(dT/dx = 0)$ show that

$$T_5 = 0.9152T_4$$

(d) Using initial values of $T_1 = 100$°C, $T_2 = 75$°C, $T_3 = 50$°C, $T_4 = 25$°C and $T_5 = 0$°C, solve the above system of equations using Gauss–Seidel iteration. Compare the results after 10 iterations with the analytical solution given in Equation (2.48):

$$\frac{T}{T_o} = \frac{\cosh m(L - x)}{\cosh mL}$$

x (mm)	0	25	50	75	100
T°C numerical	100	67.55	47.92	37.49	34.31
T°C analytical	100	67.75	48.32	37.88	34.63

(e) Using a second-order backward difference (3.41), calculate the heat transfer through the root of the spine.

[3.3 watts]

3.3 An electronics chip comprises an epoxy-based resin with $k = 0.05$ W/m K. The chip has a cross-section of $W = 10$ mm high and $L = 40$ mm wide. The base $(y = 0, 0 \leqslant x \leqslant L)$ is subjected to a constant external heat flux of $q_{in} = 200$ W/m². The sides $(x = 0, 0 \leqslant y \leqslant W$ and $x = L$, $0 \leqslant y \leqslant W$ are cooled by convection to the surrounding air where the convective heat transfer coefficient is $h_1 = 10$ W/m²K and $T_{air} = 25$°C. The top of the chip is cooled by a convective heat transfer coefficient $h_2 = 50$ W/m²K and $T_{air} = 25$°C. Electronic processes within the chip create heat, q_E (W/m²), which may be considered to take place along a line at $y = 2$ mm and $0 \leqslant x \leqslant L$.

(a) By considering an energy balance on an element of sides δx by δy show that the general equation for the interior points is given by

$$A_n T_n + A_s T_s + A_p T_p + A_w T_w + A_e T_e = -q_E/k\,\delta y \qquad (q_E = 0 \text{ when } y \neq 2\,mm)$$

where: $A_n = A_s = \dfrac{1}{\delta y^2}$

$$A_e = A_w = \dfrac{1}{\delta x^2}$$

$$A_p = -\left(\dfrac{2}{\delta x^2} + \dfrac{2}{\delta y^2}\right)$$

(b) Using second-order approximations for the derivatives, show that the finite difference equations for the boundaries are as follows:

(i) $x = 0$, $0 \leqslant y \leqslant W$

$$A_n T_n + A_s T_s + \left(A_p - \dfrac{2\,\delta x\,h_1 A_w}{k}\right)T_p$$

$$+ (A_w + A_e)T_e$$

$$= -\dfrac{q_E}{k\,\delta y} - \dfrac{2\,\delta x\,h_1 A_w T_{air}}{k}$$

(ii) $x = L/2$, $0 \leqslant y \leqslant W$

$$A_n T_n + A_s T_s + A_p T_p + (A_w + A_e)T_e$$

$$= -\dfrac{q_E}{k\delta y}$$

(iii) $0 \leqslant x \leqslant L/2$, $y = W$

$$(A_n + A_s)T_s + \left(A_p - \dfrac{2\,\delta y\,h_2 A_n}{k}\right)T_p$$

$$+ A_w T_w + A_e T_e = -\dfrac{2\,\delta y\,h_2 A_n T_{air}}{k}$$

(iv) $0 \leqslant x \leqslant L/2$, $y = 0$

$$(A_n + A_s)T_n + A_p T_p + A_w T_w + A_e T_e$$

$$= -\dfrac{2\,\delta y\,q_{in} A_w}{k}$$

And for the corners:

(v) $x = 0$, $y = 0$

$$(A_n + A_s)T_n + \left(A_p - \dfrac{2\,\delta x\,h_1 A_w}{k}\right)T_p$$

$$+ (A_e + A_w)T_e$$

$$= -\left(\dfrac{2\,\delta y\,q_{in} A_s}{k}\right) - \left(\dfrac{2\,\delta x\,h_1 A_w T_{air}}{k}\right)$$

(vi) $x = L/2$, $y = 0$

$$(A_n + A_s)T_n + A_p T_p + (A_e + A_w)T_w$$

$$= -\dfrac{2\,\delta y\,q_{in} A_w}{k}$$

(vii) $x = 0$, $y = W$

$$(A_n + A_s)T_s$$

$$+ \left(A_p - \dfrac{2\,\delta x\,h_1 A_w}{k} - \dfrac{2\,\delta y\,h_2 A_n}{k}\right)T_p$$

$$+ (A_e + A_w)T_e$$

$$= -\left(\dfrac{2\,\delta y\,q_{in} A_s}{k}\right) - \left(\dfrac{2\,\delta x\,h_1 A_w T_{air}}{k}\right)$$

$$- \left(\dfrac{2\,\delta y\,h_2 A_n T_{air}}{k}\right)$$

(viii) $x = L/2$, $y = W$

$$(A_n + A_s)T_s + \left(A_p - \dfrac{2\,\delta y h_2 A_n}{k}\right)T_p$$

$$+ (A_e + A_w)T_w = -\dfrac{2\,\delta y\,h_2 A_n T_{air}}{k}$$

(c) Using a grid of 11 nodes in the x and y directions ($\delta x = 2$ mm, $\delta y = 1$ mm), calculate the steady-state temperature distribution when $q_E = 300$ W/m^2.

3.4 An automobile disc brake is 8 mm thick and 200 mm in diameter. It is made from stainless steel with $\rho = 8055$ kg/m^3, $C = 480$ J/kg K and $k = 15.1$ W/m K. During braking from 35 m/s the average heat flux into each side of the disc brake is 0.5 MW/m^2. Using an appropriate transient finite difference scheme, calculate the temperature distribution in the disc brake after 10 s, assuming an initial temperature of 30°C and convection from both sides to surrounding air where the convective heat transfer coefficient is 500 W/m^{2}K and an ambient air temperature of 20°C.

x (mm)	T °C
0	278.2
0.4	278.7
0.8	280.1
1.2	282.4
1.6	285.7
2.0	289.9
2.4	295.1
2.8	301.1
3.2	308.1
3.6	316.0
4.0	324.7

3.5 The steady-state temperature distribution in a material with $k = 20$ W/m K and dimensions $0 \leqslant x \leqslant 0.3$ m, $0 \leqslant y \leqslant 0.4$ m obeys Laplace's equation. The three sides $x = 0, 0 \leqslant y \leqslant 0.4$ m; $0 \leqslant x \leqslant 0.3$ m, $y = 0$ and $0 \leqslant x \leqslant 0.3$ m, $y = 0.4$ m are maintained at a constant temperature of 100°C. The side $x = 0.3$ m, $0 \leqslant y \leqslant 0.4$ m exchanges heat by convection to an adjacent fluid where the convective heat transfer coefficient is 300 W/m^2 K and $T_{fluid} = 10$°C.

(a) Taking $\delta x = 0.1$ m and $\delta y = 0.2$ m, and using a grid numbering system where node 1 is located at $x = 0, y = 0$, node 2 at $x = 0$, $y = 0.2$ m, etc., show that the entire system of equations allowing the solution of all the unknown temperatures is

$$T_8 - 2.5T_5 + 150 = 0$$
$$T_5 - 2.5T_8 + T_{11} + 50 = 0$$
$$2T_8 - 5.5T_{11} + 80 = 0$$

(b) Solve for T_5, T_8 and T_{11}.

[88.69°C, 71.73°C, 40.63°C]

(c) Calculate the heat flux at node 11.

[9189 W/m^2]

Chapter four

Principles of convection

4.1 ● Introduction

Convection is the mode of heat transfer that occurs in a fluid (gas or liquid) owing to motion of the fluid itself. If this motion is created by external means, such as a pump, a fan, the wind or by relative motion of a solid surface in a stationary fluid, then it is referred to as *forced convection*. Fluid motion can also be generated by temperature differences which give rise to density variations (for example, air adjacent to a vertical heated surface will be heated, its density will decrease and as a result the air will rise). This is referred to as *free* or *natural convection*. Usually, convective heat transfer is dominated by one or the other of these two modes. When both free and forced convection effects are significant, then the term *mixed convection* is used.

Further distinctions can be made. *External convection* occurs when the fluid surrounds the surface (as, for example, in the heat transfer resulting from flow over a flat or curved surface, the external surface of a turbine blade, or a car disc brake), *internal convection* occurs when the fluid is surrounded by the surface (as, for example, in a pipe conveying steam, or the water-filled cooling passages of an internal combustion engine). Convective flows can also be described as laminar, turbulent or transitory; forced and free convection have separate criteria for delineating these regimes. The above terms are generally used together, so we would refer to the flow over a fin on a motorcycle engine at high speed as turbulent, external, forced convection (probably laminar at low speeds, and free convection when stationary). Yet another distinction can be made according to the nature of the flow: single-phase, boiling, condensing, melting or undergoing solidification. In order to convey fundamental principles without obscuring the discussion with too many complexities, much of the discussion in this book is concerned with single-phase flows. Heat transfer during a change of phase is a complex area, a discussion of the principal features of convective heat transfer during the processes of condensation and boiling is presented in Chapter 7. The reader is referred to Lock (1996) for a more extensive treatment and also coverage of both solidification and melting.

The fundamental question in convective heat transfer is to determine the surface to fluid heat transfer rate characterised by the surface *heat transfer coefficient*, *h*, defined as

$$h = \frac{q}{(T_s - T_{ref})} \tag{4.1}$$

where q is the heat flux, T_s the temperature of the surface and T_{ref} an appropriate fluid reference temperature. An axiom in the science of convective heat transfer is a statement referred to as the *no-slip condition*. A simple demonstration of this is to watch honey run off a spoon: a thin layer always remains in contact with the surface. This phenomenon occurs in the flow of air over an aircraft wing, the flow of water over the hull of a ship, in fact wherever there is relative motion between a fluid and a surface. The explanation is that a thin (molecular thickness) layer of fluid actually adheres to the surface. This simple observation has two important consequences. One is the development of boundary layer theory, which is addressed in Section 4.3. The other is that, at the surface $y = 0$, the heat flux, q, is transferred by process of conduction so using Fourier's law in Equation (4.1), we obtain

$$h = \frac{-k(\partial T / \partial y)_{y=0}}{(T_s - T_{ref})} \tag{4.2}$$

where k is the thermal conductivity of the fluid.

Equation (4.2) demonstrates that a knowledge of the temperature distribution $(\partial T / \partial y)$ in the fluid is a prerequisite to a mathematical derivation of the heat transfer coefficient. As will be shown in the following sections, the temperature distribution in a fluid is linked to its velocity distribution, which in turn is influenced by other factors such as surface geometry, pressure gradients and the presence of turbulence. Consequently, a mathematical description of convective heat transfer must be underwritten by the equations of fluid flow (the continuity and momentum equations) and the energy equation.

4.2 ● Continuity, momentum and energy equations

For clarity, and since in this book we will not discuss three-dimensional flows, the following equations are derived for a two-dimensional flow in a Cartesian coordinate system, having velocity components u in the x direction and v in the y direction. (The extension to three dimensions is relatively straightforward; it is conventional to adopt the symbol w to denote the velocity component in the z direction.)

4.2.1 Mass continuity equation

Consider the control volume shown in Figure 4.1, of dimensions Δx, Δy and unit depth in the z direction (into the paper), within a two-dimensional flow

Fig. 4.1 ●
Control volume for
mass continuity in
two-dimensional
flow

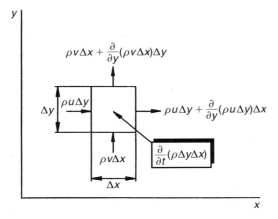

field. Over a period of time Δt, the increase in the mass of the control volume is equal to the difference between the mass flow rate entering it and that leaving it. For a fluid of density ρ, this statement can be written as:

$$\frac{\partial}{\partial t}(\rho \Delta x \Delta y) = \sum (\text{Mass flow rates in}) - \sum (\text{Mass flow rates out})$$

$$= \{(\text{Mass entering at } x) + (\text{Mass entering at } y)\} \qquad (4.3)$$
$$- \{(\text{Mass leaving at } x + \Delta x) + (\text{Mass leaving at } y + \Delta y)\}$$

where mass flow rate in or out is equal to (density) × (velocity) × (area), so

$$\text{Mass flow rate entering at } x = \rho u \Delta y \qquad (4.4a)$$

And using a Taylor series expansion,

$$\text{Mass flow rate leaving at } x + \Delta x = (\rho u \Delta y) + \frac{\partial}{\partial x}(\rho u)\Delta x \Delta y \qquad (4.4b)$$

$$\text{Mass flow rate entering at } y = \rho v \Delta x \qquad (4.4c)$$

$$\text{Mass flow rate leaving at } y + \Delta y = (\rho v \Delta x) + \frac{\partial}{\partial y}(\rho v)\Delta x \Delta y \qquad (4.4d)$$

Using (4.4a) to (4.4d) in (4.3), gives the result

$$\frac{\partial \rho}{\partial t}\Delta x \Delta y = -\frac{\partial}{\partial x}(\rho u)\Delta x \Delta y - \frac{\partial}{\partial y}(\rho v)\Delta x \Delta y$$

where after dividing through by Δx, Δy

$$\frac{\partial \rho}{\partial t} = -\frac{\partial}{\partial x}(\rho u) - \frac{\partial}{\partial y}(\rho v) \qquad (4.5)$$

If the density does not vary with time, i.e. for *steady flow*, then Equation (4.5) simplifies to

$$\frac{\partial}{\partial x}(\rho u) + \frac{\partial}{\partial y}(\rho v) = 0 \qquad (4.6)$$

If the flow can also be treated as *incompressible* (this is not the same as an incompressible fluid), then Equation (4.6) becomes

$$\frac{\partial u}{\partial x} + \frac{\partial v}{\partial y} = 0 \tag{4.7a}$$

Equation (4.5), (4.6) or (4.7a) is of little use by itself. However, as will be seen, it is frequently used to eliminate terms and simplify other equations.

The equivalent expression in cylindrical (r–z) coordinates for steady incompressible flow is

$$\frac{\partial}{\partial r}(rV_r) + \frac{\partial}{\partial z}(rV_z) = 0 \tag{4.7b}$$

where V_r and V_z are the radial and axial velocity components, respectively.

4.2.2 Momentum equations

The momentum equation for steady flow can be derived by considering the momentum changes and forces acting on the control volume of fluid shown in Figure 4.2a (x direction momentum transfer) and Figure 4.2b (x and y direction surface stress). The starting point is Newton's Second Law which, applied in the context of steady two-dimensional flow, requires that the net rate at which momentum leaves the control volume is equal to the sum of all the forces acting upon it.

For a force–momentum balance in the x direction, the rate of mass flow which crosses the surface at x is (density × area × velocity) $= \rho \Delta y u$, so the corresponding momentum flux is $(\rho \Delta y u)u = \rho u^2 \Delta y$. Similarly, the mass flow rate crossing the surface at y is $\rho \Delta x v$, and the momentum flux, since it is 'carried' by velocity, u, is $\rho \Delta x v u$. Using a first-order Taylor series expansion, the momentum flux crossing the surface at $x + \Delta x$ is

$$\rho u^2 \Delta y + \frac{\partial}{\partial x}(\rho u^2)\Delta x \Delta y$$

and the momentum flux at $y + \Delta y$ is

$$\rho u v \Delta x + \frac{\partial}{\partial y}(\rho u v)\Delta x \Delta y$$

So the net rate at which momentum leaves the control volume is

$$\frac{\partial}{\partial x}(\rho u^2)\Delta x \Delta y + \frac{\partial}{\partial y}(\rho v u)\Delta x \Delta y \tag{4.8}$$

It is important to distinguish between the two kinds of force we need consider: *body forces* which act on the entire control volume and *surface forces* which act on surface area. Buoyancy, and magnetic fields both give rise to body forces, which for the x direction we will denote as F_x per unit volume. The static pressure, p, and viscous stresses give rise to surface forces. The surface forces may be resolved into two components: a normal

Fig. 4.2 ●
Control volumes
for the momentum
equation: (a) the
transfer of
momentum in
the x direction
(b) normal and
shear stress

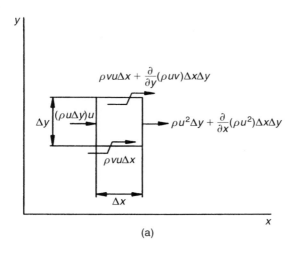

(a)

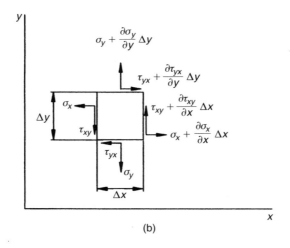

(b)

stress σ and a shear stress τ. It is usual to provide further identification of the shear stress components by using two subscripts, the first denotes the direction of the outward normal, the second the direction of the stress component. So the component τ_{yx} is uniquely identified as a viscous shear stress acting in the x direction, on a plane normal to the y direction (the normal stress components require only one subscript to be unambiguous). Remembering that a force is the product of a stress or pressure multiplied by the appropriate area, and using the Taylor series expansion, the net sum of the x direction forces acting on the control volume is

$$\left(\frac{\partial \tau_{yx}}{\partial y} + \frac{\partial \sigma_x}{\partial x} + F_x\right)\Delta x \Delta y \tag{4.9}$$

Combining Equations (4.8) and (4.9),

$$\frac{\partial}{\partial x}(\rho u^2) + \frac{\partial}{\partial y}(\rho vu) = \frac{\partial \tau_{yx}}{\partial y} + \frac{\partial \sigma_x}{\partial x} + F_x \tag{4.10}$$

The derivatives on the left-hand side of Equation (4.10) can be expanded using the product rule, and with the additional aid of the continuity equation (4.6), we obtain

$$\rho\left\{ u\frac{\partial u}{\partial x} + v\frac{\partial u}{\partial y} \right\} = \frac{\partial \tau_{yx}}{\partial y} + \frac{\partial \sigma_x}{\partial x} + F_x \tag{4.11}$$

To complete the momentum equation we must express the viscous stresses in terms of the basic variables u and v. The reasoning and mathematics that lead to Stoke's hypothesis are somewhat complicated, and the reader is referred to Schlichting (1979) for the details. For the current purpose, it is sufficient to simply quote the results, which for a Newtonian fluid relate the viscous stresses to velocity gradients, with a constant of proportionality called the dynamic viscosity μ:

$$\sigma_x = -p + 2\mu\frac{\partial u}{\partial x} - \frac{2}{3}\mu\left(\frac{\partial u}{\partial x} + \frac{\partial v}{\partial y}\right)$$

$$\sigma_y = -p + 2\mu\frac{\partial v}{\partial y} - \frac{2}{3}\mu\left(\frac{\partial u}{\partial x} + \frac{\partial v}{\partial y}\right) \tag{4.12}$$

$$\tau_{yx} = \tau_{xy} = \mu\left(\frac{\partial u}{\partial y} + \frac{\partial v}{\partial x}\right)$$

It is now a simple matter to substitute Equations (4.12) into Equation (4.11); the result is known as the Navier–Stokes equation. Of particular interest, and considerable usefulness, is the case of a flow where the viscosity can be treated as constant which, after using the mass continuity equation (4.6) to eliminate $\frac{1}{3}\mu\frac{\partial}{\partial x}\left(\frac{\partial u}{\partial x} + \frac{\partial v}{\partial y}\right)$, results in

$$\rho\left\{ u\frac{\partial u}{\partial x} + v\frac{\partial u}{\partial y} \right\} = -\frac{\partial p}{\partial x} + \left[\mu\left\{\frac{\partial^2 u}{\partial x^2} + \frac{\partial^2 u}{\partial y^2}\right\}\right] + F_x \tag{4.13}$$

It is worthwhile to offer brief comment on the physical significance of Equation (4.13). The terms on the left side are often referred to as *inertial terms*, and arise from the momentum changes. These are countered by the pressure gradient, $\partial p/\partial x$, viscous forces (in the square brackets) which always act to retard the flow, and if present, body forces. Similarly, the y direction momentum equation can be derived by considering force–momentum balances in the y direction, which result in

$$\rho\left\{ u\frac{\partial v}{\partial x} + v\frac{\partial v}{\partial y} \right\} = \frac{\partial \tau_{xy}}{\partial x} + \frac{\partial \sigma_y}{\partial y} + F_y \tag{4.14}$$

And after using relations (4.12), and Equation (4.6)

$$\rho\left\{u\frac{\partial v}{\partial x}+v\frac{\partial v}{\partial y}\right\}=-\frac{\partial p}{\partial y}+\left[\mu\left\{\frac{\partial^2 v}{\partial x^2}+\frac{\partial^2 v}{\partial y^2}\right\}\right]+F_y \qquad (4.15)$$

where F_y is the y direction body force. It is worth pointing out the symmetry between Equations (4.13) and (4.15) – the derivatives of the streamwise u velocity are exchanged for those of the transverse v velocity; the derivative of pressure follows the direction of the force balance.

For steady, incompressible, axisymmetric flow in a cylindrical coordinate system, with a body force F_z, the z direction or axial momentum equation is given by

$$\rho\left\{V_r\frac{\partial V_z}{\partial r}+V_z\frac{\partial V_z}{\partial z}\right\}=F_z-\frac{\partial p}{\partial z}+\mu\left\{\frac{\partial^2 V_z}{\partial r^2}+\frac{1}{r}\frac{\partial V_z}{\partial r}+\frac{\partial^2 V_z}{\partial z^2}\right\}$$

Similarly, the radial momentum equation is

$$\rho\left\{V_r\frac{\partial V_r}{\partial r}-\frac{V_\phi^2}{r}+V_z\frac{\partial V_r}{\partial z}\right\}=F_r-\frac{\partial p}{\partial r}+\mu\left\{\frac{\partial^2 V_r}{\partial z^2}+\frac{1}{r}\frac{\partial V_r}{\partial r}-\frac{V_r}{r^2}+\frac{\partial^2 V_r}{\partial z^2}\right\}$$

where V_ϕ is the tangential (or circumferential) velocity component.

4.2.3   Energy equation

The energy equation is derived using the first law of thermodynamics applied to a control volume, of dimensions Δx by Δy (and unit depth into the paper) and mass $\rho\Delta x\Delta y$. To maintain simplicity we will neglect heat transfer across the control volume by radiation, which is usually very small in most fluids, and assume that there are no sources of internal heat generation such as by chemical reaction. The first law of thermodynamics requires that the addition of a quantity of heat $\Delta\dot{Q}$ to the control volume in time Δt will increase its energy by ΔE and also perform work ΔW. This statement can be expressed in differential terms as

$$\frac{\mathrm{d}\dot{Q}}{\mathrm{d}t}=\frac{\mathrm{d}E}{\mathrm{d}t}+\frac{\mathrm{d}W}{\mathrm{d}t} \qquad (4.16)$$

Let us now consider the separate terms in Equation (4.16). The first two are relatively straightforward; the third, the 'work' term, is far more complex and requires additional thermodynamic relations.

Rate of heat flow, $\dfrac{\mathrm{d}\dot{Q}}{\mathrm{d}t}$

By neglecting radiative exchange, we accept that conduction is the only mechanism for heat transfer across the boundaries of the control volume. The net rate of heat flow into control volume in the x direction is

$$\frac{\mathrm{d}\dot{Q}_x}{\mathrm{d}t}\approx(q_{x+\Delta x}-q_x)\Delta y=\frac{\partial}{\partial x}q_x\Delta x\Delta y$$

which, using Fourier's law of conduction (2.1), is equal to

$$\frac{\partial}{\partial x}\left(k\frac{\partial T}{\partial x}\right)\Delta x\Delta y$$

Similarly, the net rate of heat flow into the control volume in the y direction is given by

$$\frac{d\dot{Q}_y}{dt} \approx \frac{\partial}{\partial y}\left(k\frac{\partial T}{\partial y}\right)\Delta x\Delta y$$

Hence since

$$\frac{d\dot{Q}}{dt} = \frac{d\dot{Q}_x}{dt} + \frac{d\dot{Q}_y}{dt}$$

then

$$\frac{d\dot{Q}}{dt} = \left\{\frac{\partial}{\partial x}\left(k\frac{\partial T}{\partial x}\right) + \frac{\partial}{\partial y}\left(k\frac{\partial T}{\partial y}\right)\right\}\Delta x\Delta y \tag{4.17}$$

Rate of increase of energy, $\dfrac{dE}{dt}$

The energy comprises the specific internal energy, e, and kinetic energy per unit mass, $\frac{1}{2}(u^2 + v^2)$, hence for the control volume

$$\frac{dE}{dt} = \rho\left\{\frac{de}{dt} + \frac{1}{2}\frac{d}{dt}(u^2 + v^2)\right\}\Delta x\Delta y \tag{4.18}$$

Rate at which work is done, $\dfrac{dW}{dt}$

Work is done by the control volume against the action of surface forces dW_s/dt and also the body forces dW_b/dt; dW/dt is simply the sum of these two contributions. Work is done when an applied force moves through a distance; the rate at which work is done is the product of force and distance per unit time, or force multiplied by velocity. Figure 4.3 shows the work terms we must include in the analysis, derived from the action of the surface forces. The normal components of the surface forces are shown in Figure 4.3a, the shear components in Figure 4.3b. By way of example, consider the rate at which work is done on the control volume, in the x direction by the normal stress component σ_x. Using the familiar Taylor series expansion, it follows that if the rate at which work is done at x is $-\sigma_x u\Delta y$, then the rate of work at $x + \Delta x$ is

$$\left\{\sigma_x u + \frac{\partial}{\partial x}(\sigma_x u)\Delta x\right\}\Delta y$$

The net work done by this surface force is then

$$\frac{\partial}{\partial x}(\sigma_x u)\Delta x\Delta y$$

Fig. 4.3 ●
Energy equation.
Work done on
a fluid control
volume by surface
forces: (a) normal
and (b) shear
components

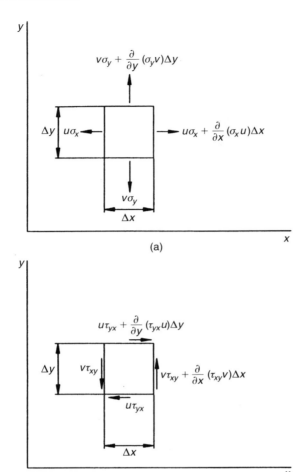

(a)

(b)

Application of similar reasoning to the other surface forces in Figure 4.3 yields the expression

$$\frac{\mathrm{d}W_s}{\mathrm{d}t} = \left\{ \frac{\partial}{\partial x}(\sigma_x u + \tau_{xy} v) + \frac{\partial}{\partial y}(\sigma_y v + \tau_{yx} u) \right\} \Delta x \Delta y \tag{4.19}$$

The work done on the control volume by the body forces, F_x and F_y, per unit mass is simply

$$\frac{\mathrm{d}W_b}{\mathrm{d}t} = (F_x u + F_y v)\Delta x \Delta y \tag{4.20}$$

We have derived in Equations (4.20) and (4.19) expressions for the work done by the forces acting on the element. The thermodynamic relation (4.16) requires the complement of this, the work done by the control volume against the forces, hence

$$\frac{\mathrm{d}W}{\mathrm{d}t} = -\left(\frac{\mathrm{d}W_s}{\mathrm{d}t} + \frac{\mathrm{d}W_b}{\mathrm{d}t}\right)$$

$$= -\left[\left\{\frac{\partial}{\partial x}(\sigma_x u + \tau_{xy} v) + \frac{\partial}{\partial y}(\sigma_y v + \tau_{yx} u)\right\} + (F_x u + F_y v)\right]\Delta x \Delta y$$

(4.21)

The terms involving the body force can be eliminated by multiplying Equation (4.11) throughout by u, and Equation (4.14) throughout by v. The resulting expressions for $F_x u$ and $F_y v$ are substituted into Equation (4.21), giving

$$-\frac{\mathrm{d}W}{\mathrm{d}t} = \rho\left[u\left(u\frac{\partial u}{\partial x} + v\frac{\partial u}{\partial y}\right) + v\left(u\frac{\partial v}{\partial x} + v\frac{\partial v}{\partial y}\right)\right]\Delta x \Delta y$$

$$+ \left[\sigma_x\frac{\partial u}{\partial x} + \tau_{xy}\frac{\partial v}{\partial x} + \sigma_y\frac{\partial v}{\partial y} + \tau_{yx}\frac{\partial u}{\partial y}\right]\Delta x \Delta y$$

(4.22)

For steady flow,

$$\frac{\mathrm{d}}{\mathrm{d}t}(u^2 + v^2) = u\frac{\partial}{\partial x}(u^2) + v\frac{\partial}{\partial y}(u^2) + v\frac{\partial}{\partial y}(v^2) + u\frac{\partial}{\partial x}(v^2)$$

so,

$$-\frac{\mathrm{d}W}{\mathrm{d}t} = \frac{1}{2}\rho\frac{\mathrm{d}}{\mathrm{d}t}(u^2 + v^2)\Delta x \Delta y$$

$$+ \left[\sigma_x\frac{\partial u}{\partial x} + \tau_{xy}\frac{\partial v}{\partial x} + \sigma_y\frac{\partial v}{\partial y} + \tau_{yx}\frac{\partial u}{\partial y}\right]\Delta x \Delta y$$

(4.23)

Using the relations for the normal and shear stresses given by Equation (4.12) in Equation (4.23), gives the result

$$-\frac{\mathrm{d}W}{\mathrm{d}t} = \frac{1}{2}\rho\frac{\mathrm{d}}{\mathrm{d}t}(u^2 + v^2)\Delta x \Delta y + \left[\mu\Phi - p\left(\frac{\partial u}{\partial x} + \frac{\partial v}{\partial y}\right)\right]\Delta x \Delta y$$

(4.24)

where Φ is the dissipation function

$$\Phi = 2\left[\left(\frac{\partial u}{\partial x}\right)^2 + \left(\frac{\partial v}{\partial y}\right)^2\right] + \left(\frac{\partial v}{\partial x} + \frac{\partial u}{\partial y}\right)^2 - \frac{2}{3}\left(\frac{\partial u}{\partial x} + \frac{\partial v}{\partial y}\right)^2$$

(4.25)

Returning now to the statement of the first law of thermodynamics given by Equation (4.16), and substituting into this the expressions we have derived for the three terms, given by Equations (4.17), (4.18) and (4.24), we get

$$\rho\frac{\mathrm{d}e}{\mathrm{d}t} = \left\{\frac{\partial}{\partial x}\left(k\frac{\partial T}{\partial x}\right) + \frac{\partial}{\partial y}\left(k\frac{\partial T}{\partial y}\right)\right\} - p\left(\frac{\partial u}{\partial x} + \frac{\partial v}{\partial y}\right) + \mu\Phi$$

(4.26)

We now focus on the left-hand side of Equation (4.26) to derive an expression for the rate of change of specific internal energy in terms of the fluid temperature T, pressure p and density ρ. The basis for much of the following discussion can be found in most thermodynamics textbooks; for example, Rogers and Mayhew (1992). Introducing the thermodynamic relationship for the specific enthalpy, i (in thermodynamics texts, enthalpy is usually given the symbol h, but to avoid confusion with the heat transfer coefficient, the symbol i is used here), defined as

$$i = e + p/\rho \tag{4.27}$$

and taking the time derivative

$$\frac{de}{dt} = \frac{di}{dt} - \frac{1}{\rho}\left(\frac{dp}{dt} - \left(\frac{p}{\rho}\right)\frac{d\rho}{dt}\right) \tag{4.28}$$

Additionally the enthalpy change is linked to the change in specific entropy, s, by

$$di = T\,ds + \frac{dp}{\rho} \tag{4.29}$$

where

$$ds = \left(\frac{\partial s}{\partial T}\right)_p dT + \left(\frac{\partial s}{\partial p}\right)_T dp \tag{4.30}$$

Using Maxwell's relations,

$$\left(\frac{\partial s}{\partial T}\right)_p = \frac{1}{T}\left(\frac{\partial i}{\partial T}\right)_p \quad \text{and} \quad \left(\frac{\partial s}{\partial p}\right)_T = \frac{1}{\rho^2}\left(\frac{\partial \rho}{\partial T}\right)_p \tag{4.31}$$

and introducing the definitions of the specific heat at constant pressure, C_p, and the volume expansion coefficient, β,

$$C_p = \left(\frac{\partial i}{\partial T}\right)_p \tag{4.32a}$$

$$\beta = -\frac{1}{\rho}\left(\frac{\partial \rho}{\partial T}\right)_p \tag{4.32b}$$

Substitution of Equations (4.30), (4.31) and (4.32) into Equation (4.29) allows the rate of change of internal energy to be expressed as

$$\frac{de}{dt} = C_p \frac{dT}{dt} - \left(\frac{\beta T}{\rho}\right)\frac{dp}{dt} + \left(\frac{p}{\rho^2}\right)\frac{d\rho}{dt} \tag{4.33}$$

The term $d\rho/dt$ in Equation (4.33) can be rewritten with the aid of mass continuity (4.6) as

$$\frac{d\rho}{dt} = \left(\frac{\partial \rho}{\partial x}\right)\left(\frac{\partial x}{\partial t}\right) + \left(\frac{\partial \rho}{\partial y}\right)\left(\frac{\partial y}{\partial t}\right)$$

$$= \left(\frac{\partial \rho}{\partial x}\right)u + \left(\frac{\partial \rho}{\partial y}\right)v \tag{4.34}$$

$$= -\rho\left(\frac{\partial u}{\partial x} + \frac{\partial v}{\partial y}\right)$$

Hence, using Equation (4.34) in Equation (4.33), we arrive at the required expression for the rate of change of internal energy:

$$\frac{de}{dt} = C_p \frac{dT}{dt} - \left(\frac{\beta T}{\rho}\right)\frac{dp}{dt} - \left\{\left(\frac{p}{\rho}\right)\left(\frac{\partial u}{\partial x} + \frac{\partial v}{\partial y}\right)\right\} \tag{4.35}$$

Finally, substituting Equation (4.35) into Equation (4.26), we get the energy equation for steady two-dimensional flow of a compressible fluid with variable properties:

$$\rho C_p \frac{dT}{dt} = (\beta T)\frac{dp}{dt} + \left\{\frac{\partial}{\partial x}\left(k\frac{\partial T}{\partial x}\right) + \frac{\partial}{\partial y}\left(k\frac{\partial T}{\partial y}\right)\right\} + \mu\Phi \tag{4.36}$$

where the dissipation function, Φ, is given by Equation (4.25).

Various simplifications can be made to Equation (4.36). For a perfect gas ($\rho = p/RT$ and, following from Equation (4.32b), $\beta = 1/T$) with a constant thermal conductivity,

$$\rho C_p \left\{u\frac{\partial T}{\partial x} + v\frac{\partial T}{\partial y}\right\} = \left\{u\frac{\partial p}{\partial x} + v\frac{\partial p}{\partial y}\right\} + k\left\{\frac{\partial^2 T}{\partial x^2} + \frac{\partial^2 T}{\partial y^2}\right\} + \mu\Phi \tag{4.37}$$

For an incompressible fluid where $\beta = 0$,

$$\rho C_p \left\{u\frac{\partial T}{\partial x} + v\frac{\partial T}{\partial y}\right\} = k\left\{\frac{\partial^2 T}{\partial x^2} + \frac{\partial^2 T}{\partial y^2}\right\} + \mu\Phi \tag{4.38a}$$

The corresponding equation in cylindrical coordinates is

$$\rho C_p \left\{V_r\frac{\partial T}{\partial r} + V_z\frac{\partial T}{\partial z}\right\} = k\left\{\frac{1}{r}\frac{\partial}{\partial r}\left(r\frac{\partial T}{\partial r}\right) + \frac{\partial^2 T}{\partial z^2}\right\} + \mu\Phi \tag{4.38b}$$

where

$$\Phi = 2\left\{\left(\frac{\partial V_r}{\partial r}\right)^2 + \left(\frac{V_r}{r}\right)^2 + \left(\frac{\partial V_z}{\partial z}\right)^2\right\} + \left\{\frac{\partial V_z}{\partial r} + \frac{\partial V_r}{\partial z}\right\}^2$$

Despite the lengthy derivation and the apparent mathematical complexity of the energy equation, the physical interpretation of Equation (4.37) is quite simple. The term on the left-hand side represents the convection of energy as a result of fluid motion. This is in equilibrium with the work done from changes in volume, conduction of energy through the fluid and, lastly, viscous dissipation. For a stationary fluid ($u = v = 0$) the energy equation becomes Laplace's equation (2.58) where, in the absence of fluid motion, conduction is the sole mechanism for energy transport, as might be expected.

Together the energy equation, the Navier–Stokes equations, the continuity equation and the relations for the fluid properties provide, in principle at least, all the information required to solve the complete flow field, the temperature field and then the heat transfer coefficient (review Equation (4.2) and accompanying comments). In practice, however, solving this system of partial differential equations presents a formidable task, even with the aid of numerical methods and high-speed computers.

In the following section, the concept of boundary layers is introduced; through the use of a few simple assumptions, this can greatly reduce the complexity of the mathematical problems encountered.

4.3 ● **Boundary layers**

At the beginning of the twentieth century, classical theory in fluid mechanics could not predict the experimental measurements made on the pressure drop in a pipe and the drag force experienced by a slender body. Classical theory treated fluids as having negligible viscosity and, considering that the most common fluids are water and air, this would initially appear to be a tempting proposition. The concept of boundary layers was first proposed by Prandtl in 1904 and provided a new theoretical model to account for the disparities between theory and experiment, due to so-called *form drag*. Since then, boundary layer theory has advanced, and to quote from the foreword to the first English edition of Schlichting (1979), it now forms 'the cornerstone of our knowledge of the flow of air and other fluids of small viscosity under circumstances of interest in many engineering applications.'

Boundary layer theory acknowledges the presence of fluid viscosity, even though experience informs us that the constant of proportionality, μ, is so small that the viscous forces should be almost negligible. To illustrate this point, let us carry out a hypothetical experiment. Imagine we are in a boat sailing across a calm sea, and we have a pitot tube built into the side of the hull under the waterline to measure the speed of the water relative to the hull. The pitot tube is an 'ideal instrument' and does not in any way disturb the flow. We make a measurement on the surface of the hull – there is no relative motion (as we would expect from the no-slip condition). We make another measurement some way out, say at 1 m. The reading is the same as the ship's speed through the water: let us call this the free stream velocity. Now we move the probe in, a little at a time. Nothing changes on the reading until we are very close to the hull (a distance of several centimetres) and then, moving closer, we note a large change in the relative velocity over a relatively short distance. This is the velocity boundary layer, and is depicted schematically in Figure 4.4a. The

Fig. 4.4 ●
Boundary layers
and nomenclature:
(a) velocity
boundary layer;
(b) thermal
boundary layer
(heated surface)

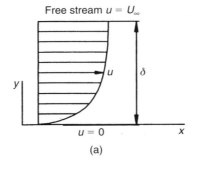
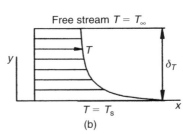

(a)

(b)

boundary layer thickness, defined as the distance at which the velocity is almost equal to the free stream velocity (99% is normal convention) and is usually denoted by the symbol δ.

Following on from this idea, we see that although the numerical value of viscosity may be small in terms of physical units, the velocity gradient $\partial u/\partial y$ is large because we experience a large change in u over a small distance δ. The shear force that is responsible for form drag is $\tau = \mu(\partial u/\partial y)_{y=0}$, so we also see how the concept of boundary layers explains the anomaly in classical fluid mechanics. The existence of boundary layers is not limited to velocity, or momentum, boundary layers. We can carry out a similar experiment measuring the fluid temperature between a heated surface and a cold flow over it. And we would find a thin layer of thickness δ_T in which there are large temperature changes from the free stream value of the cold fluid to the surface value on the heated plate (see Figure 4.4b). We can do the same with a concentration of species such as water vapour in air and once again find a boundary layer – the concentration boundary layer. Since convective heat transfer concerns itself with the interaction between a fluid and solid surface, we recognise that boundary layer theory can have tremendous implications.

By virtue of its nature this section can offer only a brief view of boundary layer theory and how it aids the study of convective heat transfer. The classic work in this field is Schlichting (1979); the more recent book by Bejan (1995) contains some excellent descriptive text and relatively simple mathematical arguments.

Let us now explore the implications of boundary layer theory to the flow over a solid surface and the associated convective heat transfer. For simplicity, we consider an incompressible flow with constant properties and ignore body forces (Equations (4.7), (4.13) and (4.15)). For this flow, the velocity u changes from $u = U_\infty$, the free stream value at the edge of the boundary layer, to $u = 0$ on the surface. The boundary layer thickness we denote by δ, and the length in the direction of flow is L. The changes in x, y and u are therefore of the same order of magnitude (indicated by the symbol $\sim$) as L, δ and U_∞, or

$$x \sim L, \quad y \sim \delta \quad \text{and} \quad u \sim U_\infty$$

From the continuity equation (4.7a) we have

$$\frac{\partial u}{\partial x} + \frac{\partial v}{\partial y} = 0$$

$$\text{or} \quad \left[\frac{U_\infty}{L}\right] \sim \left[\frac{v}{\delta}\right]$$

So we find that the ratio U_∞/L is of the same order of magnitude as v/δ, hence

$$v \sim U_\infty \frac{\delta}{L} \tag{4.39}$$

From the x direction momentum equation (4.13) and using the above result for v, we obtain

$$u\frac{\partial u}{\partial x} + v\frac{\partial u}{\partial y} = -\frac{1}{\rho}\frac{\partial p}{\partial x} + v\left\{\frac{\partial^2 u}{\partial x^2} + \frac{\partial^2 u}{\partial y^2}\right\} \tag{4.40}$$

$$\left\{\left[\frac{U_\infty^2}{L}\right],\left[\frac{U_\infty^2}{L}\right]\right\} \sim \left[\frac{p}{\rho L}\right]\left\{\left[\frac{\mu U_\infty}{\rho L^2}\right],\left[\frac{\mu U_\infty}{\rho \delta^2}\right]\right\}$$

Since $\delta \ll L$, the term representing $\partial^2 u/\partial x^2$ in the x direction momentum equation, $\mu U_\infty/\rho L^2$, must be small in relation to $\mu U_\infty/\rho \delta^2$, so we are justified in neglecting $\partial^2 u/\partial x^2$. The same argument can be applied to the y direction momentum equation, $\partial^2 v/\partial y^2$ is retained and $\partial^2 v/\partial x^2$ is omitted. So, modified in light of our boundary layer argument that $\delta \ll L$, the momentum equations become

$$u\frac{\partial u}{\partial x} + v\frac{\partial u}{\partial y} = -\frac{1}{\rho}\frac{\partial p}{\partial x} + \frac{\mu}{\rho}\frac{\partial^2 u}{\partial y^2} \tag{4.41}$$

$$u\frac{\partial v}{\partial x} + v\frac{\partial v}{\partial y} = -\frac{1}{\rho}\frac{\partial p}{\partial y} + \frac{\mu}{\rho}\frac{\partial^2 v}{\partial y^2} \tag{4.42}$$

The next step is to simplify the mathematical statement of the pressure gradient. Intuitively, we may speculate that the pressure change in the y direction is small in relation to the pressure change in the x direction. To justify this, consider the pressure at any point which, as it depends on x and y, then

$$\frac{dp}{dx} = \frac{\partial p}{\partial x} + \frac{\partial p}{\partial y}\frac{dy}{dx} \tag{4.43}$$

Equations (4.41) and (4.42) represent a balance between either inertia or friction and pressure forces. For arguments sake, consider a balance between inertia and pressure (although the same result is obtained with a friction–pressure balance); Equations (4.41) and (4.42) can be written in order of magnitude terms as

$$\frac{\partial p}{\partial x} \sim \frac{\rho U_\infty^2}{L} \quad \text{and} \quad \frac{\partial p}{\partial y} \sim \frac{\rho U_\infty^2 \delta}{L^2}$$

The ratio of the first and second terms of the right side in Equation (4.43) is

$$\frac{\partial p/\partial x}{(\partial p/\partial y)\,(dy/dx)} = \left(\frac{L}{\delta}\right)^2 \gg 1 \tag{4.44}$$

The implication of Equation (4.44) in the context of Equation (4.43) is that our intuitive reasoning is correct and mathematically we are able to replace $\partial p/\partial x$ with the ordinary derivative dp/dx and omit $\partial p/\partial y$. Furthermore, since the pressure now depends only on x, the pressure inside the boundary layer is equal to that outside it. Pressure is therefore not a boundary layer variable, but a known quantity imposed upon the boundary layer. To show this mathematically, if we apply Equation (4.41) at the outer edge of the boundary layer where $u = U_\infty$, and as by definition there are no variations of u or v with coordinate y, then

$$U_\infty \frac{dU_\infty}{dx} = -\frac{1}{\rho}\frac{dp}{dx}$$

Integrating we obtain Bernoulli's equation

$$p + \frac{1}{2}\rho U_\infty^2 = \text{a constant} \qquad (4.45)$$

Thus, providing U_∞ is known, we need only two equations, since there are only two unknowns, u and v. One equation will be provided by mass continuity, the other is formally known as the 'boundary layer momentum equations'. For steady flow of an incompressible fluid, and replacing the body force term, the two equations become

$$\frac{\partial u}{\partial x} + \frac{\partial v}{\partial y} = 0 \qquad (4.46)$$

$$\rho\left[u\frac{\partial u}{\partial x} + v\frac{\partial u}{\partial y}\right] = F_x - \frac{dp}{dx} + \mu\frac{\partial^2 u}{\partial y^2} \qquad (4.47)$$

Similar arguments can be applied to the energy equation (4.38) and the dissipation function (4.25). This will produce a scaling relation involving terms with the ratio δ/δ_T. Strictly, the thermal boundary layer and velocity boundary layer do not have the same thickness, but a consequence of boundary layer theory is that (δ/δ_T) varies with v/α, the Prandtl number Pr. Fluids with $Pr \ll 1$ (liquid metals) and $Pr \gg 1$ (oils) require special attention. However, gases have a Prandtl number of order unity, as do most common liquids (oils and liquid metals being notable exceptions). The constraint that δ and δ_T are of the same order of magnitude, does not severely limit the usefulness of the boundary layer energy equation, which for fluids with moderate values of Prandtl number is

$$\rho C_p\left[u\frac{\partial T}{\partial x} + v\frac{\partial T}{\partial y}\right] = k\frac{\partial^2 T}{\partial y^2} + \mu\left(\frac{\partial u}{\partial y}\right)^2 \qquad (4.48)$$

Compared with the equations for the entire flow field (Equations (4.7), (4.13), (4.15) and (4.38)), Equations (4.46), (4.47) and (4.48) for the boundary layer represent a tremendous reduction in complexity.

4.4 ● Dimensionless groups for convection

The basic non-dimensional groups that characterise convective heat transfer can be derived from dimensional analysis using Buckingham's Π theorem and the method of indices. Free and forced convection are generated by different physical mechanisms, so these will be treated separately.

4.4.1 Forced convection

Consider flow over a heated flat plate, neglecting the pressure gradient. From equations (4.46), (4.47) and (4.48) we expect the heat flux, q, to be a function of the coordinate x, the free stream velocity U_∞, the temperature difference $T_s - T_\infty$ and fluid properties ρ, μ, C_p and k. Since by definition, $h = q/(T_s - T_\infty)$, this can be expressed in the following functional relationship:

$$h = h(x, U_\infty, \rho, \mu, C_p, k) \tag{4.49}$$

The heat transfer coefficient has dimensions of $H/L^2\Theta t$, where H represents the dimensions of heat (joules), Θ represents temperature, L represents length, and t represents time. Similarly, heat H appears as a dimensional entity in two other physical quantities, k and C_p. The thermal conductivity k has dimensions $H/L\Theta t$; C_p has dimensions $H/M\Theta$ (where M represents mass). These are set out in Table 4.1.

Inspection of Table 4.1 reveals that the quantities H and Θ do not appear independently, but as the group H/Θ. Consequently, there are only four primary dimensions: H/Θ, L, M and t. And as there are seven variables, there will be $7 - 4 = 3$ independent dimensionless groups, Π_1, Π_2, Π_3. Taking x, U_∞, k and μ as the basic variables:

$$\Pi_1 = h(x)^{a_1}(U_\infty)^{b_1}(k)^{c_1}(\mu)^{d_1} \tag{4.50a}$$

$$\Pi_2 = C_p(x)^{a_2}(U_\infty)^{b_2}(k)^{c_2}(\mu)^{d_2} \tag{4.50b}$$

$$\Pi_3 = \rho(x)^{a_3}(U_\infty)^{b_3}(k)^{c_3}(\mu)^{d_3} \tag{4.50c}$$

Firstly consider the group Π_1,

$$\Pi_1 = h(x)^{a_1}(U_\infty)^{b_1}(k)^{c_1}(\mu)^{d_1}$$

Table 4.1 ●
Dimensional groups for forced convention

Quantity	Dimensions	Units
h	$\dfrac{H}{L^2\Theta t}$	W/m²K
x	L	m
U_∞	$\dfrac{L}{t}$	m/s
k	$\dfrac{H}{L\Theta t}$	W/m K
μ	$\dfrac{M}{Lt}$	kg/m s
C_p	$\dfrac{H}{\Theta M}$	J/kg K
ρ	$\dfrac{M}{L^3}$	kg/m³

In dimensional quantities, since Π_1 is dimensionless, the left side is equal to zero, so

$$0 = \left(\frac{H}{L^2 \Theta t}\right)(L)^{a_1}\left(\frac{L}{t}\right)^{b_1}\left(\frac{H}{L\Theta t}\right)^{c_1}\left(\frac{M}{Lt}\right)^{d_1} \tag{4.51}$$

Equating powers of

length, L: $\quad -2 + a_1 + b_1 - c_1 - d_1 = 0$

H/Θ: $\quad 1 + c_1 = 0$

mass, M: $\quad d_1 = 0$ $\qquad\qquad\qquad\qquad$ (4.52)

time, t: $\quad -1 - b_1 - c_1 - d_1 = 0$

From which (four equations with four unknowns, a_1, b_1, c_1 and d_1) we get $d_1 = 0$; $c1 = -1$; $b1 = 0$ and $a1 = 1$. Hence $\Pi_1 = hx/k = Nu_x$ which is known as the Nusselt number. This represents a dimensionless heat transfer coefficient and can be thought of as the ratio of the heat transfer by convection to that of conduction through the fluid. A value of $Nu \approx 1$ implies that convective effects are either very weak or not present. The subscript x denotes that the local coordinate x is used for the characteristic dimension. If the value of the heat transfer coefficient at x is also used in the Nusselt number, then it is termed a *local Nusselt number*. Other forms used are Nu_L (local value based on plate length L) or, for a cylinder, Nu_D (value based on diameter). *Average Nusselt numbers* are denoted by Nu_{av} or $\overline{Nu}$.

Continuing in a similar fashion, we find $\Pi_2 = \mu C_p/k = Pr$, the *Prandtl number*, a dimensionless fluid property. The physical interpretation of the Prandtl number is easier to recognise when rewritten as (since $\mu = \rho v$ and $\alpha = k/\rho C_p$) $Pr = v/\alpha$, the ratio of momentum to thermal diffusivities. As mentioned in Section 4.3, and further recalled in the next two chapters, the Prandtl number has considerable significance as a measure of the relative thickness of the velocity and thermal boundary layers.

Finally we find that $\Pi_3 = \rho U_\infty x/\mu = Re_x$, which is the well-known *Reynolds number*. Again use is made of a subscript to denote the characteristic length: x for a local value at distance x, or L for an overall value based on the length L. A Reynolds number for the flow over a cylinder or through a pipe would use the cylinder or pipe diameter D, hence Re_D. For a surface that rotates at Ω radians per second, we would replace the velocity U_∞ with the tangential velocity of the surface Ωr, and use the radius r as the characteristic dimension, leading to the definition of a 'rotational' Reynolds number as $Re_\Omega = \rho \Omega r^2/\mu$.

From the original statement we find that for forced convection

$$Nu_x = f(Re_x, Pr) \tag{4.53}$$

At high speed, frictional heating can be significant and a further group, the Eckert number Ec, is also used:

$$Ec = U_\infty^2/C_p(T_s - T_\infty) \tag{4.54}$$

Dimensional analysis does not provide any further information other than the existence of the relevant dimensionless groups. The functional relation f is unspecified and can be found through experimentation or by solution of the boundary layer equations. This is covered in Chapter 5.

4.4.2 Free convection

In free convection, U_∞ is zero and motion is generated through a buoyancy force. For example, the fluid adjacent to a hot surface will be heated, its density will decrease and the heated air will move in the opposite direction to the gravitational field (other force fields, for example centrifugal force, also create free convection motion), to a new position where thermal equilibrium is established. In the case of gravitationally induced free convection, the buoyancy force per unit volume B ($F_x = -\rho g$ in Equation (4.47) together with the hydrostatic pressure gradient $dp/dx = -\rho_\infty g$) is

$$B = g(\rho_\infty - \rho) \tag{4.55}$$

where the subscript ∞ applies to conditions outside the boundary layer. From the definition of the volume expansion coefficient (4.32b), we have:

$$\beta = -\frac{1}{\rho}\left(\frac{\partial \rho}{\partial T}\right)_p$$

$$\approx -\frac{1}{\rho}\frac{(\rho_\infty - \rho)}{(T_\infty - T)} \tag{4.56}$$

Equation (4.56) is known as the *Boussinesq approximation*, and from Equations (4.55) and (4.56)

$$B \approx \rho\beta g(T - T_\infty) \tag{4.57a}$$

Considering the variation across the whole boundary layer where the temperature ranges from T_s at the surface to T_∞ at the free stream,

$$B \approx \rho\beta g(T_s - T_\infty) \tag{4.57b}$$

which has dimensions of $M/L^2 t^2$. So, formulating the functional relationship in terms of the relevant dimensional quantities gives

$$h = h(B, x, \rho, \mu, C_p, k) \tag{4.58}$$

Since there are seven variables and four primary dimensions, there will be three dimensionless groups. Selecting h, C_p and B as the basic variables and proceeding as above for forced convection, we find

$$\Pi_1 = h(\rho)^{a_1}(\mu)^{b_1}(x)^{c_1}(k)^{d_1} = \frac{hx}{k} = Nu_x \tag{4.59a}$$

$$\Pi_2 = C_p(\rho)^{a_2}(\mu)^{b_2}(x)^{c_2}(k)^{d_2} = \frac{\mu C_p}{k} = Pr \tag{4.59b}$$

$$\Pi_3 = B(\rho)^{a_3}(\mu)^{b_3}(x)^{c_3}(k)^{d_3} = \frac{\rho x^3 B}{\mu^2} = \rho^2 x^3 \beta g \frac{(T_s - T_\infty)}{\mu^2} = Gr_x \tag{4.59c}$$

The groups Π_1 and Π_2, the Nusselt and Prandtl numbers, were brought to light in the dimensional analysis of forced convection. The group Π_3, which in free convection replaces the forced convection parameter of the Reynolds number, is known as the *Grashof number*. Consequently, for free convection $Nu_x = f(Gr_x Pr)$ implies that the Nusselt number depends on the surface to free stream temperature difference $(T_s - T_\infty)$, whereas in forced convection it does not. The product $(Gr_x Pr)$ is frequently used in free convection; it is known as the Rayleigh number and given the symbol Ra_x. The relationships between the Nusselt number and the Grashof and Prandtl numbers are covered in detail in Chapter 6.

4.4.3 Average and local values

Engineers are likely to be concerned with the average heat transfer from a surface. A brief discussion on the average Nusselt number requires some attention, as in the author's teaching experience this is often a source of confusion. The average Nusselt number, Nu_{av}, is defined as

$$Nu_{av} = \frac{q_{av} L}{\Delta T_{av} k} \tag{4.60}$$

where L is an appropriate characteristic length, and q_{av} and ΔT_{av} are the area-averaged heat flux and surface-to-fluid reference temperature difference, respectively. These quantities are evaluated using

$$q_{av} = \frac{1}{A} \int q \, dA \tag{4.61}$$

$$\Delta T_{av} = \frac{1}{A} \int \Delta T \, dA \tag{4.62}$$

For a rectangle of length L in the x direction and constant width W, $A = LW$ and $dA = W \, dx$. For a circular disc of outer radius $r = b$, $A = \pi b^2$ and $dA = 2\pi r \, dr$. In many circumstances the surface temperature to fluid reference temperature difference does not vary, so $\Delta T_{av} = \Delta T$. Where ΔT does vary with the streamwise coordinate x, it is convenient to express this variation as a power law, i.e. $\Delta T = \Delta T_{max} (x/L)^m$ with m assigned a value: $m = 2$ indicates a square law variation of surface temperature with distance; $m = 1$ indicates a linear variation; $m = 0$ indicates a constant surface to free stream temperature difference and so on. Equation (4.62) can then be integrated. The integrand in Equation (4.61) can be obtained from the dimensionless relationship for Nu_x, which can usually be expressed as

$$Nu_x = C_1(Pr^a Re_x^b)$$

for forced convection and

$$Nu_x = C_2(Pr^d Gr_x^e)$$

for free convection, where C_1, C_2, a, b, d and e are known constants. For forced convection from a flat plate, with a constant free stream velocity (see for example Equations (5.67) and (5.83)) and constant values of μ, C_p, k and ρ, we would obtain $q = D_1 x^{b+m-1}$ where D_1 is shorthand for the compounded constant

$$D_1 = k\Delta T_{\max} C_1 \left(\frac{\mu C_p}{k}\right)^a \left(\frac{\rho U_\infty}{\mu}\right)^b \tag{4.63}$$

After carrying out the integrals it is easy to verify that

$$Nu_{av} = \frac{q_{av} L}{\Delta T_{av} k}$$

$$= \frac{C_1 (Pr^a Re_L^b)(m+1)}{(m+b)} \tag{4.64a}$$

which for a constant surface temperature ($m = 0$) simplifies to

$$Nu_{av} = \frac{C_1 (Pr^a Re_L^b)}{b} \tag{4.64b}$$

since $b < 1$ the division by this constant actually makes the average Nusselt number greater than the maximum local Nusselt number at $x = L$.

4.4.4 Choice of reference temperature and temperature for evaluating fluid properties

For external flows, fluid properties such as ρ, C_p, k and μ are generally evaluated at the *mean film temperature*, $T_m = \frac{1}{2}(T_s + T_\infty)$. For internal flows, it is usual to use a mean of the bulk mean inlet and outlet temperatures, $T_m = T_{b,in} + T_{b,out}$ (see Equation (5.24) for a definition of T_b). Sometimes a correlation will specify some other temperature; for internal flow it may be the mean of T_m and the pipe wall temperature. If the temperature differences (surface to fluid, inlet to outlet) are small enough that changes in the fluid properties are also small, then the choice is relatively unimportant, providing consistency is maintained.

To account for the effect of frictional heating, the Nusselt number should be defined using the adiabatic wall temperature, $T_{s,ad}$, hence T_∞ in the denominator of the definition of Nu is replaced by

$$T_{s,ad} = T_\infty + \frac{\mathcal{R} U_\infty^2}{2C_p}$$

where $\mathcal{R}$ is a recovery factor ($\mathcal{R} = Pr^{1/2}$ for moderate Pr and $\mathcal{R} = 1.9 Pr^{1/3}$ for large Pr). In flows where $Ec \ll 1$, the temperature difference $T_{s,ad} - T_\infty$ will be small and T_∞ may be used.

4.5 ● Turbulence and the time-averaged equations

4.5.1 The nature of turbulence

In the 1880s, Osborne Reynolds carried out his historic visualisation studies of flow in a pipe. He observed that well-ordered laminar flow degenerated into a 'sinuous' motion when the velocity of the flow in a pipe reached a certain value. In laminar flow, a dye injected into the moving fluid stream exhibits a well-ordered structure with parallel streamlines (not unlike the grain in wood) highlighted by the presence of the dye. If the flow rate, or to be more precise, the Reynolds number is increased, then at some point this orderly structure of laminae will lose its identity, giving rise to a flow structure characterised by large-scale disturbances called eddies. Examples of turbulent flows can be seen in the development of a smoke plume rising from a cigarette and water flowing from a tap (both of which display an initial laminar section which further downstream breaks off into turbulent flow), and the formation of cumulus clouds on a summer's day.

A schematic diagram of a boundary layer undergoing transition from laminar to turbulent flow is shown in Figure 4.5. Transition from laminar to turbulent boundary layer flow depends on a number of factors, for example the flow Reynolds number, surface roughness, pressure gradient and turbulence in the free stream itself. For forced convection, the onset of transition is characterised by the Reynolds number. For Reynolds numbers less than the transition value ($Re < Re_{\text{trans}}$), disturbances initiated for example by surface irregularities decay owing to the damping action of viscosity. Above the transition value of Reynolds number, the action of viscosity is insufficient to damp out these disturbances, which then become amplified and propagate through the fluid, causing the main flow to break down into eddy motion.

For flow in smooth pipes, transition occurs at

$$Re_{\text{trans}} = \frac{\rho U_{\text{b}} D}{\mu} \approx 2000$$

where U_{b} is the bulk average velocity of the fluid and D the pipe diameter. For external flows, transition occurs at a different value of Re_{trans}, and for flow over a flat plate,

$$Re_{\text{trans}} = \frac{\rho U_{\infty} x}{\mu} \approx 3 \times 10^5 \text{ to } 1 \times 10^6$$

Fig. 4.5 ●
Transition to turbulence in a boundary layer for forced convection over a flat plate

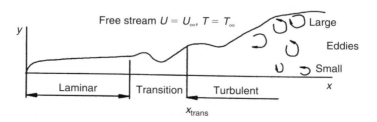

Free stream $U = U_\infty$, $T = T_\infty$

Large

Eddies

Small

Laminar | Transition | Turbulent

x_{trans}

where U_∞ is the velocity in the free stream and x is the distance from the beginning of the boundary layer at the leading edge of the plate. The different value of Re_{trans} for pipe flow and external flow is sometimes a source of confusion, the reason being that the two Reynolds numbers are defined differently: one uses a pipe diameter (in effect one half of the boundary layer thickness in developed flow), the other the streamwise coordinate.

For free convection, transition is characterised by the dimensionless groups Gr and Pr. It has generally been accepted that transition from laminar to turbulent free convection occurs when $(Gr \cdot Pr)_{\text{trans}} \approx 10^9$. However, the criterion has been revised by Bejan and Large (1990) who suggest that transition is independent of the Prandtl number and that $Gr_{\text{trans}} \approx 10^9$ is more appropriate.

Turbulence is of fundamental interest to engineers because most flows encountered in engineering are turbulent. This is not necessarily owing to chance but more a result of design. Convective heat transfer is considerably enhanced by the action of turbulent eddies which create a mixing effect, taking heat from the surface and transporting it into the free stream. Compared with laminar flow, the presence of turbulence also increases the skin friction – for an aircraft wing, golf ball, compressor or turbine blade this implies that separation of the flow from the surface will be delayed, resulting in better aerodynamic efficiency. The nature of turbulence and the phenomenon of transition are vast subject areas in their own right. The aim of this section is therefore to present some of the main features of classical turbulence theory and the implications of this to convective heat transfer. There are many excellent monographs on the subject of turbulence (Hinze, 1975; McComb, 1994). Bradshaw (1978) provides a comprehensive review and Schlichting (1979) is also a useful source for further reading.

4.5.2 Time-averaged equations

Figure 4.6 shows the characteristic unsteady output from a fast-response thermocouple probe inserted into a turbulent boundary layer flow. At time t, the instantaneous temperature T has two components, a mean value $\bar{T}$ and a randomly fluctuating or eddy component T', hence

$$T = \bar{T} + T' \tag{4.65a}$$

Similarly, each velocity component comprises steady and eddy or fluctuating contributions, so for two-dimensional flow

$$u = \bar{u} + u' \text{ and } v = \bar{v} + v' \tag{4.65b}$$

Likewise for the pressure and density

$$p = \bar{p} + p' \text{ and } \rho = \bar{\rho} + \rho' \tag{4.65c}$$

The reader will no doubt by now have grasped that the instantaneous picture of turbulence, obtained by substituting Equations (4.65) into

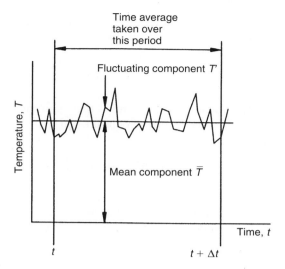

Fig. 4.6 ●
Thermocouple in turbulent flow: unsteady output

Time average taken over this period

Fluctuating component T'

Mean component $\bar{T}$

Temperature, T

Time, t

t $t + \Delta t$

Equations (4.46) to (4.48), leads to a set of equations which are far outnumbered by the unknowns. This picture can be simplified if we consider the average behaviour by integrating over some suitable time period, Δt say, so that the mean value of the relevant variable is time-independent (see Figure 4.6). Then we obtain

$$\bar{T} = \frac{1}{\Delta t} \int_t^{t+\Delta t} T \, dt \quad \text{and} \quad \bar{u} = \frac{1}{\Delta t} \int_t^{t+\Delta t} u \, dt \quad \text{etc.} \tag{4.66}$$

Using Equations (4.65) and (4.66) it follows that, since they are random fluctuations, the time average of the fluctuating components must be zero:

$$\overline{T'} = \overline{u'} = \overline{\rho'} = \overline{v'} = \overline{\rho'} = 0 \tag{4.67}$$

It therefore follows that

$$\bar{\bar{u}} = \bar{u}$$

$$\overline{u + v} = \bar{u} + \bar{v}$$

$$\overline{uv} = \bar{u}\bar{v} + \overline{u'v'} \tag{4.68}$$

$$\overline{\frac{\partial \bar{u}}{\partial x}} = \frac{\partial \bar{u}}{\partial x}$$

and identical relations exist for the other variables.

Using Equations (4.65), the continuity equation (4.46) for constant density flow can be written out in terms of the steady and fluctuating components:

$$\frac{\partial u}{\partial x} + \frac{\partial v}{\partial y} = \left(\frac{\partial u'}{\partial x} + \frac{\partial v'}{\partial y} \right) + \left(\frac{\partial \bar{u}}{\partial x} + \frac{\partial \bar{v}}{\partial y} \right) = 0 \tag{4.69}$$

By taking a time average, the terms in the left-hand bracket become equal to zero, as in (4.67), so it is apparent that

$$\left(\frac{\partial \bar{u}}{\partial x} + \frac{\partial \bar{v}}{\partial y}\right) = 0 \tag{4.70}$$

Ignoring body forces, and using the continuity equation, the incompressible boundary layer momentum equation (4.47) can be written as follows:

$$\rho\left(\frac{\partial u^2}{\partial x} + \frac{\partial (uv)}{\partial y}\right) = -\frac{dp}{dx} + \mu\frac{\partial^2 u}{\partial y^2} \tag{4.71}$$

Substituting in the mean and fluctuating components, taking the time average and using Equations (4.68) gives, after some manipulation

$$\rho\left(\bar{u}\frac{\partial \bar{u}}{\partial x} + \bar{v}\frac{\partial \bar{u}}{\partial y}\right)$$

$$= -\frac{d\bar{p}}{dx} - \rho\frac{\partial}{\partial x}\overline{u'^2} + \frac{\partial}{\partial y}\left(\mu\frac{\partial \bar{u}}{\partial y} - \rho\overline{u'v'}\right) \tag{4.72}$$

In boundary layer flows, where $\partial u/\partial y \gg \partial u/\partial x$, the term $\partial/\partial x(\overline{u'^2})$ can be neglected and Equation (4.72) becomes

$$\rho\left(\bar{u}\frac{\partial \bar{u}}{\partial x} + \bar{v}\frac{\partial \bar{u}}{\partial y}\right) = -\frac{d\bar{p}}{dx} + \frac{\partial \tau}{\partial y} \tag{4.73}$$

where the shear stress, τ, is made up from the laminar and turbulent contributions:

$$\tau = \tau_{\text{lam}} + \tau_{\text{turb}} = \mu\frac{\partial \bar{u}}{\partial y} - \rho\overline{u'v'} \tag{4.74}$$

The turbulent contribution represented by the term, $\rho\overline{u'v'}$ is known as the eddy or *Reynolds stress*; it is zero for laminar flow, and it tends to zero near a solid wall for turbulent flow as the eddies decay. Since u' and v' are the velocity fluctuations due to a circular-like eddy as it moves along with the mean flow, then they are of comparable magnitudes. Also, for a flow with $\partial \bar{u}/\partial y > 0$ it follows that fluid particles moving upwards in the boundary layer ($v' > 0$) arrive in a region of larger mean velocity; and as the original mean velocity is maintained, they give rise to a negative value of u'. It also follows that a negative value of v' aids to increase u', so the term $(-\rho\overline{u'v'})$ is always positive and can greatly increase the shear stress. For a flow with $\partial \bar{u}/\partial y < 0$ the term $(-\rho\overline{u'v'})$ is negative, but so is the laminar shear stress. Thus, irrespective of the sign of $\partial \bar{u}/\partial y$, the turbulent contribution always enhances the laminar shear stress.

As a crude illustration of the magnitude of the turbulent shear stress, consider a flow of air with $U_\infty = 40\,\text{m/s}$ at 1 bar and $20°\text{C}$ ($\rho = 1.2\,\text{kg/m}^3$ and $\mu = 1.8 \times 10^{-5}\,\text{kg/m s}$), having a 3% fluctuation ($u' = 1.2\,\text{m/s}$) and for a boundary layer thickness of $\delta = 4\,\text{mm}$:

$$\mu\frac{\partial u}{\partial y} \approx \frac{\mu U_\infty}{\delta}$$

$$= \frac{1.8 \times 10^{-5} \times 40}{4 \times 10^{-3}} = 0.18\,\text{Pa}$$

$$|\rho\overline{u'v'}| \approx \rho u'^2 = 1.2 \times 1.2 \times 1.2 = 1.73\,\text{Pa}$$

so there is roughly a tenfold increase in the apparent shear stress due to the presence of turbulence.

Note the value of the shear stress at the wall, τ_s, will not be increased by this amount, since u' and v' are zero at the wall, hence $\tau_s = \mu(\partial u/\partial y)_{y=0}$. Nonetheless, there will be an increase in τ_s in turbulent flow because the effect of the increase in the shear stress in the bulk of the flow is to increase also the velocity gradient at the wall.

Similar arguments can be applied to the boundary layer energy equation (4.48) where for incompressible flow, neglecting dissipation and applying the procedure of separating the instantaneous velocities and temperatures into mean and fluctuating components, then time averaging, we get the following equation:

$$\rho C_p\left\{\bar{u}\frac{\partial\bar{T}}{\partial x} + \bar{v}\frac{\partial\bar{T}}{\partial y}\right\} = \frac{\partial q}{\partial y} \tag{4.75}$$

where the heat flux is made up from laminar and turbulent (eddy) contributions

$$q = q_{\text{lam}} + q_{\text{turb}} = -\left(k\frac{\partial\bar{T}}{\partial y} - \rho C_p\overline{T'v'}\right) \tag{4.76}$$

As with the shear stress in the momentum equation, the heat flux q in the turbulent energy equation is greatly enhanced by the term $\rho C_p\overline{T'v'}$.

It is convenient to adopt the following definitions which introduce a *momentum eddy diffusivity* ε_M and an *eddy thermal diffusivity* ε_H. Since these parameters involve only flow variables such as u', T' and v', they are not fluid properties but flow parameters.

$$\tau_{\text{turb}} = -\rho\overline{u'v'} = \rho\varepsilon_M\frac{\partial\bar{u}}{\partial y} \tag{4.77}$$

$$q_{\text{turb}} = -\rho C_p\overline{T'v'} = \rho C_p\varepsilon_H\frac{\partial\bar{T}}{\partial y} \tag{4.78}$$

Introducing Equations (4.77) and (4.78) into Equations (4.73) and (4.75):

$$\bar{u}\frac{\partial\bar{u}}{\partial x} + \bar{v}\frac{\partial\bar{u}}{\partial y} = -\frac{1}{\rho}\frac{d\bar{p}}{dx} + \frac{\partial}{\partial y}\left[(\nu + \varepsilon_M)\frac{\partial\bar{u}}{\partial y}\right] \tag{4.79}$$

$$\bar{u}\frac{\partial\bar{T}}{\partial x} + \bar{v}\frac{\partial\bar{T}}{\partial y} = \frac{\partial}{\partial y}\left[(\alpha + \varepsilon_H)\frac{\partial\bar{T}}{\partial y}\right] \tag{4.80}$$

In laminar flow, the turbulent shear stress and heat flux terms do not arise, so we have three equations (continuity, momentum and energy) and three unknowns (u, v and T), which in principle at least can lead to a closed form of solution. For turbulent flow there are two extra unknowns (ε_M and ε_H), resulting in three equations with five unknowns. So there is little chance of solving the time-averaged boundary layer equations without recourse to experimental data. Turbulence modelling is the study which aims to attain closure of the turbulent flow problem. This is achieved by using experimental data to obtain expressions for the turbulent shear stress and heat flux. There are various degrees of sophistication with turbulence models, often referred to as zero-order, first-order and second-order turbulence models. The simplest is the zero-order mixing length model, attributed to Prandtl.

4.5.3 Prandtl's mixing length model of turbulence

An order of magnitude estimate of the momentum eddy diffusivity ε_M can be obtained from the arguments originally proposed by Prandtl. Imagine an eddy located at a distance y from the wall where the mean longitudinal velocity is $\bar{u}_{(y)}$. This eddy now moves to a new position, at $y - L$, where the mean velocity is $\bar{u}_{(y-L)}$, and retains its original identity. The distance L is known as the mixing length. A physical interpretation of this quantity is the transverse distance travelled by a group of fluid particles such that the change in average velocity is equal to the mean transverse fluctuation of the transverse component in turbulent flow. As longitudinal momentum is conserved in this transition, the change in velocity, or velocity fluctuation, u' is given by the Taylor series expansion

$$u' = \bar{u}_{(y)} - \bar{u}_{(y-L)} \approx L\left(\frac{\partial \bar{u}}{\partial y}\right) \tag{4.81a}$$

Since, as previously noted, the transverse velocity fluctuation, v', is of the same order as the longitudinal fluctuation, u', then

$$|v'| \approx L\left|\left(\frac{\partial \bar{u}}{\partial y}\right)\right| \tag{4.81b}$$

And since the product $\overline{u'v'}$ is negative when $\partial\bar{u}/\partial y > 0$, we can approximate the time average of the product of the fluctuating components as

$$-\overline{u'v'} \approx L^2\left(\frac{\partial \bar{u}}{\partial y}\right)^2$$

Using the definition of ε_M (4.77) it follows that

$$\varepsilon_M = -\frac{\overline{u'v'}}{\partial\bar{u}/\partial y} = \frac{L^2(\partial\bar{u}/\partial y)^2}{\partial\bar{u}/\partial y} = L^2\frac{\partial \bar{u}}{\partial y} \tag{4.82a}$$

Also the turbulent shear stress must be of opposite sign to the velocity gradient which from (4.77) implies that ε_M is always postive, hence

$$\varepsilon_M = L^2 \left| \frac{\partial \bar{u}}{\partial y} \right| \tag{4.82b}$$

The mixing length is determined by the distance from the wall, so it is reasonable to assume that, physically,

$$L = \kappa y \tag{4.83}$$

where κ is a constant. A value of $\kappa = 0.4$ is consistent with experimental evidence in simple boundary layer flows, although the numerical value will vary with different flow configurations.

   Prandtl's mixing length hypothesis is a commonly and successfully used turbulence model. Compared with the more sophisticated first- and second-order models, it is relatively easy to program, and does not impose too much of a computational overhead. It does, however, have its limitations. First, as noted above, the mixing length 'constant' is not a universal constant, but varies with flow configuration. Secondly, the concept of the mixing length as a fraction of the distance to the closest wall loses meaning in the proximity of corners for two-dimensional flows. Thirdly, the formulation of the mixing length is difficult to extend to the case of three-dimensional flows. And finally, in flows where separation occurs, the results give poor agreement with experimental data.

4.5.4 Analogies between heat and momentum

Intuitively we may suspect a link between friction and heat transfer. After all, the mechanism of turbulence increases both friction and heat transfer, owing to eddies creating transverse mixing between adjacent layers of fluid. Turbulence promoters, such as twisted tape inserts, are used in heat exchanger tubes to enhance the heat transfer and as a penalty they incur a larger pressure drop, through surface friction, than a smooth tube. This casual observation can be formalised mathematically and applied to a variety of turbulent flows, allowing the heat transfer rate to be obtained from wall friction correlations.

   For forced convection over a flat surface, of a flow with constant density, with zero pressure gradient, and neglecting buoyancy and viscous dissipation, the time-averaged boundary layer momentum and energy equations (4.79) and (4.80) become ($v = \mu/\rho$, $\alpha = k/\rho C_p$)

$$\bar{u}\frac{\partial \bar{u}}{\partial x} + \bar{v}\frac{\partial \bar{u}}{\partial y} = \frac{\partial}{\partial y}\left\{ (v + \varepsilon_M)\frac{\partial \bar{u}}{\partial y}\right\} = \frac{\partial}{\partial y}\left(\frac{\tau}{\rho}\right) \tag{4.84}$$

$$\bar{u}\frac{\partial \bar{T}}{\partial x} + \bar{v}\frac{\partial \bar{T}}{\partial y} = \frac{\partial}{\partial y}\left\{ (\alpha + \varepsilon_H)\frac{\partial \bar{T}}{\partial y}\right\} = \frac{\partial}{\partial y}\left(\frac{-q}{\rho C_p}\right) \tag{4.85}$$

where all the variables are time-averaged.

In the region close to the wall, the velocity components $\bar{u}$, $\bar{v}$ approach zero as the distance from the wall approaches zero. It is therefore clear that the inertial terms (the left-hand side of Equations (4.84) and (4.85)) also approach zero. Since the gradient of a constant is zero, we infer that the region close to the wall is characterised by a constant heat flux rate and constant shear stress. This supposition is in fact supported by a large amount of experimental evidence. The name given to this region is the *viscous sublayer*, and it is generally accepted that the boundary occurs at values of the non-dimensional coordinate

$$y^+ = \frac{y}{v}\left(\frac{\tau_s}{\rho}\right)^{1/2} < 5$$

From Equations (4.84) and (4.85) the ratio τ/q can be expressed as

$$-\frac{\tau}{q} = \frac{1}{C_p}\left(\frac{Pr + \alpha^{-1}\varepsilon_H}{1 + \alpha^{-1}\varepsilon_M}\right)\frac{d\bar{u}}{d\bar{T}} \tag{4.86}$$

The ratio $\varepsilon_M/\varepsilon_H$ is known as the *turbulent Prandtl number*, Pr_T. From the separate definitions of ε_M and ε_H, it is easy to see that Pr_T is not a fluid property but more a flow property. The turbulent Prandtl number can be measured, albeit with difficulty, by simultaneous measurements of the fluctuating components u', v' and T' and the gradients of the time-averaged quantities $\partial\bar{u}/\partial y$ and $\partial\bar{T}/\partial y$. Although, as previously noted, Pr_T should depend on the circumstances of the flow, a value of order unity is virtually general to a wide range of flows. For a fluid with $Pr = 1$ (an idealised case but, as we shall see later, one that it is possible to accommodate for fluids where the 'laminar' Prandtl number differs from unity) and $Pr_T \approx 1$, the quantity in the brackets in Equation (4.86) is also around unity, hence

$$-\frac{\tau}{q} = \frac{1}{C_p}\frac{d\bar{u}}{d\bar{T}} \tag{4.87}$$

Integrating from the wall, where $\tau = \tau_s$ and $q = q_s$; $\bar{T} = T_s$ and $\bar{u} = 0$, to the free stream where $\tau = 0$ and $q = 0$; $\bar{T} = T_\infty$ and $\bar{u} = U_\infty$:

$$\frac{\tau_s}{q_s} = \frac{U_\infty}{(T_s - T_\infty)C_p} \tag{4.88}$$

From the definitions

$$Nu_x = \frac{q_s x}{(T_s - T_\infty)k}$$

$$Pr = \frac{\mu C_p}{k}$$

$$Re_x = \frac{\rho U_\infty x}{\mu}$$

$$C_{f,x} = \frac{\tau_s}{\frac{1}{2}\rho U_\infty^2} \qquad \text{(friction coefficient)}$$

and after recalling that we have imposed the condition $Pr = 1$, Equation (4.88) becomes

$$Nu_x = \tfrac{1}{2} Re_x C_{f,x} \tag{4.89}$$

where the subscript x denotes a value obtained at distance x from the start of the boundary layer. Equation (4.89) is known as the *Reynolds analogy* and forms a link between the shear stress and the heat transfer. An identical expression to Equation (4.89) is obtained for laminar flow, since $Pr_T = 0$.

Although there appears to be a consensus on the value of the turbulent Prandtl number, just about all fluids have a value of Pr which differs from unity, Equation (4.89) is often modified to allow for this by including the empirical factor $Pr^{1/3}$:

$$Nu_x = \tfrac{1}{2} Re_x C_{f,x} Pr^{1/3} \tag{4.90}$$

which is known as the *Colburn analogy* and can be applied for $0.6 < Pr < 60$.

4.6 ● Integral equations

A complete solution of the boundary layer equations (4.47) and (4.48) is, even in the relatively simple case of laminar flow, a formidable task, and apart from a few special cases, it requires the aid of a computer. Approximate solutions are desirable as they can give some physical insight into a particular problem. For most design purposes they are sufficiently accurate and they can also help with the analysis of experimental data. One such method is the integral method attributed to von Kármán (1921) and Pohlhausen (1921). The method is referred to as an integral method because the partial differential equations are integrated across the boundary layer thickness and in doing so the partial derivatives are transformed into ordinary differentials. This process brings about a loss of information in the behaviour of velocity and temperature at specific points in the fluid since the differential equations are not satisfied for every fluid particle. What is important, however, is the behaviour at the wall and the edge of the boundary layer, and this information is retained through the use of appropriate boundary and compatibility conditions.

First, taking the boundary layer momentum equation (Equation (4.47) with the aid of Equation (4.57) for the buoyancy force) and integrating this across a distance Δ which exceeds the thickness of the velocity boundary layer thickness δ and the thermal boundary layer thickness δ_T:

$$\int_0^\Delta \rho\left(u\frac{\partial u}{\partial x} + v\frac{\partial u}{\partial y}\right)dy = \int_0^\Delta -\frac{dp}{dx}dy + \int_0^\Delta \rho\beta g(T - T_\infty)\,dy + \int_0^\Delta \frac{\partial \tau}{\partial y}dy \tag{4.91}$$

Since at this stage no assumption has been made about the shear stress, the above equation is valid for both laminar and turbulent flow (for clarity the overbars have been omitted). We can also express the pressure gradient in terms of the free stream velocity gradient (a form of Bernoulli's equation), where

$$-\frac{1}{\rho}\frac{dp}{dx} = U_\infty \frac{dU_\infty}{dx} \tag{4.92}$$

so

$$\int_0^\Delta \left(u\frac{\partial u}{\partial x} + v\frac{\partial u}{\partial y} \right) dy$$
$$= \int_0^\Delta U_\infty \frac{dU_\infty}{dx}\,dy + \int_0^\Delta \beta g(T - T_\infty)\,dy + \int_0^\Delta \frac{1}{\rho}\frac{\partial \tau}{\partial y}\,dy \tag{4.93}$$

Using continuity (4.46), $\partial u/\partial x + \partial v/\partial y = 0$, the term in braces on the left-hand side of Equation (4.93) can be rewritten with the aid of the rule for integrating a product:

$$\int_0^\Delta v\frac{\partial u}{\partial y}\,dy = \Big|_0^\Delta uv - \int_0^\Delta u\frac{\partial v}{\partial y}\,dy$$

Since at $y = 0$, $u = v = 0$ and at $y = \Delta$, $u = U_\infty$ then

$$\Big|_0^\Delta uv = U_\infty(v)_{y=\Delta}$$

and since from continuity $v = \int_0^\Delta (\partial v/\partial y)\,dy = -\int_0^\Delta (\partial u/\partial x)\,dy$ then

$$\int_0^\Delta v\frac{\partial u}{\partial y}\,dy = -U_\infty \int_0^\Delta \frac{\partial u}{\partial x}\,dy + \int_0^\Delta u\frac{\partial u}{\partial x}\,dy \tag{4.94}$$

Equation (4.94) is substituted into Equation (4.93) and the transverse component of velocity, v (which is normal to the surface and less easy to specify than u), and also the derivatives with respect to y are eliminated. After some rearrangement, and noting that at $y = 0$, $\tau = \tau_s$ (subscript 's' is used to denote value at the surface) and at $y = \Delta$, $\tau = 0$, Equation (4.93) becomes

$$\frac{d}{dx}\left\{ \int_0^\Delta u(U_\infty - u)\,dy \right\} + \frac{dU_\infty}{dx}\left\{ \int_0^\Delta (U_\infty - u)\,dy \right\}$$
$$= \frac{\tau_s}{\rho} - \int_0^\Delta \beta g(T - T_\infty)\,dy \tag{4.95}$$

Equation (4.95) is the *momentum integral* equation for steady, incompressible, two-dimensional laminar or turbulent flow. Note that the partial derivatives have been replaced by ordinary derivatives because all variables are now expressed as functions of the streamwise coordinate, x. Further simplification is possible: for example, in free convection, $U_\infty = 0$; in forced convection, the body force term B (4.57) may usually be neglected and if the free stream velocity is constant then $dU_\infty/dx = 0$.

To solve the momentum integral equation for a particular physical situation, U_∞ is usually known and all that needs to be specified is the shear stress and $u = u(y)$. For laminar flow, the velocity profile can be deduced using compatibility conditions which result in a parabolic velocity profile. For turbulent flow, $u(y)$ can be obtained from experimental data, a common example being the one-seventh power law distribution. For laminar flow, the shear stress at the surface is known in terms of the velocity gradient, $\tau = \mu \partial u/\partial y$; for turbulent flow empirical data is used.

The integral energy equation may be derived in a similar way by integrating the boundary layer energy equation (4.48), neglecting dissipation, in the cross-stream direction across a thickness Δ, which again is taken to be the maximum of either the velocity thermal boundary layer thicknesses, i.e. $\Delta = \max(\delta, \delta_T)$:

$$\int_0^\Delta \rho C_p \left\{ u \frac{\partial T}{\partial x} + v \frac{\partial T}{\partial y} \right\} dy = -\int_0^\Delta \frac{\partial q}{\partial y}\, dy \tag{4.96}$$

Once again the unknown velocity, v, may be eliminated by using the continuity equation. Integrating by parts and noting that at $y = 0$, $v = 0$ and at $y = \Delta$, $T = T_\infty$, gives

$$\int_0^\Delta v \frac{\partial T}{\partial y}\, dy = \left. vT \right|_0^\Delta - \int_0^\Delta T \frac{\partial v}{\partial y}\, dy$$

$$= -\int_0^\Delta T_\infty \frac{\partial u}{\partial x}\, dy + \int_0^\Delta T \frac{\partial u}{\partial x}\, dy \tag{4.97}$$

Substitution of Equation (4.97) into Equation (4.96), with $q = q_s$ at $y = 0$ and $q = 0$ at $y = \Delta$, this gives

$$\int_0^\Delta \rho C_p \left\{ u \frac{\partial T}{\partial x} + \frac{\partial u}{\partial x}(T - T_\infty) \right\} dy = q_s \tag{4.98}$$

For many flows, T_∞ is constant and Equation (4.98) can be rewritten as

$$\frac{d}{dx} \left\{ \int_0^\Delta u(T - T_\infty)\, dy \right\} = \frac{q_s}{\rho C_p} \tag{4.99}$$

which is the *energy integral* equation and is valid for steady, laminar or turbulent, incompressible, two-dimensional boundary layer flow with an isothermal free stream temperature. To solve this, and the momentum integral equation equation, distributions of u and T ($u = u(y)$ and $T = T(y)$) are assumed, resulting in a pair of ordinary differential equations, one for δ and the other for δ_T. It is worth noting that errors in the calculation of the heat flux, caused by differences between asumed and 'real' distributions of u and T, are not usually significant. Although the assumed distributions are chosen to match conditions only at the surface and the free stream, when the equations are integrated the errors tend to average out. Since the integrals are performed over the boundary layer, the integrands in Equations (4.95) and (4.99) containing the terms $(u - U_\infty)$ and $(T - T_\infty)$ vanish for values of $y > \delta$ and δ_T. Hence it is necessary to integrate only over the physical extent of the boundary layer.

The integral equations are a useful simplification to the boundary layer equations, which are themselves a simplification of the Navier–Stokes and energy equations. As is shown in Chapters 5 and 6, the integral equations are also particularly amenable to analytical solution. Although this is straightforward in the case of forced convection, requiring only the solution of two separate ordinary differential equations; for free convection, where the variable T appears in both momentum and energy equations, (4.95) and (4.99), it requires the solution of two *simultaneous* first-order differential equations.

4.7 ● Scale analysis

A further stage of simplification is possible by using scale analysis. This was applied in Section 4.3 to examine the relative significance of terms in the momentum equation. In this section we apply those same arguments to the boundary layer region to estimate not only the thickness of the velocity and thermal boundary layers but also the heat transfer coefficient. In relation to its simplicity, this method, promoted by Bejan (1995), gives remarkably informative results.

In the velocity boundary layer, y varies from 0 to δ, and from 0 to δ_T in the thermal boundary layer. The corresponding changes in u and T are 0 to U_∞ and $\Delta T (= T_s - T_{ref})$ to 0, respectively. The streamwise coordinate, x, is characterised by a length scale, L. Hence in terms of scale, the fundamental question in convective heat transfer posed by Equation (4.2),

$$h = \frac{-k(\partial T/\partial y)_{y=0}}{(T_s - T_{ref})}$$

becomes

$$h \sim \frac{k\Delta T}{\delta_T \Delta T} = \frac{k}{\delta_T} \tag{4.100}$$

and since $Nu = hL/k$,

$$Nu \sim \frac{L}{\delta_T} \tag{4.101}$$

So, to estimate Nu we need to know the thermal boundary layer thickness, which in turn requires knowledge of the velocity boundary layer thickness. This we are able to do by examining the order of magnitude value of terms in the continuity, and boundary layer momentum and energy equations. For incompressible flow, (4.46), (4.47) and (4.48), neglecting viscous dissipation:

$$\frac{\partial u}{\partial x} + \frac{\partial v}{\partial y} = 0 \tag{4.102}$$

$$\rho\left(u\frac{\partial u}{\partial x} + v\frac{\partial u}{\partial y}\right) = F_x - \frac{\mathrm{d}p}{\mathrm{d}x} + \mu\frac{\partial^2 \mu}{\partial y^2} \tag{4.103}$$

$$\rho\,C_{\mathrm{p}}\left(u\frac{\partial T}{\partial x} + v\frac{\partial T}{\partial y}\right) = k\frac{\partial^2 T}{\partial y^2} \tag{4.104}$$

4.7.1 Laminar forced convection

A simple yet relevant case is that of flow over a flat plate $(\mathrm{d}p/\mathrm{d}x = 0)$. From Equation (4.102) we are able to estimate the scale of the transverse velocity component v as

$$v \sim U_\infty \delta/L \tag{4.105}$$

Since forced convection implies that F_x is zero, Equation (4.103) becomes in terms of scale

$$\{[U_\infty^2/L], [vU_\infty/\delta]\} \sim (\mu/\rho)[U_\infty/\delta^2] \tag{4.106}$$

Substituting the estimate of v from Equation (4.105), we obtain the result

$$\{[U_\infty^2/L], [U_\infty^2/L]\} \sim (\mu/\rho)[U_\infty/\delta^2] \tag{4.107}$$

Equation (4.107) tells us that the inertial terms in the momentum equation are of comparable magnitude, hence

$$[U_\infty^2/L] \sim (\mu/\rho)[U_\infty/\delta^2] \tag{4.108}$$

or, after rearranging,

$$\delta/L \sim (\mu/\rho U_\infty L)^{1/2} = Re_L^{-1/2} \tag{4.109}$$

Equation (4.109) predicts the exponent of the Reynolds number derived from more 'exact' analysis. As we have examined orders of magnitude, a leading constant, which would be obtained from more sophisticated methods, is not present.

The extension to the thermal boundary layer depends on the ratio $\delta/\delta_{\mathrm{T}}$.

Case 1: $\delta_{\mathrm{T}} \gg \delta$ (or Pr $\ll$ 1, for example liquid metals)
In the thermal boundary layer, $u \sim U_\infty$ and, from Equation (4.105), $v \sim U_\infty\delta/L$. The order of magnitude of the terms in the boundary layer energy equation becomes

$$\{[U_\infty\Delta T/L], [U_\infty\delta\Delta T/\delta_{\mathrm{T}}L]\} \sim \alpha[\Delta T/\delta_{\mathrm{T}}^2] \tag{4.110}$$

The second term in the left-hand bracket is small in relation to the first since we have imposed the condition that $\delta_{\mathrm{T}} \gg \delta$, so

$$[U_\infty\Delta T/L] \sim \alpha[\Delta T/\delta_{\mathrm{T}}^2] \tag{4.111}$$

Rearranging then dividing and multiplying by the viscosity, μ:

$$\left(\frac{\delta_T}{L}\right) \sim \left(\frac{\alpha}{U_\infty}L\right)^{1/2} = \left\{\left(\frac{k}{\rho}C_p\right)\left(\frac{\mu}{\mu}\right)\left(\frac{1}{U_\infty}L\right)\right\}^{1/2}$$

$$= (Re_L \cdot Pr)^{-1/2} \tag{4.112}$$

The Nusselt number from Equation (4.101) is $Nu \sim L/\delta_T$, the inverse of Equation (4.112), so for a relatively thick thermal boundary layer

$$Nu \sim (Re_L\,Pr)^{1/2} \tag{4.113}$$

Case 2: $\delta \gg \delta_T$ and $\delta \approx \delta_T$ (or $Pr \gg 1$ and $Pr \approx 1$, for example oils and gases, respectively)

For a relatively thin thermal boundary layer, the velocity u does not reach its free stream value in the distance δ_T. As an estimate, we may consider u to vary from zero at the surface to $U_\infty(\delta_T/\delta)$ at the limit of the thermal boundary layer. From applying continuity, with this scale (4.102), we find that $v \sim U_\infty \delta_T^2/\delta L$. The order of magnitude of the terms in the boundary layer energy equation becomes

$$\{[U_\infty \Delta T \delta_T/\delta L], [U_\infty \Delta T \delta_T/\delta L]\} \sim \alpha[\Delta T/\delta_T^2] \tag{4.114}$$

In this case the two terms in the left-hand bracket are of comparable magnitude, and so

$$[U_\infty \delta_T/\delta L] \sim \alpha/\delta_T^2 \tag{4.115}$$

From which

$$\delta_T^3 \sim \frac{\alpha \delta L}{U_\infty} \tag{4.116}$$

Rearranging then dividing and multiplying by the viscosity, μ, and using Equation (4.109)

$$\left(\frac{\delta_T}{L}\right) \sim \left(\frac{\alpha \delta}{U_\infty L^2}\right)^{1/3} = \left\{\left(\frac{\delta}{L}\right)(Pr \cdot Re_L)^{-1}\right\}^{1/3}$$

$$= Re_L^{-1/2}Pr^{-1/3} \tag{4.117}$$

The Nusselt number from Equation (4.101) is still $Nu \sim L/\delta_T$, the inverse of Equation (4.117), so for a relatively thin thermal boundary layer

$$Nu \sim Re_L^{1/2}Pr^{1/3} \tag{4.118}$$

4.7.2 Laminar free convection

In free convection the velocity in the thermal boundary layer is zero at both $y = 0$ and $y = \delta$, so there does not appear to be a simple and obvious choice of scale for u. Applying our scaling relations to the continuity equation across the thermal boundary layer, we find that $v \sim u\delta_T/L$. Using this result to examine the terms in the energy equation

$$\{[u\Delta T/L], [u\Delta T/L]\} \sim \alpha[\Delta T/\delta_T^2] \qquad (4.119)$$

we are able to deduce the appropriate scale for u as

$$u \sim \alpha\frac{L}{\delta_T^2} \qquad (4.120)$$

As explained in Section 4.4.2, the body force arising from buoyancy and the hydrostatic pressure gradient are combined in the momentum equation as

$$F_x - \frac{dp}{dx} = \rho\beta g\Delta T$$

The terms in the momentum equation are then

$$\{[u^2/L], [u^2/L]\} \sim (\mu/\rho)[u/\delta_T^2], [\beta g\Delta T] \qquad (4.121)$$

Since buoyancy is the motivating force in free convection, we are unable to ignore it, so it is useful to compare the size of the remaining terms, arising from inertia (on the left-hand side) and friction (on the right-hand side), with buoyancy represented by unity. After dividing by $\beta g\Delta T$, and using the scale for u as defined in Equation (4.120),

$$[(L/\delta_T)^4 Ra_L^{-1}Pr^{-1}] \sim [(L/\delta_T)^4 Ra_L^{-1}], 1 \qquad (4.122)$$

where Ra_L is the Rayleigh number $(Ra_L = Gr_L Pr = \rho^2 g\beta\Delta T L^3 C_p/k\mu)$. Inspection of Equation (4.122) tells us that the difference between a flow dominated by a balance of inertial to buoyancy forces and one dominated by a balance of frictional to buoyancy forces depends on the value of the Prandtl number.

Case 1: Pr ≫ 1 and Pr ≈ 1
In this case the inertial term is small in relation to friction and Equation (4.122) provides

$$\frac{\delta_T}{L} \sim Ra_L^{-1/4} \qquad (4.123)$$

And from Equation (4.101)

$$Nu \sim Ra_L^{1/4} \qquad (4.124)$$

Case 2: Pr ≪ 1
There is now a balance between inertial and buoyancy forces, from which

$$\frac{\delta_T}{L} \sim (Ra_L Pr)^{-1/4} \qquad (4.125)$$

And from Equation (4.101)

$$Nu \sim (Ra_L Pr)^{1/4} \qquad (4.126)$$

4.8 ● Closing comments

Convection is the mode of heat transfer that occurs due to fluid motion. Convective heat transfer may be subdivided according to a number of categories:

1. The physical cause of motion: free or natural convection, forced convection, mixed convection.
2. The flow configuration: internal or external flow.
3. The presence of turbulence: laminar, transitory or turbulent flow.
4. The nature of flow: single-phase, boiling, condensation, melting, solidifying.

Knowledge of the convective heat transfer coefficient, h, allows us to calculate heat transfer rates by convection. However, h is not a physical constant (such as the thermal conductivity) but varies from one flow to another and from one part of a flow to another. Dimensional analysis leads to the result that the heat transfer coefficient can be represented by the Nusselt number, a dimensionless group. For forced convection, the Nusselt number depends on the Reynolds and Prandtl numbers. For free convection, the Nusselt number depends on the Grashof and Prandtl numbers. To be able to *calculate* the heat transfer coefficient for any given flow, we must calculate the temperature distribution in the fluid, which in turn is influenced by the velocity distribution. The complete mathematical description of convective heat transfer is therefore complex, and it normally requires considerable computing power to solve the governing simultaneous partial differential equations.

The concept of boundary layer theory greatly reduces this complexity. By recognising that the most significant changes in temperature and velocity components occur in a thin (boundary) layer close to the surface, we are able to eliminate pressure as a variable and ignore certain derivatives in the terms arising from the viscous stress. The resulting boundary layer equations are more manageable, but some form of numerical (as opposed to analytical) solution is almost always necessary.

Integral equations represent a further step in the direction of simplification, and are amenable to direct analytical solution. A further step that provides much useful information with relatively little effort is scale analysis.

Laminar flow is, by its very nature, ordered. However, turbulent convection is probably more common on the scale of engineering applications. By separating the instantaneous values of velocity, temperature, pressure and density into mean and fluctuating components, and time averaging, we are able to obtain time-averaged boundary layer equations. The physical significance of these is that the shear stress and heat flux are greatly enhanced in turbulent flow, owing to the inherent eddy motion. However, two new unknowns are introduced: the eddy momentum and thermal diffusivities. Consequently it is necessary to use some form of turbulence model which links these with velocity and temperature distributions. The analogy between heat and momentum is a useful means of obtaining heat transfer rates from established skin friction data.

This chapter has presented the fundamental equations, various simplifications and dimensionless groups relevant to convective heat transfer. In Chapters 5 and 6 we apply these relations to examine heat transfer in forced and free convection, respectively. The subject of heat transfer during the processes of condensation and boiling is discussed in Chapter 7.

4.9 ● References

Bejan, A. (1995). *Convective Heat Transfer*. New York: Wiley

Bejan, A. and Large, J.L. (1990). The Prandtl number effect on the transition in natural convection along a vertical surface. *J. Heat Transfer*, **112**, 787–90

Bradshaw, P., ed. (1978). *Topics in Applied Physics*. Vol. 12: *Turbulence* 2nd edn. Berlin: Springer-Verlag

Hinze, J.O. (1975). *Turbulence* 2nd edn. New York: McGraw-Hill

Lock, G.S.H. (1996). *Latent Heat Transfer*. Oxford: Oxford Science Publications

McComb, W.D. (1994). *The Physics of Fluid Turbulence*. Oxford: Oxford Science Publications/Clarendon Press

Pohlhausen, K (1921). Zur näherungsweisen Integration der Differentialgleichung der laminaren Reibungsschicht. *ZAMM*, **1**, 252–68

Rogers, G.F.C. and Mayhew, Y.R. (1992). *Engineering Thermodynamics, Work and Heat Transfer* 4th edn. Harlow: Longman

Schlichting, H. (1979). *Boundary Layer Theory* 7th edn. Maidenhead: McGraw-Hill

Von Kármán, T. (1921). Uber Laminare und Turbulente Reibung. *ZAMM*, **1**, 233–52

4.10 ● End of chapter questions

4.1 Calculate the Prandtl number ($Pr = \mu C_p/k$) for the following.

(a) Water at 20°C: $\mu = 1.002 \times 10^{-3}$ kg/m s, $C_p = 4.183$ kJ/kg K and $k = 0.603$ W/m K.

[6.95]

(b) Water at 90°C: $\rho = 965$ kg/m³, $v = 3.22 \times 10^{-7}$ m²/s, $C_p = 4208$ J/kg K and $k = 0.676$ W/m K.

[1.93]

(c) Air at 20°C and 1 bar: $R = 287$ J/kg K, $v = 1.563 \times 10^{-5}$ m²/s, $C_p = 1005$ J/kg K and $k = 0.026\,24$ W/m K.

[0.719]

(d) Air at 100°C: $\mu = \dfrac{1.46 \times 10^{-6} T^{3/2}}{(110 + T)}$ kg/m s

$C_p = 0.917 + 2.58 \times 10^{-4} T - 3.98 \times 10^{-8} T^2$ kJ/kg K (where T is the absolute temperature in K) and $k = 0.031\,86$ W/m K.

[0.689]

(e) Mercury at 20°C: $\mu = 1520 \times 10^{-6}$ kg/m s, $C_p = 0.139$ kJ/kg K and $k = 0.0081$ kW/m K.

[0.0261]

(f) Liquid sodium at 400 K: $\mu = 420 \times 10^{-6}$ kg/m s, $C_p = 1369$ J/kg K and $k = 86$ W/m K.

[0.0067]

(g) Engine oil at 60°C: $\mu = 8.36 \times 10^{-2}$ kg/m s, $C_p = 2035$ J/kg K and $k = 0.141$ W/m K.

[1207]

4.2 Calculate the appropriate Reynolds numbers and state if the flow is laminar or turbulent for the following.

(a) A 10 m (waterline length) long yacht sailing at 13 km/h in seawater $\rho = 1000\,kg/m^3$ and $\mu = 1.3 \times 10^{-3}\,kg/m\,s$.

$$[2.78 \times 10^6, \text{turbulent}]$$

(b) A compressor disc of radius 0.3 m rotating at 15 000 rev/min in air at 5 bar and 400°C and

$$\mu = \frac{1.46 \times 10^{-6}T^{3/2}}{(110 + T)}\,kg/m\,s$$

$$[1.12 \times 10^7, \text{turbulent}]$$

(c) 0.05 kg/s of CO_2 gas at 400 K flowing in a 20 mm dia. pipe. For the viscosity take

$$\mu = \frac{1.56 \times 10^{-6}T^{3/2}}{(233 + T)}\,kg/m\,s$$

$$[1.6 \times 10^5, \text{turbulent}]$$

(d) The roof of a coach 6 m long, travelling at 100 km/h in air ($\rho = 1.2\,kg/m^3$ and $\mu = 1.8 \times 10^{-5}\,kg/m\,s$).

$$[11.1 \times 10^6, \text{turbulent}]$$

(e) The flow of exhaust gas (p = 1.1 bar, $T = 500°C$, $R = 287\,J/kg\,K$ and $\mu = 3.56 \times 10^{-5}\,kg/m\,s$) over a valve guide of diameter 10 mm in a 1.6 litre, four-cylinder, four-stroke engine running at 3000 rev/min (assume 100% volumetric efficiency, an inlet density of 1.2 kg/m³ and an exhaust port diameter of 25 mm).

$$[6869, \text{laminar}]$$

4.3 Calculate the appropriate Grashof numbers and state if the flow is laminar or turbulent for the following.

(a) A central heating radiator, 0.6 m high with a surface temperature of 75°C in a room at 18°C ($\rho = 1.2\,kg/m^3$, $Pr = 0.72$ and $\mu = 1.8 \times 10^{-5}\,kg/ms$).

$$[1.84 \times 10^9, \text{mostly laminar}]$$

(b) A horizontal oil sump, with a surface temperature of 40°C, 0.4 m long and 0.2 m wide containing oil at 75°C ($\rho = 854\,kg/m^3$, $Pr = 546$, $\beta = 0.7 \times 10^{-3}\,K^{-1}$ and $\mu = 3.56 \times 10^{-2}\,kg/m\,s$).

$$[4.1 \times 10^4, \text{laminar}]$$

(c) The external surface of a heating coil, 30 mm dia., having a surface temperature of 80°C in water at 20°C ($\rho = 1000\,kg/m^3$, $Pr = 6.95$, $\beta = 0.227 \times 10^{-3}\,K^{-1}$ and $\mu = 1.00 \times 10^{-3}\,kg/m\,s$.

$$[3.6 \times 10^6, \text{laminar}]$$

(d) Air at 20°C ($\rho = 1.2\,kg/m^3$, $Pr = 0.72$ and $\mu = 1.8 \times 10^{-5}\,kg/m\,s$) adjacent to a 60 mm dia. horizontal light bulb with a surface temperature of 90°C.

$$[3.5 \times 10^4, \text{laminar}]$$

4.4 Calculate the Nusselt numbers for the following.

(a) A gas flow ($Pr = 0.71$, $\mu = 4.63 \times 10^{-5}\,kg/m\,s$ and $C_p = 1175\,J/kg\,K$) over a turbine blade of chord length 20 mm, where the average heat transfer coefficient is 1000 W/m²K.

$$[261]$$

(b) A horizontal electronics component with a surface temperature of 35°C, 5 mm wide and 10 mm long, dissipating 0.1 W by free convection from one side into air where the temperature is 20°C and $k = 0.026\,W/m\,K$.

$$[8.5]$$

(c) A 1 kW central heating radiator 1.5 m long and 0.6 m high with a surface temperature of 80°C dissipating heat by radiation and convection into a room at 20°C ($k = 0.026\,W/m\,K$, assume black body radiation and $\sigma = 56.7 \times 10^{-9}\,W/m\,K^4$).

$$[249]$$

(d) Air at 4°C ($k = 0.024\,W/m\,K$) adjacent to a wall 3 m high and 0.15 m thick made of brick with $k = 0.3\,W/m\,K$, the inside temperature of the wall is 18°C, the outside wall temperature 12°C.

$$[188]$$

4.5 The local Nusselt number for flow over a triangular plate of length L and base width W is given by the expression $Nu_x = C \cdot Re_x^n$, where x is the distance from the apex. For a plate, with a constant surface temperature and with the apex at the leading edge of the flow, show that the average Nusselt number, Nu_{av}, is given by

$$Nu_{av} = \frac{2C}{(n + 1)}Re_L^n$$

4.6 In forced convection for flow over a flat plate the local Nusselt number can be represented by the general expression $Nu_x = C_1 Re_x^n$. In free convection from a vertical surface the local Nusselt number is represented by $Nu_x = C_2 Gr_x^m$, where for a fixed value of Pr, C_1, C_2, n and m are constants.

(a) Show that the local heat transfer coefficient is independent of the surface-to-air temperature difference in forced convection, whereas in free convection h depends upon $(T_s - T_\infty)^m$.

(b) In turbulent free convection, it is generally recognised that $m = \frac{1}{3}$. Show that the local heat transfer coefficient does not vary with coordinate x.

4.7 In a boundary layer flow over a flat plate with zero external pressure gradient, there is a balance between viscous and inertial forces. Using this argument show that the boundary layer thickness δ, at a distance x downstream, is approximated by

$$\frac{\delta}{x} \approx Re_x^{-1/2}$$

4.8 Consider an annular control volume of fluid of axial length Δz and radial thickness Δr with velocity components V_z and V_r in the axial and radial directions, respecitvely. Show that for steady, axisymmetric and incompressible flow the continuity equation is

$$\frac{\partial}{\partial r}(r\,V_r) + \frac{\partial}{\partial z}(r\,V_z) = 0$$

4.9 Following on from Question 4.8, show that the z direction momentum equation for steady, incompressible, axisymmetric flow in a cylindrical coordinate system, with a body force F_z is given by

$$\rho\left\{V_r \frac{\partial V_z}{\partial r} + V_z \frac{\partial V_z}{\partial z}\right\} = F_z - \frac{\partial p}{\partial z}$$

$$+ \mu\left\{\frac{\partial^2 V_z}{\partial r^2} + \frac{1}{r}\frac{\partial V_z}{\partial r} + \frac{\partial^2 V_z}{\partial z^2}\right\}$$

Assume the following relations for the stress components:

$$\sigma_z = -p + 2\mu\frac{\partial V_z}{\partial z}$$

$$\sigma_r = -p + 2\mu\frac{\partial V_r}{\partial r}$$

$$\tau_{r,z} = \mu\left\{\frac{\partial V_r}{\partial z} + \frac{\partial V_z}{\partial r}\right\}$$

4.10 (a) Draw a labelled schematic diagram showing the features of a velocity boundary layer for forced convection over a flat plate. Also show in the diagram the essential differences between the laminar and turbulent parts of the boundary layer.

(b) Results from a series of heat transfer tests on printed circuit boards (PCBs) indicate that the local Nusselt numbers are correlated by

$$Nu_x = 0.3Re_x^{0.5} \quad \text{for } Re_x \leqslant 2 \times 10^4$$
$$Nu_x = 0.02Re_x^{0.9} \quad \text{for } Re_x > 2 \times 10^4$$

Show that the average Nusselt number for a board of length L, of constant surface temperature, for which $Re_x \gg 2 \times 10^4$ is given by the expression

$$Nu_{av,L} = 0.0222Re_L^{0.9} - 80.2$$

(c) Four PCBs, each dissipating 60 W of electrical power, are contained within a cabinet. Each PCB is 125 mm high and 30 mm long. The spacing between adjacent boards is 40 mm and that between the walls of the cabinet and the PCB at each end is 20 mm.

Use the above equation to select from the table below a cooling fan for the cabinet which will maintain the surface temperature of the PCBs at 50°C when the ambient air temperature is 25°C (for air take $\rho = 1.2$ kg/m³, $k = 0.026$ W/m K and $\mu = 1.8 \times 10^{-5}$ kg/m s).

Fan	Flow rate (m³/s)
a	0.015
b	0.07
c	0.28
d	7.0

[0.068 m³/s, fan b]

4.11 The heat transfer by free convection from the casing of a jet engine of 2 m dia. and 3 m length is to be investigated using a laboratory-scale model in water. The full-size engine casing has a steady-state operating temperature of 120°C when the external air temperature is 18°C and the model is to be heated to 60°C and immersed in water at 20°C.

(a) At what temperature should the fluid properties for air and water be evaluated?

[air 69°C; water 40°C]

(b) What should be the size of the model?

[0.3169 m dia., 0.475 m long]

(c) During a test on the model it is found that 17.4 kW are required to balance the losses by convection from the cylindrical surface. What would be the heat loss from the full-size engine?

For air at 69°C take these values:
$\mu = 2.04 \times 10^{-5}$ kg/m s, $k = 0.0293$ W/m K, $\rho = 1.02$ kg/m³ and $C_p = 1007$ J/kg K.

For water at 40°C take these values:
$\mu = 651 \times 10^{-6}$ kg/m s, $k = 0.632$ W/m K, $\rho = 1000$ kg/m³, $C_p = 4179$ J/kg K and $\beta = 381 \times 10^{-6}$ K⁻¹.

[13 kW]

4.12 The data listed below give measurements of the pressure drop, Δp, against the bulk mean velocity, U_m, for a test carried out on fully developed flow in a heat exchanger pipe with a turbulence promoter. The test section is 1 m long with an effective pipe diameter of 40 mm, and water ($\rho = 1000$ kg/m³ and $\mu = 1 \times 10^{-3}$ kg/m s) was used as the test fluid.

U_m (m/s)	Δp (kN/m²)
1	6
2	24
5	150
10	600

(a) Consider a force balance on an element in fully developed pipe flow and derive a relationship between the friction factor, f, and the skin friction coefficient, C_f.

[$f = 4C_f$]

(b) Plot the relationship between U_m and Δp, and using the result for part (a) and the Colburn analogy (4.90), show that

$$Nu_D = 0.06 Re_D Pr^{1/3}$$

(c) Estimate the surface heat transfer coefficient when a geometrically similar pipe of 0.01 m effective diameter is used to convey 1.8×10^{-5} m³/s of liquid refrigerant. ($\rho = 669$ kg/m³ and $\mu = 0.205 \times 10^{-3}$ kg/m s, $C_p = 4.52$ kJ/kg K and $k = 0.547$ W/m K.)

[29 000 W/m²K]

Chapter five

Forced convection

5.1 ● Introduction

In this section we shall apply the principles outlined in Chapter 4 to forced convection. As mentioned in the introductory remarks to that chapter, treating a flow as forced convection implies that the buoyancy force is negligible. Forced convective heat transfer is created by flow that is driven directly from the action of a fan or a pump, or due to relative motion such as that between a moving surface in a stationary fluid. Idealised examples considered in this chapter are the heat transfer from a flat plate, through a duct, a rotating disc, objects such as spheres and cylinders in a forced crossflow and the heat transfer due to an impinging jet. In some simple cases, such as the flat plate and a circular duct, the derivation of an analytical solution is given. Correlations are cited for cases where the flow or geometry is more complex. Although idealised, these geometries have wide application to design of thermal systems. Typical applications to engineering hardware are given in the text where appropriate.

In addition to the Nusselt number, the other relevant dimensionless groups in forced convection are the Reynolds number, the Prandtl number and, for high-speed flows, the Eckert number. An important consequence of the distinctions laminar/turbulent, internal/external flows is the combination of these dimensionless groups and the numerical value of constants such as C_1, a and b in Equation (4.64b).

5.2 ● Parallel flow

Figure 5.1 shows a schematic representation of parallel flow between two plates having different velocities and separated by a sufficiently small distance, s, to avoid the formation of separate boundary layers on each plate. In parallel flow only one of the velocity components is different from zero; in this example the streamwise velocity, $u \neq 0$. The case of parallel flow between two plates (known as Couette flow if one of the plates is moving and Poiseuille flow if both plates are stationary) is one of the few cases where there is an exact solution to the boundary layer

Fig. 5.1 ●

Parallel flow
between two
plates

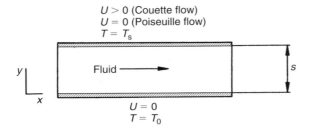

$U > 0$ (Couette flow)
$U = 0$ (Poiseuille flow)
$T = T_s$

Fluid

s

y

x

$U = 0$
$T = T_0$

equations. Parallel flow occurs for example in the narrow gap between a lubricated bearing and its journal.

Starting with the two-dimensional continuity, and boundary layer momentum and energy equations for constant density, steady flow (4.46) to (4.48), and omitting the body force term ($F_x = 0$),

$$\frac{\partial u}{\partial x} + \frac{\partial v}{\partial y} = 0 \tag{5.1}$$

$$\rho\left[u\frac{\partial u}{\partial x} + v\frac{\partial u}{\partial y}\right] = -\frac{\mathrm{d}p}{\mathrm{d}x} + \mu\frac{\partial^2 u}{\partial y^2} \tag{5.2}$$

$$\rho C_p\left[u\frac{\partial T}{\partial x} + v\frac{\partial T}{\partial y}\right] = k\frac{\partial^2 T}{\partial y^2} + \mu\left(\frac{\partial u}{\partial y}\right)^2 \tag{5.3}$$

Using the condition that $v = 0$, Equations (5.1) and (5.2) become

$$\frac{\mathrm{d}u}{\mathrm{d}x} = 0 \tag{5.4}$$

$$0 = -\frac{\mathrm{d}p}{\mathrm{d}x} + \mu\frac{\mathrm{d}^2 u}{\mathrm{d}y^2} \tag{5.5}$$

Each plate is at a uniform temperature, so $\partial T/\partial x = 0$, and with $v = 0$ the boundary layer energy equation (5.3) becomes

$$0 = k\frac{\mathrm{d}^2 T}{\mathrm{d}y^2} + \mu\left(\frac{\mathrm{d}u}{\mathrm{d}y}\right)^2 \tag{5.6}$$

Note the partial derivatives in the original equations have been replaced with ordinary derivatives since u and T are dependent only on y.

For *Couette flow*, the plate at $y = 0$ is stationary and the plate at $y = s$ is moving with a velocity U. Equation (5.5) can be integrated to give

$$u = \left(\frac{y}{s}\right)\left\{U - \left(\frac{s^2}{2\mu}\right)\left(\frac{\mathrm{d}p}{\mathrm{d}x}\right)\left[\left(1 - \frac{y}{s}\right)\right]\right\} \tag{5.7}$$

In Couette flow, fluid motion is created not by the pressure gradient but by the movement of one plate relative to the other, hence $\mathrm{d}p/\mathrm{d}x = 0$ and

$$u = \frac{Uy}{s} \tag{5.8}$$

Likewise, and using the velocity distribution given by Equation (5.8), the simplified energy equation (5.6) may be integrated to give

$$T = T_0 + \frac{\mu U^2}{2k}\left[\frac{y}{s} - \left(\frac{y}{s}\right)^2\right] + (T_s - T_0)\frac{y}{s} \tag{5.9}$$

Equation (5.9) shows us something common to all problems in convection, namely the interdependence between the velocity and temperature fields.

Example 5.1

The journal bearing shown in Figure 5.2 has a radius of 30 mm, runs at 3000 rev/min and is lubricated using oil with $\mu = 0.3\,\text{kg/m s}$ and $k = 0.15\,\text{W/m K}$. The respective temperatures of the outer stationary and inner moving surfaces are 20°C and 40°C, and the radial clearance between them is 0.1 mm. Calculate:

(a) The temperature distribution and the maximum temperature in the lubricating oil.

(b) The surface heat fluxes.

Solution **(a)** Although the journal bearing is circular, the clearance, s, is much less than the radius, so we can neglect curvature effects and undertake an analysis in Cartesian coordinates. Further, if we define the coordinate y

Fig. 5.2 ●
Parallel flow: a journal bearing (Example 5.1)

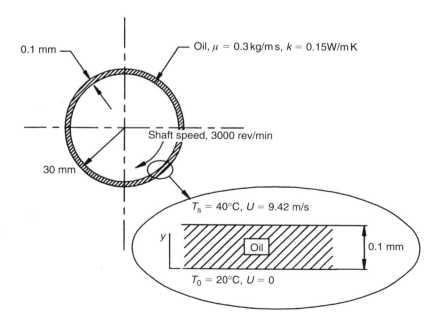

having its origin on the stationary surface, then the temperature distribution in the lubricating oil will be given by Equation (5.9):

$$T = T_0 + \frac{\mu U^2}{2k}\left[\frac{y}{s} - \left(\frac{y}{s}\right)^2\right] + (T_s - T_0)\frac{y}{s}$$

where $U = \Omega r = 3000 \times 2 \times \pi \times 0.03/60 = 9.42\,\text{m/s}$ and $s = 10^{-4}\,\text{m}$.

Inserting numerical values for T_0, T_s, μ, k and U gives

$$T = 20 + 108.8 \times 10^4 y - 88.8 \times 10^8 y^2 \tag{5.10}$$

The maximum temperature occurs when $\mathrm{d}T/\mathrm{d}y = 0$, and by differentiating Equation (5.10), it is found to occur at

$$y_{(T_{max})} = \frac{108.8 \times 10^{-4}}{177.6} = 0.061\,261\,\text{mm}$$

At this value of y the temperature is

$$T = 20 + 108.8 \times 10^4 \times (0.061261 \times 10^{-3}) - 88.8 \times 10^8 \times (0.061261 \times 10^{-3})^2$$

$$T = 53.33°\text{C}$$

(b) From Fourier's law, $q = -k(\mathrm{d}T/\mathrm{d}y)_{y=\text{surface}}$, and from differentiation of Equation (5.9)

$$-k\left(\frac{\mathrm{d}T}{\mathrm{d}y}\right)_{y=0} = -k\left[\frac{\mu U^2}{2ks} + \frac{(T_s - T_0)}{s}\right]$$

$$= -0.15\left[\left(\frac{0.3 \times 9.42^2}{0.3 \times 0.1 \times 10^{-3}}\right) + \left(\frac{20}{0.1 \times 10^{-3}}\right)\right]$$

$$= -163 \times 10^3\,\text{W/m}^2 \qquad \text{(heat flux from the oil to the stationary surface)}$$

For the surface at $y = s$

$$-k\left(\frac{\mathrm{d}T}{\mathrm{d}y}\right)_{y=s} = -k\left[\frac{-\mu U^2}{2ks} + \frac{(T_s - T_0)}{s}\right]$$

$$= 103 \times 10^3\,\text{W/m}^2 \qquad \text{(heat flux from the oil to the moving surface)}$$

Comment The maximum temperature found in the oil exceeds the surface temperature of either bearing surface owing to the action of viscous dissipation in the oil. Using the definitions of $Nu = qs/(T_s - T_0)k$; $Ec = U^2/C_p(T_s - T_0)$ and $Pr = \mu C_p/k$, it is quite simple to verify that the above relations may be expressed non-dimensionally as $Nu = -1 \pm \frac{1}{2}Ec \cdot Pr$, where the negative sign applies to the surface at $y = 0$ and the positive sign to the surface at $y = s$.

5.3 ● Laminar flow in a pipe

For two-dimensional pipe flow, it is more convenient to work in axial and radial velocity components, denoted by V_z and V_r respectively, and a cylindrical coordinate system with its origin on the centreline of the pipe. For incompressible pipe flow, the continuity equation and the (axial direction) momentum equation, neglecting the body force term, are

$$\frac{\partial V_r}{\partial r} + \frac{V_r}{r} + \frac{\partial V_z}{\partial z} = 0 \tag{5.11}$$

$$\rho\left[V_r\frac{\partial V_r}{\partial r} + V_z\frac{\partial V_z}{\partial z}\right] = -\frac{dp}{dz} + \mu\left[\frac{\partial^2 V_z}{\partial r^2} + \frac{1}{r}\frac{\partial V_z}{\partial r} + \frac{\partial^2 V_z}{\partial z^2}\right] \tag{5.12}$$

The corresponding form for the energy equation, neglecting viscous dissipation ($\Phi = 0$) is

$$\rho C_p\left[V_r\frac{\partial T}{\partial r} + V_z\frac{\partial T}{\partial z}\right] = k\left[\frac{\partial^2 T}{\partial r^2} + \frac{1}{r}\frac{\partial T}{\partial r} + \frac{\partial^2 T}{\partial z^2}\right] \tag{5.13}$$

Equations (5.11), (5.12) and (5.13) bear obvious similarities to their Cartesian counterparts (4.7a), (4.15) and (4.38a); the recurring presence of the factor, $1/r$, accounts for the variation in area with radial coordinate in a cylindrical coordinate system.

A schematic diagram of the boundary layer development in pipe flow is shown in Figure 5.3. Flow enters the pipe (or duct) with a uniform velocity profile. Downstream of the pipe entrance, the action of viscous forces retards the fluid in the region of the pipe wall and gives rise to a sheared velocity profile. In the entry region, the boundary layer develops over the entire surface of the pipe wall, leaving an inviscid 'core' region. After some distance downstream, the boundary layers forming on opposing sides of the pipe wall merge and fill the entire pipe, thus eliminating the core region. When this occurs the flow is said to be *fully developed* and, as a consequence, both V_z and the non-dimensionalised temperature become

Fig. 5.3 ●
How laminar hydrodynamic boundary layers develop in a circular tube

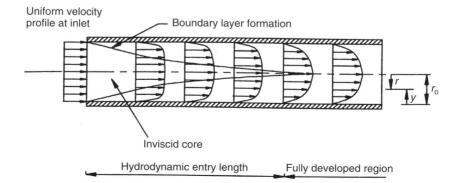

Uniform velocity profile at inlet

Boundary layer formation

Inviscid core

Hydrodynamic entry length   Fully developed region

independent of axial distance, z. The distance downstream along the pipe before the flow becomes fully developed is known as the *hydrodynamic entry length*, $L_{h,\,entry}$, and for laminar pipe flow $Re_D \lesssim 2000$. It is generally accepted that

$$\frac{L_{h,\,entry}}{D} \approx 0.05 Re_D \tag{5.14}$$

For fully developed pipe flow, where $\partial V_z / \partial z = 0$, Equation (5.11) simplifies to,

$$\frac{\partial V_r}{\partial r} + \frac{V_r}{r} = \frac{\partial}{\partial r}(rV_r) = 0 \tag{5.15}$$

Since $V_r = 0$ at $r = r_o$ (the pipe wall), the implication of Equation (5.15) is that $V_r = 0$ everywhere. Using this together with the condition of fully developed flow that $\partial V_z / \partial z = 0$, Equation (5.12) becomes

$$0 = -\frac{dp}{dz} + \mu \left(\frac{d^2 V_z}{dr^2} + \frac{1}{r}\frac{dV_z}{dr} \right)$$

which after rearranging gives

$$\frac{d}{dr}\left(r\frac{dV_z}{dr} \right) = \frac{r}{\mu}\frac{dp}{dz} \tag{5.16}$$

The same equation can be derived directly from considering a force–momentum balance on an element of fluid in fully developed pipe flow. Integrating Equation (5.16) and using the conditions that $V_z = 0$ at $r = r_o$ and $dV_z / dr = 0$ at $r = 0$ (axis of symmetry), provides the velocity distribution for fully developed laminar pipe flow:

$$V_z = \frac{-r_o^2}{4\mu}\frac{dp}{dz}\left\{ 1 - \left(\frac{r}{r_o} \right)^2 \right\} \tag{5.17}$$

It is more meaningful to relate the axial velocity to the *bulk average velocity* U_b, a quantity that is easily measured or obtained from mass continuity $\dot{m} = \rho U_b A$, where

$$U_b = \frac{1}{\pi\, r_o^2} \int_0^{r_o} 2\pi\, rV_z\, dr \tag{5.18}$$

Using Equation (5.17) for the axial velocity in Equation (5.18)

$$U_b = -\left(\frac{r_o^2}{8\mu} \right)\frac{dp}{dz} \tag{5.19}$$

and taking the ratio of Equation (5.17) to (5.19), to eliminate the pressure gradient, gives

$$\frac{V_z}{U_b} = 2\left[1 - \left(\frac{r}{r_o} \right)^2 \right] \tag{5.20}$$

For pipe flow, the friction factor, f, is defined as

$$f = \frac{-2r_o(dp/dz)}{\frac{1}{2}\rho U_b^2} \tag{5.21}$$

Using (5.19) in (5.21) gives the result for laminar flow:

$$f = \frac{64}{Re_D} \tag{5.22}$$

In general, the thermal and hydrodynamic entry lengths are not of course equal. For laminar flow, the *thermal entry length*, $L_{t,\,entry}$, is given by Kays and Crawford (1993) as

$$\frac{L_{t,\,entry}}{D} \approx 0.05 Re_D Pr \tag{5.23}$$

Comparison with Equation (5.14) indicates the presence of the Prandtl number in Equation (5.23). As noted in Chapter 4, the value of the Prandtl number determines the ratio δ/δ_T, so it is not surprising that the Prandtl number appears in Equation (5.23) and also determines the ratio of the thermal to hydrodynamic entry lengths for laminar pipe flow.

The implications of a fully developed temperature profile are slightly more difficult to envisage than the fully developed velocity profile. To begin, we define a non-dimensional temperature ratio $\theta = (T - T_s)/(T_b - T_s)$ where T is the temperature anywhere in the fluid, and T_s the surface temperature of the pipe walls. The *bulk temperature* of the fluid in the pipe, T_b, is calculated in a similar way (as an enthalpy average) to the bulk velocity:

$$T_b = \frac{1}{\pi U_b r_o^2} \int_0^{r_o} 2\pi r V_z T \, dr \tag{5.24}$$

If we postulate that the temperature ratio, θ, is independent of axial distance, z, then it is solely a function of radial coordinate, hence $\theta = \theta(r)$. And since T_b and T_s do not depend on radial coordinate, it follows from the definition of θ that

$$\left(\frac{\partial \theta}{\partial r}\right)_{r=r_o} = \text{constant} = \frac{1}{T_b - T_s}\left(\frac{\partial T}{\partial r}\right)_{r=r_o} \tag{5.25}$$

The heat transfer coefficient, h, is based on the temperature difference $T_s - T_b$, and the heat flux at the pipe wall is given by the two relations

$$q_s = h(T_s - T_b) \tag{5.26}$$

$$q_s = -k\left(\frac{\partial T}{\partial r}\right)_{r=r_o} \tag{5.27}$$

Taking the ratio of these two:

$$\frac{h}{k} = -\frac{1}{T_s - T_b}\left(\frac{\partial T}{\partial r}\right)_{r=r_o} = \text{constant and invariant with } z \qquad (5.28)$$

For the case of a pipe with a constant heat flux at the walls (a boundary condition with important engineering applications such as electrical resistance heating, radiant heating and nuclear heating), $(\partial T/\partial r)_{r=r_o} = \text{constant}$. Hence from Equation (5.28) the quantity $(T_s - T_b)$ is also constant.

Taking a heat balance over an elemental pipe length Δz,

$$2\pi r_o q_s \Delta z = \frac{d}{dz}\left[C_p \int_0^{r_o} 2\pi \rho r V_z T \, dr\right]\Delta z$$

or using Equation (5.24) for T_b,

$$q_s = \frac{1}{2} r_o \rho C_p U_b \frac{dT_b}{dz} \qquad (5.29)$$

implies, for a constant heat flux, that $dT_b/dz = \text{constant}$.

A dimensionless temperature profile that is invariant with z tells us that $(\partial\theta/\partial z) = 0$. Rearranging the definition of θ we obtain

$$(T - T_s) = \theta(T_b - T_s)$$

and differentiating,

$$\frac{\partial}{\partial z}(T - T_s) = \frac{(T - T_s)}{(T_b - T_s)}\frac{d}{dz}(T_b - T_s)$$

from which

$$\frac{\partial T}{\partial z}(T_b - T_s) = \frac{dT_s}{dz}(T_b - T) + \frac{dT_b}{dz}(T - T_s) \qquad (5.30)$$

From our previous findings, that $(T_b - T_s) = \text{constant}$ and $dT_b/dz = \text{constant}$, it is obvious that $dT_b/dz = dT_s/dz$. It also follows from Equation (5.30) that for a constant heat flux at the pipe wall

$$\frac{\partial T}{\partial z} = \frac{dT_s}{dz} = \frac{dT_b}{dz} = \text{constant} \qquad (5.31)$$

Applying the conditions that $V_r = 0$ and $\partial^2 T/\partial z^2 = 0$ to the energy equation (5.13) gives

$$\rho C_p\left[V_z\frac{\partial T}{\partial z}\right] = k\left[\frac{\partial^2 T}{\partial r^2} + \frac{1}{r}\frac{\partial T}{\partial r}\right] = \frac{k}{r}\frac{\partial}{\partial r}\left(r\frac{\partial T}{\partial r}\right) \qquad (5.32)$$

Using our previous results for the temperature profile (5.31), it is legitimate to replace $\partial T/\partial z$ in (5.32) with the ordinary derivative dT_b/dz, which leads to the following equation:

$$\frac{\partial}{\partial r}\left(r\frac{\partial T}{\partial r}\right) = \frac{r\rho C_p V_z}{k}\frac{dT_b}{dz} \qquad (5.33a)$$

Integrating Equation (5.33a) once, with the condition that $\partial T/\partial r = 0$ at $r = 0$, and using Equation (5.20) to express the axial velocity in terms of the bulk average velocity gives

$$r\frac{\partial T}{\partial r} = \frac{2U_b\rho C_p}{k}\frac{\mathrm{d}T_b}{\mathrm{d}z}\left[\frac{r^2}{2} - \frac{r^4}{4r_o^2}\right] \tag{5.33b}$$

and integrating again, with the condition that $T = T_s$ at $r = r_o$

$$T = T_s - \frac{2U_b\rho C_p}{k}\frac{\mathrm{d}T_b}{\mathrm{d}z}\left[\frac{3r_o^2}{16} - \frac{r^2}{4} + \frac{r^4}{16r_o^2}\right] \tag{5.33c}$$

Using Equation (5.29), i.e. $q_s = \frac{1}{2}r_o\rho C_p U_b(\mathrm{d}T_b/\mathrm{d}z)$ in Equation (5.33c) together with Equation (5.20) for V_z in Equation (5.24) and performing the integration yields the result

$$T_s - T_b = \frac{11}{24}\frac{q_s r_o}{k} \tag{5.34}$$

Since the Nusselt number for pipe flow is defined using the diameter of the pipe $(D = 2r_o)$, then

$$Nu_D = \frac{2q_s r_o}{(T_s - T_b)k} = \frac{48}{11} \approx 4.364 \tag{5.35a}$$

The Nusselt number for fully developed laminar pipe flow with a *constant heat flux* at the surface is therefore constant and independent of both z and Re_D. This may at first seem surprising, but the key word here is 'fully developed'. In the entry section, the Nusselt number is Reynolds number dependent. The exact relationship depends on the different possibilities such as a fully developed hydrodynamic flow with a developing thermal entry length; developing hydrodynamic and thermal entry lengths; and so on. The reader is referred to Kays and Crawford (1993) for further details.

For the case of a pipe with a *constant surface temperature*, and with fully developed hydrodynamic and thermal flow, an iterative solution is required, giving the result

$$Nu_D \approx 3.658 \tag{5.35b}$$

The above Nusselt number relations may also be applied to annular passages and ducts of non-circular cross-section, providing the mean hydraulic diameter D_h ($D_h = 4 \times$ cross-sectional area/wetted perimeter) is used in place of the pipe diameter for a circular tube. A useful review of various correlations in stationary and rotating annular passages is given by Childs and Long (1996).

5.4 ● Turbulent flow in a pipe

The case of fully developed laminar pipe flow discussed in Section 5.3 has limited application to engineering problems; the reader need only insert a few numbers: for water at 20°C with a bulk average velocity of 0.1 m/s in a pipe of 15 mm dia., the flow is just laminar ($Re = 1500$) and has a hydrodynamic entry length of 75 diameters (1.125 m). In water-to-gas heat exchangers the overall heat transfer coefficient is controlled by the gas-side heat transfer coefficient, so in order to reduce the water-side pressure drop and cause only a minor reduction in the overall heat transfer coefficient, the flow in these tubes may be laminar. However, turbulent flows as a whole are more common in engineering practice than laminar flows. The principles outlined in this section of applying turbulent flow data to predict the heat transfer in pipes have more general use and can be applied to turbulent flow over plates, etc. (Section 5.6).

As is the case with laminar flow, in *turbulent pipe flow* there are also hydrodynamic and thermal entry lengths. Fully developed velocity profiles in turbulent flow occur after about 15 to 40 pipe diameters, and can be estimated using the relation

$$\frac{L_{h,\text{entry}}}{D} = 4.4 Re_D^{1/6} \tag{5.36}$$

The thermal entry length depends on the Prandtl number, and with the exception of fluids having a Prandtl number much less than unity (i.e. liquid metals), fully developed temperature profiles are established within 15 to 40 pipe diameters. These values contrast with those for laminar flow (Equations (5.14) and (5.23)), where the entry lengths are considerably greater.

A starting point for the analysis of heat transfer in fully developed turbulent pipe flow is the analogy between heat and momentum discussed in Section 4.5. For a fluid of non-unity Prandtl number, and using the diameter as the characteristic dimension, the Colburn modification (4.90) to the Reynolds analogy tells us that

$$Nu_D = \frac{1}{2} C_f Re_D Pr^{1/3} \tag{5.37}$$

where C_f is the skin friction coefficient used in external flows and defined in terms of the surface shear stress τ_s and free stream velocity U_∞ as

$$C_f = \frac{\tau_s}{\frac{1}{2}\rho U_\infty^2}$$

It is relatively straightforward to relate C_f to the friction factor, f, since

$$\tau_s = -\mu \left(\frac{\partial V_z}{\partial r} \right)_{r=r_o}$$

and from Equation (5.17) and the definition of the friction factor (5.21) we have

$$C_f = \frac{f}{4} \tag{5.38}$$

Hence, using Equation (5.38) in Equation (5.37), the desired relation in pipe flow variables is

$$Nu_D = \frac{f}{8}\, Re_D\, Pr^{1/3} \tag{5.39}$$

To obtain the Nusselt number, we require an explicit formula for the friction factor in turbulent pipe flow. One such formula that has been used with success over the years is the *Blasius resistance formula* (Blasius, 1913), compiled from a large range of experimental data from smooth pipes and valid for the Reynolds number range $2300 \leqslant Re_D \leqslant 10^5$. The formula states that

$$f = 0.3164 Re_D^{-0.25} \tag{5.40}$$

Substituting the Blasius resistance formula into Equation (5.39)

$$Nu_D = 0.03955 Re_D^{3/4}\, Pr^{1/3} \tag{5.41}$$

Other similar equations exist for the Nusselt number in developed turbulent pipe flow, the *Dittus–Boelter equation* (Dittus and Boelter, 1930) accounts for both heating of the fluid ($T_s > T_b$) and cooling ($T_s < T_b$) and has the form

$$Nu_D = 0.023 Re_D^{0.8}\, Pr^n \tag{5.42}$$

where $n = 0.4$ for heating or $n = 0.3$ for cooling. This particular correlation has been validated by experiment over the range of conditions $0.7 \leqslant Pr \leqslant 160$; $Re_D \geqslant 10^4$; $L/D \geqslant 10$. The fluid property values in Equations (5.41) and (5.42) should be calculated at the bulk mean temperature, T_b. In circumstances where there are large variations in the properties, the *Sieder–Tate equation* (Sieder and Tate, 1936) should be used, where all property values with the exception of μ_s are evaluated at T_b. The equation is given as

$$Nu_D = 0.027 Re_D^{0.8} Pr^{1/3}\, [\mu/\mu_s]^{0.14} \tag{5.43}$$

Equations (5.41) to (5.43) are simple to apply. However, their range of applica-bility is somewhat limited and can result in significant errors. A formula having greater range of validity ($10^4 \leqslant Re_D \leqslant 5 \times 10^6$ and $0.5 \leqslant Pr \leqslant 2000$), at the expense of a slight increase in complexity is given by Gnielinski (1976) as

$$Nu_D = \frac{(f/8)\,\{Re_D - 1000\}Pr}{1 + \{12.7(f/8)^{1/2}(Pr^{2/3} - 1)\}} \tag{5.44}$$

The effects of surface roughness may be incorporated either by using the Moody chart (Figure 5.4) or the Colebrook (1939) formula to evaluate the friction factor, where

$$f^{-1/2} = 1.74 - 2\log_{10}\left\{ \left(\frac{e}{r_o}\right) + \frac{18.7}{(Re_D\sqrt{f})} \right\} \tag{5.45}$$

Fig. 5.4 ●
Variation of skin
friction coefficient,
C_f, with pipe
Reynolds number,
Re_D, for laminar
and turbulent flow
in smooth and
rough pipes

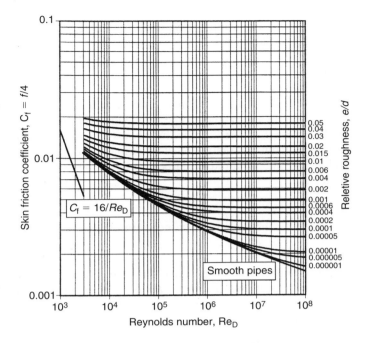

and e is the mean height of the roughness. Equation (5.45) can then be used either with the Colburn analogy (4.90) or Equation (5.44) to obtain the Nusselt number.

Liquid metals which have Prandtl numbers in the range $0.1 \geqslant Pr \geqslant 0.001$ deserve a special mention. The large value of thermal conductivity (which gives rise to the low value of Prandtl number) makes conduction a significant heat transport mechanism, even in the fully turbulent part of the flow field. A correlation for pipes with a constant wall heat flux (Notter and Sleicher, 1972) gives good agreement to available experimental data for liquid metals:

$$Nu_D = 6.3 + (0.0167 Re_D^{0.85} Pr^{0.93})$$ (5.46)

Example 5.2

A water heater uses 15 mm dia. copper pipe, with a mean roughness height of 0.075 mm, which is heated electrically to a constant surface temperature of 95°C. Water enters the pipe at 0.1 kg/s at a temperature of 10°C and leaves at a temperature of 75°C. What length of pipe is required to achieve this? Compare this with the length required for a perfectly smooth pipe.

Solution   Consider, as shown in Figure 5.5, a heat balance on an element of fluid in the pipe. The heat transferred from the wall is equal to the increase in enthalpy of the fluid, hence, with $\bar{T} = T_b - T_s$

Fig. 5.5 ●
Heat balance on a
fluid element in a
pipe (Example 5.2)

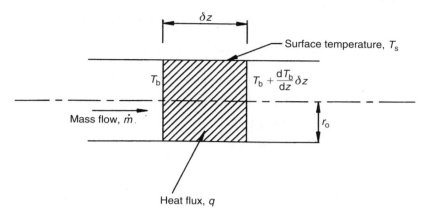

$$2\pi r_o q \; \delta z = -\dot{m} \; C_p \frac{d\bar{T}}{dz} \delta z$$

and so

$$\frac{d\bar{T}}{dz} = \frac{-2\pi r_o q}{\dot{m} C_p} \tag{5.47}$$

Using the definitions of

$$Pr = \frac{\mu C_p}{k}$$

$$Re_D = \frac{\rho U_b D}{\mu} = \frac{4\dot{m}}{\pi D \mu}$$

$$Nu_D = \frac{qD}{(T_s - T_b)k}$$

Equation (5.47) may be rewritten as

$$d\bar{T}/dz = -4 \; Nu_D \bar{T}/(Re_D Pr D)$$

Rearranging and integrating between inlet (where $\bar{T} = T_s - T_{b,in}$) and outlet
(where $\bar{T} = T_s - T_{b,out}$) gives

$$\frac{(T_s - T_{b,out})}{(T_s - T_{b,in})} = \exp(-\lambda) \tag{5.48}$$

where $\lambda = 4Nu_D L/(Re_D Pr D)$
  For the water heater

$$\dot{m} = 0.1 \text{ kg/s}$$

$$T_s = 95°C$$

$$T_{b,out} = 75°C$$

$$T_{b,in} = 10°C$$

Evaluating properties at the mean of the bulk average fluid temperatures,

$$T_m = \frac{(T_{b,\text{out}} + T_{b,\text{in}})}{2} = 42.5°C$$

From tables, for water at 42.5°C

$$\mu = 6.28 \times 10^{-4} \, \text{kg/m s}$$

$$C_p = 4180 \, \text{J/kg K}$$

$$k = 0.63 \, \text{W/m K}$$

$$Pr = 4.17$$

so we have

$$Re_D = \frac{4\dot{m}}{\pi D \mu}$$

$$= \frac{0.4}{(\pi \times 0.015 \times 6.28 \times 10^{-4})}$$

$$= 0.135 \times 10^5$$

From Equation (5.45)

$$f^{-1/2} = 1.74 - 2\log_{10}\left\{ \left(\frac{e}{r_o}\right) + \frac{18.7}{(Re_D\sqrt{f})} \right\}$$

For $Re_D = 0.135 \times 10^5$ and $e/r_o = 0.01$, by trial and error iteration, $f = 0.036$.
Using Equation (5.44),

$$Nu_D = \frac{(f/8)\{Re_D - 1000\}Pr}{1 + \{12.7(f/8)^{1/2}(Pr^{2/3} - 1)}$$

$$= 96.8$$

And from Equation (5.48)

$$\frac{(95 - 75)}{(95 - 10)} = \exp(-\lambda) = 0.235$$

From which, $\lambda = \ln(1/0.235) = 1.447$. Since $\lambda = 4Nu_D L/Re_D PrD$, then

$$L = \frac{\lambda Re_D PrD}{4 Nu_D}$$

$$= \frac{1.447 \times 1.35 \times 10^4 \times 4.17 \times 0.015}{4 \times 96.8}$$

$$= 3.15 \, \text{m}$$

For a smooth tube, using the Blasius resistance formula (5.40),

$$f = 0.3164 Re_D^{-1/4} = 0.0294$$

From Equation (5.44), $Nu_D = 90.2$, hence using Equation (5.48),

$$L = 3.38 \, \text{m}$$

Comment The difference in lengths is comparatively small since the Reynolds number is in the range where the roughness (of $0.075/15 = 0.005$) does not have a large effect on the friction factor. As shown in the Moody chart (Figure 5.4) for values of $Re_D > 10^6$, there is a marked increase in the friction factor with roughness and consequently a significant difference between the lengths required for rough and smooth pipes.

5.5 ● Laminar flow over a flat plate

The simple geometry of the flat plate is a convenient starting point and useful to many of the more complex configurations classified as external forced convection. In this and the following section we will examine laminar and turbulent flow and the associated heat transfer from a flat plate, with a constant free stream velocity ($dp/dx = 0$) and a constant free stream temperature. This somewhat idealised configuration may appear to have limited use in the thermal modelling of engineering components which may need to include such effects as surface curvature, free stream temperature and pressure variations, surface roughness and so on. Although these effects can be modelled mathematically, this is inevitably complex and beyond the scope of this text. However, in considering the heat transfer characteristics of the flat plate, we will expose fundamental behaviour that is common to the more complex flows.

The problem is to obtain an analytical expression for the heat transfer coefficient (expressed non-dimensionally as a Nusselt number) for the case of a laminar boundary layer with a constant free stream velocity U_∞, constant free stream temperature T_∞, flowing over a heated (or cooled) plate having a constant surface temperature T_s. Although exact solutions to the boundary layer equations exist, for example Bejan (1993), an approximate method, based on the integral momentum and energy equations (4.95) and (4.99) is used here, since it can also be employed in turbulent flow and in other external flow configurations. The method was pioneered by von Kármán (1921) and first applied to laminar flow over a flat plate by Pohlhausen (1921).

Since we are considering forced convection, the buoyancy force may be neglected. For constant density ρ, constant free stream velocity U_∞, and a constant free stream temperature ($\rho, U_\infty, T_\infty = $ constant), Equations (4.95) and (4.99) may be written as

$$\frac{d}{dx}\left\{\int_0^\delta u(U_\infty - u)\,dy\right\} = \frac{\tau_s}{\rho}$$

$$= v\left(\frac{\partial u}{\partial y}\right)_{y=0} \tag{5.49}$$

and

$$\frac{d}{dx}\left\{\int_0^{\delta_T} u(T - T_\infty)\,dy\right\} = \frac{q_s}{\rho C_p}$$

$$= -\alpha\left(\frac{\partial T}{\partial y}\right)_{y=0} \tag{5.50}$$

where $v = \mu/\rho$ and $\alpha = k/\rho C_p$ are, respectively, the kinematic viscosity and thermal diffusivity (both have units of m^2/s). As a result of neglecting the buoyancy force, u appears alone in Equation (5.49); the integral equations are then said to be uncoupled. However, in Equation (5.50) u and T appear together, so it is first necessary to solve Equation (5.49). The integral momentum and energy equations will be applied to a free convection problem in Chapter 6, leading to a set of coupled equations which have to be solved simultaneously.

The strategy used to solve Equation (5.49) is to assume a defined functional relation for the velocity distribution, $u(y)$, substitution of this into the momentum integral equation forms a first-order ordinary differential equation for the boundary layer thickness, δ.

If we make the velocity ratio u/U_∞ equal to a function of the distance y/δ, then

$$u/U_\infty = f(\eta)$$

where $\eta = y/\delta$ and f denotes some, as yet unspecified, function. It is therefore apparent that

$$dy = \delta\,d\eta$$

$$u = U_\infty f$$

$$\frac{\partial u}{\partial y} = \left(\frac{\partial f}{\partial \eta}\right)\left(\frac{\partial \eta}{\partial y}\right)\left(\frac{\partial u}{\partial f}\right)$$

$$= \left(\frac{\partial f}{\partial \eta}\right)\frac{U_\infty}{\delta} \tag{5.51}$$

Using Equations (5.51) in Equation (5.49) we get, where the limit on the integral is now $\eta = \delta/\delta = 1$,

$$\frac{d}{dx}\left\{\delta\int_0^1 f(\eta)\,(f(\eta) - 1)\,d\eta\right\} = -\left\{\frac{v}{\delta U_\infty}\right\}\left(\frac{\partial f(\eta)}{\partial \eta}\right)_{\eta=0} \tag{5.52}$$

A fairly accurate solution to Equation (5.52) can be obtained from assuming a purely linear relationship, as in Couette flow (5.8), but this solution $(f(\eta) = u/U_\infty = y/\delta)$ contradicts the boundary layer principle that at $y = \delta$, $\partial u/\partial y = 0$ and must be rejected on physical grounds. An improvement is to assume a polynomial for the velocity distribution in the boundary layer:

$$f(\eta) = a + b\eta + c\eta^2 + d\eta^3 + \ldots$$

where a, b, c, d, etc., are constants evaluated using known physical boundary conditions. The series can be extended beyond a cubic, but a cubic gives sufficiently accurate results and complies with the physical boundary conditions.

From inspection of Figure 5.6a it is evident that:

at $\quad \eta = 1$ (i.e. $y = \delta$), $u = U_\infty$

at $\quad \eta = 1$, $\dfrac{\partial u}{\partial y} = 0$

at $\quad \eta = 0$ $(y = 0)$, $u = 0$

So this gives us three boundary conditions to find three of the coefficients. The fourth is obtained from the compatibility condition, which from Equation (4.47),

$$\rho\left(u\frac{\partial u}{\partial x} + v\frac{\partial u}{\partial y} \right) = -\frac{dp}{dx} + \mu\frac{\partial^2 u}{\partial y^2}$$

tells us that, since u and v approach zero as y approaches zero, then at $y = 0$ the left-hand side of this equation equals zero. We are ignoring the pressure gradient, $dp/dx = 0$, and so the fourth condition is therefore

$$\left(\frac{\partial^2 u}{\partial y^2} \right)_{y=0} = 0$$

Fig. 5.6 ●
Boundary layer
development for
laminar forced
convection over
a flat plate:
(a) velocity
boundary layer;
(b) thermal
boundary layer;
(c) unheated
starting length

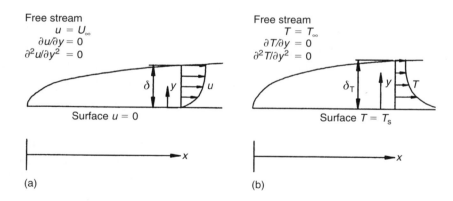

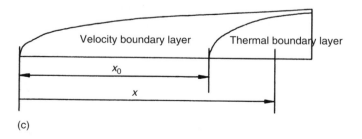

Using these four conditions it is easily shown that, $a = c = 0$, $b = \frac{3}{2}$, $d = -\frac{1}{2}$ and our polynomial representing the velocity profile for laminar flow over a flat plate is

$$\frac{u}{U_\infty} = f(\eta) = \frac{3}{2}\eta - \frac{1}{2}\eta^3 \qquad (5.53)$$

Substituting the velocity profile given by Equation (5.53) into Equation (5.52) allows the momentum integral equation to be rewritten in terms of one unknown, δ, the boundary layer thickness. Differentiation of Equation (5.53) gives

$$\left(\frac{\partial f}{\partial \eta}\right)_{\eta = 0} = \frac{3}{2}$$

Using this result, Equation (5.52) becomes, after evaluating the integral,

$$\delta \frac{d\delta}{dx} = \frac{140\nu}{13U_\infty} \qquad (5.54)$$

which is now a first-order ordinary differential equation for δ. Noting that

$$\frac{d}{dx}y^2 = y\frac{dy}{dx} + y\frac{dy}{dx} = 2y\frac{dy}{dx}$$

Equation (5.54) may be rewritten as

$$\frac{d}{dx}\left(\frac{1}{2}\delta^2\right) = \frac{140\nu}{13U_\infty} \qquad (5.55)$$

Integrating Equation (5.55) and imposing the condition that the velocity boundary layer starts at $x = 0$ ($\delta = 0$ at $x = 0$) provides the equation for the growth of a laminar boundary layer for flow over a flat plate:

$$\frac{\delta}{x} = 4.64 Re_x^{-1/2} \qquad (5.56)$$

where Re_x is the familiar Reynolds number ($Re_x = \rho U_\infty x/\mu = U_\infty x/\nu$).

Physically, Equation (5.56) demonstrates that the boundary layer thickness grows with $x^{1/2}$ and also with $(\nu/U_\infty)^{1/2}$. In other words, the more viscous the fluid, the thicker the boundary layer for a given velocity. At this stage it is instructive to make a comparison between the approximate integral solution and an exact solution of the boundary layer equations by comparing values of the skin friction coefficient,

$$C_f = \frac{2\tau_s}{\rho U_\infty^2}$$

and

$$\tau_s = \mu\left(\frac{\partial u}{\partial y}\right)_{y=0} = \frac{\mu U_\infty}{\delta}\left(\frac{\partial f}{\partial \eta}\right)_{\eta=0}$$

So, using Equations (5.53) for f and (5.56) for δ, we get

$$C_f = 0.647 Re_x^{-1/2} \qquad (5.57)$$

which compares well with a value of $C_f = 0.664 Re_x^{-1/2}$ obtained from the exact solution. Since C_f depends on the velocity gradient at the wall, the good agreement (2.5% error in the leading coefficient and identical value of exponent) serves to illustrate the value of the integral technique. It is also useful to compare the result expressed by Equation (5.46) with that derived using scale analysis (4.109). The Reynolds number dependence of the boundary layer growth is confirmed from the simpler scale analysis; the additional complexity involved in the integral analysis leads to a value of the constant of proportionality.

Before proceeding further, it is instructive to look at the rationale for the next step. If we are able to determine the heat flux, then the values of heat transfer coefficient and Nusselt number are a simple matter of arithmetic. At the surface $q = -k(\partial T/\partial y)$, so if we have a functional relation for $T(y)$ then differentiation of this expression to obtain the heat flux will involve the thermal boundary layer thickness, δ_T. For generality, it will be assumed that δ and δ_T are not equal, so we define a new variable, η_T, the non-dimensional thermal boundary layer thickness and represent the temperature non-dimensionally as $g(\eta_T)$, where

$$g(\eta_T) = \frac{(T - T_s)}{(T_\infty - T_s)}$$

and

$$\eta_T = \frac{y}{\delta_T}$$

Continuing in a way similar to that for the velocity profile, with a cubic polynomial

$$g(\eta_T) = a + b\eta_T + c\eta_T^2 + d\eta_T^3 \qquad (5.58)$$

and the following conditions (see Figure 5.6b):

at $\quad \eta_T = 0, \, g = 0$

at $\quad \eta_T = 1, \, g = 1$

at $\quad \eta_T = 1, \, \dfrac{\partial g}{\partial \eta_T} = 0$

The fourth condition follows in an analogous way to that for the velocity boundary layer, but by using the boundary layer energy equation (4.48):

at $\quad \eta_T = 0, \, \dfrac{\partial^2 g}{\partial \eta_T^2} = 0$

Hence,

$$g(\eta_T) = \frac{3}{2}\eta_T - \frac{1}{2}\eta_T^3 \qquad (5.59)$$

The integral energy equation (5.50) may now be written as

$$\frac{d}{dx}\left\{\delta_T\int_0^1 f(\eta)\{1-g(\eta_T)\}d\eta_T\right\} = \frac{\alpha}{U_\infty\delta_T}\left(\frac{\partial g(\eta_T)}{\partial \eta_T}\right)_{\eta_T=0} \tag{5.60}$$

Equation (5.60) can be further simplified by introducing the ratio ζ, where

$$\zeta = \frac{\delta_T}{\delta} \tag{5.61}$$

Substituting for $f(\eta)$ from (5.53) and $g(\eta_T)$ from (5.58), Equation (5.60) gives a first-order, ordinary differential equation in δ_T:

$$\frac{d}{dx}\left[\zeta\,\delta_T\left(\frac{3}{20}-\frac{3\zeta^2}{280}\right)\right] = \frac{3}{2}\frac{\alpha}{U_\infty\delta_T} \tag{5.62}$$

As previously noted, the velocity boundary layer thickness varies with the diffusion of momentum (kinematic viscosity). The thermal boundary layer thickness depends on the diffusion of heat, characterised by the thermal diffusivity, α. So $\zeta(\zeta = \delta_T/\delta) \sim \alpha/\nu$, and for moderate values of Prandtl number, i.e. Pr of order unity, we can neglect the term with ζ^2. Recalling that $\delta_T = \zeta\delta$, Equation (5.62) then becomes

$$\frac{d}{dx}(\delta\zeta^2) = \frac{10\alpha}{\delta\zeta U_\infty} \tag{5.63}$$

If both the thermal and velocity boundary layers begin at $x = 0$, then ζ is independent of x. As already shown in Equation (5.54),

$$\delta\frac{d\delta}{dx} = \frac{140\nu}{13U_\infty}$$

Substitution of this result into Equation (5.63) gives

$$\zeta = \frac{\delta_T}{\delta}$$

$$= \left(\frac{13\alpha}{14\nu}\right)^{1/3}$$

$$= 0.976Pr^{-1/3} \tag{5.64}$$

This result is valid for Prandtl number of order unity, which is the case for most gases and also water. It is not valid when the velocity and thermal boundary layers do not both begin at $x = 0$, for example a plate with an initial section that is unheated, often referred to as an unheated starting length (Figure 5.6c).

It is now a simple matter to obtain the local heat transfer coefficient and local Nusselt number Nu_x. The local heat transfer coefficient, h, is defined as

$$h = \frac{q_s}{(T_s - T_\infty)k} \tag{5.65}$$

where

$$q_s = -k \left(\frac{\partial T}{\partial y} \right)_{y=0} = k \left(\frac{\partial g}{\partial \eta_T} \right)_{\eta_T=0} \frac{(T_s - T_\infty)}{\delta_T}$$

So, using Equation (5.59) for $g(\eta_T)$,

$$h = \frac{3k}{2\,\delta_T} \tag{5.66a}$$

and, from Equations (5.64) and (5.56),

$$\delta_T = \zeta\delta = 0.976 Pr^{-1/3} \times 4.64 x Re_x^{-1/2} \tag{5.66b}$$

From the definition of the local Nusselt number, $Nu_x = hx/k$, it follows from Equations (5.66a) and (5.66b) that

$$Nu_x = 0.331 Pr^{1/3} Re_x^{1/2} \tag{5.67}$$

and from Equations (4.60) to (4.62) the *average* Nusselt number for a plate of length L is then given by

$$Nu_{av} = 0.662 Pr^{1/3} Re_L^{1/2} \tag{5.68}$$

Equation (5.67) compares extremely well (0.3% error) with the exact solution of the differential boundary layer equations for moderate values of Pr, $Nu_x = 0.332 Pr^{1/3} Re_x^{1/2}$. Also, Equation (5.67) shows that $h \sim x^{-1/2}$, illustrating that the heat transfer coefficient decreases in the streamwise direction as the boundary layer develops and thickens. Again it is useful to compare this with the result derived from the more economical scale analysis, where the $Pr^{1/3} Re_L^{1/2}$ behaviour is confirmed. The extra effort required for the integral analysis leads to a value for the constant of proportionality.

For an unheated starting length with laminar flow over a flat plate where the velocity boundary layer begins at $x = 0$ and the thermal boundary layer at $x = x_0$,

$$Nu_x = \frac{0.331 Pr^{1/3} Re_x^{1/2}}{\left(1 - (x_0/x)^{3/4}\right)^{1/3}} \tag{5.69}$$

5.6 ● Turbulent flow over a flat plate

Transition from laminar to turbulent flow over a flat plate generally takes place at $Re_x \approx 5 \times 10^5$. However, the value of the transition Reynolds number can vary by up to an order of magnitude each way depending on the presence of disturbances in the freestream. For turbulent flow we can apply the Colburn analogy (4.90), using a relationship for the shear stress derived from empirical measurements. It is possible to use the Blasius resistance formula (5.40), obtained for pipe flow, where the pipe friction factor, f, is equal to $0.3164 Re_D^{-1/4}$, by modifying this to suit the coordinate system for flow over a flat plate.

First we note that from Equation (5.38) $f = 4C_f = 8\tau_s/\rho U_b^2$, so the surface shear stress is given by

$$\tau_s = \frac{\rho U_b^2 f}{8}$$

$$= 0.039\,55\rho U_b^2 Re_D^{-1/4}$$

and with $D = 2r_o$,

$$\tau_s = 0.033\,25\rho U_b^{7/4}\left(\frac{\nu}{r_o}\right)^{1/4} \tag{5.70}$$

Defining a friction velocity $U^* = (\tau_s/\rho)^{1/2}$, Equation (5.70) can be rearranged as

$$\frac{U_b}{U^*} = 6.993\left(\frac{U^* r_o}{\nu}\right)^{1/7} \tag{5.71}$$

Experimental measurements of the time-averaged velocity profiles in fully developed ($z/D > 20$) turbulent pipe flow have demonstrated that the local velocity at distance y from the pipe wall can be related to that on the centreline U_c, using a power law relationship of the form

$$\frac{u}{U_c} = \left(\frac{y}{r_o}\right)^{1/n} \qquad (n = 7, 8, 9 \text{ or } 10) \tag{5.72}$$

The most appropriate value for n increases with the pipe Reynolds number, Re_D. For moderate values of Re_D ($Re_D < 10^5$), a value of $n = 7$ is used, and is commonly referred to as the 'one-seventh power law'. Pipe flow velocities are, however, characterised not by U_c, but by the bulk average velocity U_b, where

$$U_b = \frac{1}{\pi r_o^2}\int_0^{r_o} 2\pi r u\, dr \tag{5.73}$$

Integrating Equation (5.73) using Equation (5.72) with $n = 7$ and changing coordinates such that $r = (r_o - y)$ and $dr = -dy$, the relation between average and centreline velocities is

$$U_b \approx 0.8 U_c \tag{5.74}$$

From Equations (5.74) and (5.71),

$$\frac{U_c}{U^*} = 8.74\left(\frac{U^* r_o}{\nu}\right)^{1/7} \tag{5.75}$$

By analogy with pipe flow, it is assumed that a similar relation holds for turbulent flow over a flat plate, and by replacing U_c with U_∞ and r_o with δ:

$$\frac{U_\infty}{U^*} = 8.74\left(\frac{U^* \delta}{\nu}\right)^{1/7} \tag{5.76a}$$

and at a general point, y, in the boundary layer:

$$\frac{u}{U^*} = 8.74 \left(\frac{U^* y}{v}\right)^{1/7} \tag{5.76b}$$

and dividing Equation (5.76b) by (5.76a)

$$\frac{u}{U_\infty} = \left(\frac{y}{\delta}\right)^{1/7} \tag{5.77}$$

Since $U^* = (\tau_s/\rho)^{1/2}$, then rearrangement of Equation (5.76a) gives

$$\frac{\tau_s}{\rho} = 0.0225\, U_\infty^{7/4} \left(\frac{v}{\delta}\right)^{1/4} \tag{5.78}$$

To complete the solution we need information about the growth of δ, which may be obtained from solving the momentum integral equation (5.49):

$$\frac{\mathrm{d}}{\mathrm{d}x} \int_0^\delta u(U_\infty - u)\, \mathrm{d}y = \frac{\tau_s}{\rho} \tag{5.79}$$

Using (5.77) for $u(y)$ and (5.78) for the shear stress in Equation (5.79) gives the first-order ordinary differential equation for δ:

$$\frac{7}{72}\frac{\mathrm{d}\delta}{\mathrm{d}x} = 0.0225 \left(\frac{v}{U_\infty \delta}\right)^{1/4} \tag{5.80}$$

and as $\delta = 0$ when $x = 0$:

$$\frac{\delta}{x} = 0.37 Re_x^{-1/5} \tag{5.81}$$

Introducing the local skin friction coefficient $C_{f,x} (= \tau_s / \frac{1}{2}\rho U_\infty^2)$ and using Equation (5.78) for τ_s and Equation (5.81) for δ, gives

$$C_{f,x} = 0.0576 Re_x^{-1/5} \tag{5.82}$$

Finally, using the Colburn analogy (4.90) and Equation (5.82) allows us to obtain an expression for the local Nusselt number for turbulent flow over a flat plate as

$$Nu_x = 0.0288 Re_x^{0.8} Pr^{1/3} \tag{5.83}$$

and the average Nusselt number for a plate of length L as

$$Nu_{av} = 0.036 Re_L^{0.8} Pr^{1/3} \tag{5.84}$$

For an unheated starting length (Figure 5.6c) in turbulent flow with a thermal boundary layer that begins at $x = x_0$,

$$Nu_x = \frac{0.0288 Re_x^{0.8} Pr^{1/3}}{\left(1 - (x_0/x)^{9/10}\right)^{1/9}} \tag{5.85}$$

Example 5.3

For forced convection from a flat plate, the local Nusselt number is given by the following correlations for laminar and turbulent flow:

laminar: $\quad Nu_x = 0.331\,Re_x^{1/2}\,Pr^{1/3}$

turbulent: $\quad Nu_x = 0.0288\,Re_x^{0.8}\,Pr^{1/3}$

The fins on an air-cooled motorcycle engine may be considered as individual flat plates of length $L = 0.2\,\text{m}$. Owing to disturbances in the free stream, transition occurs at $Re_{x,\,\text{trans}} = 2 \times 10^5$. Calculate, for a roadspeed of 140 kph, the average heat transfer coefficient from the fin surface allowing for separate laminar and turbulent sections, and compare this with the result obtained assuming purely turbulent flow. Take $\rho = 1.1\ \text{kg/m}^3$, $\mu = 1.7 \times 10^{-5}\,\text{kg/m s}$, $k = 0.026\ \text{W/m K}$ and $Pr = 0.7$.

Solution Consider a plate of length L, and having a constant surface temperature T_s. A forced convection boundary layer flow begins as laminar and then becomes turbulent at a distance $x = x_{\text{trans}}$. From the definition of the average Nusselt number,

$$Nu_{\text{av}} = \frac{q_{\text{av}} L}{(T_s - T_\infty)k}$$

where the average heat flux is obtained by integrating over the laminar and turbulent sections of the plate:

$$q_{\text{av}} = \frac{1}{L}\left\{\int_0^{x_{\text{trans}}} q_{\text{lam}}\,\text{d}x + \int_{x_{\text{trans}}}^L q_{\text{turb}}\,\text{d}x\right\}$$

From the Nu_x–Re_x relations given above, and also Equations (5.67) and (5.83)

$$q_{\text{lam}} = 0.331(T_s - T_\infty)\,k\left(\frac{\rho U_\infty}{\mu}\right)^{1/2} Pr^{1/3}x^{-1/2}$$

$$q_{\text{turb}} = 0.0288(T_s - T_\infty)\,k\left(\frac{\rho U_\infty}{\mu}\right)^{0.8} Pr^{1/3}x^{-0.2}$$

From which

$$Nu_{\text{av}} = \frac{q_{\text{av}} L}{(T_s - T_\infty)k}$$

$$= \left\{0.662 Re_{x,\,\text{trans}}^{1/2} + 0.036\left[Re_L^{0.8} - Re_{x,\,\text{trans}}^{0.8}\right]\right\} Pr^{1/3}$$

At 140 kph, $U_\infty = 38.89\,\text{m/s}$ and

$$Re_L = \frac{\rho U_\infty L}{\mu}$$

$$= \frac{1.1 \times 38.89 \times 0.2}{1.7 \times 10^{-5}} = 5 \times 10^5$$

So

$$Nu_{av} = \left\{ 0.662 \times (2 \times 10^5)^{1/2} + 0.036 \times \left[(5 \times 10^5)^{0.8} - (2 \times 10^5)^{0.8} \right] \right\} \times 0.7^{1/3}$$

$$= 865$$

$$h_{av} = \frac{Nu_{av}k}{L}$$

$$= \frac{865 \times 0.026}{0.2} = 112 \, \text{W/m}^2\text{K}$$

For purely turbulent flow,

$$Nu_{av} = 0.036 Re_L^{0.8} Pr^{1/3}$$

$$= 0.036 (5 \times 10^5)^{0.8} \times (0.7)^{1/3}$$

$$= 1158$$

$$h_{av} = \frac{Nu_{av}k}{L}$$

$$= \frac{1158 \times 0.026}{0.2} = 150 \, \text{W/m}^2\text{K}$$

Comment In this example transition occurs at 40% of the fin length, so it is therefore necessary to include laminar and turbulent portions in the analysis to find the average heat transfer coefficient.

5.7 ● Heat transfer from a rotating disc

The heat transfer from rotating surfaces has particular relevance to the design of compressor and turbine discs, automotive components such as brake assemblies and clutches, and electronic components such as disc drives.

Figure 5.7 shows the flow induced by a disc, rotating at an angular velocity Ω (rad/s) in an infinite fluid at rest. On the surface of the disc the no-slip condition applies and, owing to the action of viscosity and centrifugal force, fluid is thrown radially outwards. As a result, a stream of fluid is drawn towards the disc to balance this flow.

The subject of flow and heat transfer in rotating disc systems is as rich and diverse as any of the other major fields of heat transfer. Examples of relevant configurations are two discs rotating at different speeds, one disc with a superposed radial flow of cooling air, an enclosed 'cavity' comprising two discs and a peripheral 'shroud', and so on. All of these geometries have application to gas turbine engines and the interested reader is advised to consult the two texts by Owen and Rogers (1989, 1995) for a thorough overview of the subject area.

Fig. 5.7 ●
Flow induced by a
rotating disc in a
quiescent
environment

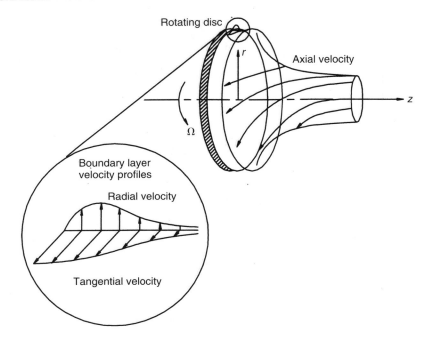

Fig. 5.7 Flow induced by a rotating disc in a quiescent environment

In so far as the single disc rotating in a quiescent environment is concerned, the heat transfer depends on the Prandtl number and the rotational Reynolds number Re_Ω. The latter non-dimensional group is defined by replacing U_∞ in the conventional (linear) Reynolds number with the local tangential velocity Ωr, and replacing the streamwise coordinate x with the local radius r. Hence

$$Re_\Omega = \frac{\rho \Omega r^2}{\mu} \qquad (5.86)$$

At values of $Re_\Omega < 2 \times 10^5$, the flow is generally agreed to be laminar; above this value transition occurs, and the flow becomes fully turbulent at $Re_\Omega > 3 \times 10^5$.

For a disc of outer radius b, with a surface-to-fluid temperature variation of the form

$$\frac{(T_{s,r} - T_\infty)}{(T_{s,b} - T_\infty)} = \left(\frac{r}{b}\right)^n$$

where $T_{s,r}$ is the surface temperature at radius r, $T_{s,b}$ the surface temperature at the outer radius, and n a known constant (see Section 4.4.3), the local Nusselt number for laminar flow is given by Dorfman (1963) as

$$Nu = \frac{qr}{(T_{s,r} - T_\infty)k} = 0.308(n+2)^{0.5} Re_\Omega^{0.5} Pr^{0.5} \qquad (5.87)$$

and for turbulent flow as

$$Nu = 0.0197(n + 2.6)^{0.2} Re_\Omega^{0.8} Pr^{0.6} \tag{5.88}$$

Note that the dependence on the Reynolds number for laminar and turbulent flow is the same as for flow over a flat plate.

Example 5.4

The local Nusselt number at radius r for a rotating disc of outer radius b, in conditions of turbulent flow over the entire surface, and a surface-to-air temperature difference given by the power law variation $(T_{s,r} - T_\infty)/(T_{s,b} - T_\infty) = (r/b)^n$, is given by Equation (5.88) as

$$Nu = 0.0197(n + 2.6)^{0.2} Re_\Omega^{0.8} Pr^{0.6}$$

(a) Derive an expression for the average Nusselt number.

(b) Calculate the heat flow from one side of a compressor disc of radius $b = 0.2\,\mathrm{m}$ which rotates at 19 000 rev/min in air at 8 bar and 550°C. The disc has a linear surface temperature distribution and the maximum surface temperature at $r = b$ is 650°C.

Solution (a) From the definitions of Re_Ω and Nu,

$$q = 0.0197(n + 2.6)^{0.2}(\rho\Omega/\mu)^{0.8} Pr^{0.6} k\,(T_{s,b} - T_\infty)\,b^{-n}\,\{r^{1.6+n-1}\}$$
$$= C_0\{r^{0.6+n}\}$$

where C_0 is a constant. The local surface temperature is given by

$$(T_{s,r} - T_\infty) = (T_{s,b} - T_\infty)\left(\frac{r}{b}\right)^n$$

From Equations (4.60), (4.61) and (4.62), integrating between $r = 0$ and $r = b$, with $dA = 2\pi r\,dr$ and $A = \pi b^2$,

$$q_{av} = \frac{2C_0 b^{n+0.6}}{n + 2.6}$$

$$(T_{s,r} - T_\infty)_{av} = \frac{2(T_{s,b} - T_\infty)}{n + 2}$$

Since $Nu_{av} = q_{av}b/(T_{s,r} - T_\infty)_{av}\,k$, substituting the relations found for the average heat flux and temperature difference gives

$$Nu_{av} = \frac{C_0(n + 2)b^{n+1.6}}{(T_{s,b} - T_\infty)(n + 2.6)}$$

After substituting for the constant C_0, we get

$$Nu_{av} = 0.0197(n + 2)(n + 2.6)^{-0.8} Re_{\Omega,b}^{0.8} Pr^{0.6}$$

(b) First we must establish a value of the mean film temperature to evaluate the fluid properties, in this case an average of the average disc temperature and air temperature:

$$(T_{s,r} - T_\infty)_{av} = \frac{2(T_{s,b} - T_\infty)}{(n+2)}$$

and for a linear temperature distribution, $n = 1$,

$$(T_{s,r} - T_\infty)_{av} = \frac{2(650 - 550)}{3} = 66.7°C$$

hence

$$T_m = \frac{\{550 + (550 + 66.7)\}}{2} = 583.3°C \quad (856.4\,K)$$

For air at 8 bar and 856.4 K,

$$\rho = \frac{p}{RT} = 8 \times 10^5 / (287 \times 856.4) = 3.25\,kg/m^3$$

$$\mu = 3.763 \times 10^{-5}\,kg/m\,s$$

$$k = 0.0603\,W/m\,K$$

$$C_p = 1101\,J/kg\,K$$

$$Pr = \frac{\mu C_p}{k} = 3.763 \times 10^{-5} \times 1101/0.0603 = 0.687$$

$$\Omega = \frac{19\,000 \times 2\pi}{60} = 1990\,rad/s$$

$$Re_{\Omega,b} = \frac{\rho\Omega b^2}{\mu} = \frac{3.25 \times 1990 \times 0.2^2}{3.763 \times 10^{-5}} = 6.87 \times 10^6$$

$$Nu_{av} = 0.0197(n+2)(n+2.6)^{-0.8} Re_{\Omega,b}^{0.8} Pr^{0.6}$$

$$Nu_{av} = 0.0197 \times (1+2) \times (1+2.6)^{-0.8} \times (6.87 \times 10^6)^{0.8} \times 0.687^{0.6} = 4992$$

From the definition of Nu_{av},

$$q_{av} = \frac{Nu_{av}(T_{s,r} - T_\infty)_{av} k}{b} = \frac{4992 \times 66.7 \times 0.0603}{0.2} = 100.4\,kW/m^2$$

The heat flow, $\dot{Q}$, is therefore

$$\dot{Q} = q_{av}\pi b^2 = 100.4 \times \pi \times 0.2^2 = 12.61\,kW$$

Comment Note the use of area integration to obtain the average heat flux and average temperature difference. Assuming a transitional Reynolds number of 3×10^5, at $Re_{\Omega,b} = 6.87 \times 10^6$, transition from laminar to turbulent flow will occur at $r = 0.042\,m$, roughly 20% of the disc radius, or 4.5% of the disc surface area. The error in treating this as turbulent is therefore small.

5.8 ● **Impinging jets, cylinders in crossflow and spheres**

5.8.1 **Impinging jets**

Jet impingement, either by single jets or an array of many jets, is commonly used in situations which demand a high heat transfer coefficient. Gas torches, electric heat guns, heating of plastic sheets prior to forming and the cooling of metal plates, are all applications using impinging jets.

A comprehensive review on the flow and heat transfer from a single (round and slot) jet and arrays of jets has been carried out by Martin (1977). As shown in Figure 5.8a, the flow due to the jet comprises a stagnation (or impingement) region immediately under the jet, followed by the formation of a wall jet on the surface. As shown qualitatively in Figure 5.8b, the local Nusselt number distribution is generally a bell-shaped curve, with a maximum in the stagnation region, and decreasing with radius r from the stagnation point. If the distance from the jet exit to the surface, H, is less than about five nozzle diameters then a second maximum in the local Nusselt number distribution will occur.

Fig. 5.8 ●
Impinging jet flow:
(a) flow structure;
(b) some typical
distributions for
Nu, the Nusselt
number

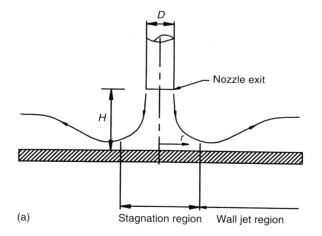

(a)   Stagnation region   Wall jet region

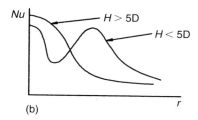

(b)

The average Nusselt number due to a single circular jet of diameter D impinging on a circular surface of radius r and located a distance H above the surface is correlated by the following expression due to Martin:

$$Nu_{D,\text{av}} = F(Re)\, G(r/D,\, H/D)\, Pr^{0.42} \tag{5.89}$$

where $F(Re) = 2Re^{1/2}(1 + 0.005\, Re^{0.55})^{1/2}$

$$G(r/D,\, H/D) = \frac{D}{r}\left\{1 - 1.1\left(\frac{D}{r}\right)\right\} \Big/ \left\{1 + 0.1\left(\frac{D}{r}\right)\left(\frac{H}{D} - 6\right)\right\}$$

$$Re = U_{b,\text{exit}}\, D/\nu$$

and $U_{b,\text{exit}}$ is the bulk mean velocity at the jet exit.

Equation (5.89) is valid for $2 \times 10^3 \leqslant Re \leqslant 4 \times 10^5$; $2 \leqslant H/D \leqslant 12$ and $2.5 \leqslant r/D \leqslant 7.5$. Martin also presents correlations for a single-slot jet and arrays of round and slot jets. Space does not permit these to be given here since a fairly involved description of the geometry is required. The interested reader is referred to Martin's review for further details.

Example 5.5

A mass $0.0212\,\text{kg/s}$ of air at $20°C$ ($C_p = 1007\,\text{J/kg K}$, $k = 0.026\,\text{W/m K}$ and $\mu = 1.8 \times 10^{-5}\,\text{kg/m s}$) is to be passed through a $15\,\text{mm}$ exit dia. jet to cool a $180\,\text{mm}$ dia. and $4\,\text{mm}$ thick titanium plate ($\rho = 4540\,\text{kg/m}^3$, $k = 10\,\text{W/m K}$ and $C = 523\,\text{J/kg K}$) previously heated to $400°C$. The exit from the jet is located $45\,\text{mm}$ above the plate, and the surface of the plate not exposed to the impinging jet is assumed to be adiabatic.

(a) Find the average heat transfer coefficient.

(b) Estimate the time taken for the plate to be cooled to $40°C$.

Solution **(a)** $Re = \dfrac{4\dot{m}}{\pi D \mu} = \dfrac{4 \times 0.0212}{\pi \times 0.015 \times 1.8 \times 10^{-5}} = 1 \times 10^5$

$Pr = \dfrac{\mu C_p}{k} = \dfrac{1.8 \times 10^{-5} \times 1007}{0.026} = 0.697$

$\dfrac{r}{D} = \dfrac{90}{15} = 6$

$\dfrac{H}{D} = \dfrac{45}{15} = 3$

$\dfrac{D}{r} = \dfrac{15}{90} = 0.1667$

From Equation (5.89)

$$Nu_{D,\,av} = F(Re)\,G(r/D,\,H/D)\,Pr^{0.42}$$

$$F(Re) = 2Re^{1/2}(1 + 0.005Re^{0.55})^{1/2} = 1235$$

$$G(r/D,\,H/D) = \frac{D}{r}\left\{1 - 1.1\left(\frac{D}{r}\right)\right\}\Big/\left\{1 + 0.1\left(\frac{D}{r}\right)\left(\frac{H}{D} - 6\right)\right\}$$

$$= 0.143$$

Hence

$$Nu_{D,\,av} = 1235 \times 0.143 \times (0.697)^{0.42}$$

$$= 152$$

so

$$h_{av} = Nu_{D,\,av}\,\frac{k}{D} = \frac{152 \times 0.026}{0.015} = 264\,\text{W/m}^2\text{K}$$

(b) Since $Bi < 1$ $(Bi = hL/k = 264 \times 0.004/10 = 0.1)$, we can use the lumped mass approximation (2.84a) where

$$\frac{(T - T_\infty)}{(T_i - T_\infty)} = \exp(-\lambda h_{av}t) \quad \text{and} \quad \lambda = \frac{A}{mC}$$

For the titanium disc, the mass, $m = 4540 \times \pi \times 0.09 \times 0.09 \times 0.004$
$= 0.462\,\text{kg}$, so

$$A = \pi \times 0.09 \times 0.09$$

$$= 0.0254\,\text{m}^2$$

At the required condition of $T = 40°\text{C}$,

$$\frac{40 - 20}{400 - 20} = \frac{1}{19}$$

hence

$$t = \ln\frac{(1/19)}{-\lambda h_{av}}$$

$$= \ln\frac{(1/19)}{-(264 \times 0.0254/0.462 \times 523)}$$

$$= 106\,\text{s}$$

Comment Compared with flow over a flat plate (Example 5.3, $h_{av} = 112\,\text{W/m}^2\text{K}$ at $Re_L = 5 \times 10^5$), an impinging jet gives a relatively high heat transfer coefficient.

5.8.2 Cylinders in crossflow

The heat transfer from a single cylinder, or banks of cylinders, exposed to a fluid stream directed perpendicular to their axis has obvious applications in heat exchanger design. Another application, albeit on a much smaller scale is the design of a hot-wire anemometer, a device consisting of a fine (typically $\approx 20\,\mu m$ dia.) wire used to determine local flow velocities from measurements of the current flowing through the wire (see Example 5.6).

A schematic diagram of the formation of the boundary layer over a single cylinder, of diameter D in crossflow is shown in Figure 5.9. The approach velocity of the flow is U_∞, which is not the same as the local freestream velocity u_∞, owing to the bending of the streamlines. The local value of u_∞ is zero at the stagnation point, at which point the pressure is a maximum. Downstream of the stagnation point the pressure decreases $(dp/dx < 0)$ causing the flow to be accelerated. At some point, the pressure reaches a minimum and the pressure gradient changes sign, $dp/dx > 0$. Further downstream of this point, the fluid velocity is decelerated and eventually the fluid near the surface has insufficient momentum to overcome the pressure gradient and so the flow separates from the surface. The accepted criterion for transition from laminar to turbulent flow is $Re_D > 2 \times 10^5$ ($Re_D = U_\infty D/v$). The separation point, given in terms of the angle θ from the stagnation point is different for laminar ($\theta \approx 80°$) and turbulent ($\theta \approx 140°$) flows.

The foregoing brief description serves to illustrate that an analytical treatment of the heat transfer behaviour is complex. Several attempts have been made to correlate data for the heat transfer from single cylinders. The most recent is due to Churchill and Bernstein (1977) valid for $Re_D Pr > 0.2$:

$$Nu_{D,\,av} = 0.3 + \left(\frac{A}{B}\right) C^{4/5} \tag{5.90}$$

Fig. 5.9 ●
Flow and boundary layer development over a cylinder n crossflow

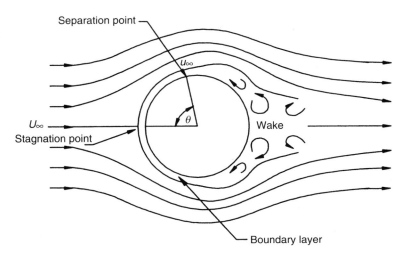

Separation point

u_∞

Wake

U_∞

Stagnation point

θ

Boundary layer

where $A = 0.62 \, Re_D^{1/2} Pr^{1/3}$

$$B = \left[1 + \left(\frac{0.4}{Pr}\right)^{2/3}\right]^{1/4}$$

$$C = 1 + \left(\frac{Re_D}{282\,000}\right)^{5/8}$$

where the properties for $Nu_{D,\mathrm{av}}$, Re_D and Pr are evaluated at $\frac{1}{2}(T_\infty + T_w)$.

Correlations for aligned and staggered arrays of cylinders in crossflow are presented in the review by Zukauskas (1987). Again, lack of space precludes their replication here. In brief they are of the form

$$Nu_{D,\mathrm{av}} = C_1 C_n Re_D^n Pr^m (Pr/Pr_w)^{1/4} \tag{5.91}$$

where the values of the constants C_1, n and m depend on the value of Re_D and the value of the constant C_n depends on the number of rows of cylinders in the array. The term (Pr/Pr_w) is the ratio of the Prandtl number evaluated at the mean fluid temperature to that evaluated at the cylinder wall temperature.

Example 5.6

A hot-wire probe comprises a 5 mm length of $10 \, \mu m$ dia. wire with an electrical resistance of 140 ohms/metre. The wire is maintained at a constant temperature. Estimate the current required to maintain the wire at 22°C in a 10 m/s airstream at 1.5 bar absolute pressure and 20°C.

Solution   Obtain fluid properties at the mean film temperature of 21°C:

$$\rho = \frac{p}{RT} = \frac{1.5 \times 10^5}{287 \times 294} = 1.778 \, \text{kg/m}^3$$

$\mu = 1.82 \times 10^{-5} \, \text{kg/m s}$ (tables and interpolation)

$k = 0.02577 \, \text{W/m K}$ (tables and interpolation)

$C_p = 1005 \, \text{J/kg K}$

$$Re_D = \frac{\rho U_\infty D}{\mu} = \frac{1.778 \times 10 \times 10 \times 10^{-6}}{1.82 \times 10^{-5}} = 9.77$$

$$Pr = \frac{\mu C_p}{k} = \frac{1.82 \times 10^{-5} \times 1005}{0.0257} = 0.712$$

$Re_D \, Pr = 9.77 \times 0.712 = 6.95$ (i.e. > 0.2, so we can use (5.90))

From Equation (5.90)

$$A = 0.62 \, Re_D^{1/2} Pr^{1/3} = 0.62 \times 9.77^{1/2} \times 0.717^{1/3} = 1.735$$

$$B = \left[1 + (0.4/Pr)^{2/3}\right]^{1/4} = \left[1 + (0.4/0.717)^{2/3}\right]^{1/4} = 1.138$$

$$C = 1 + \left(\frac{Re_D}{282\,000}\right)^{5/8} = 1 + \left(\frac{9.77}{282\,000}\right)^{5/8} = 1.0016$$

$$Nu_{D,\,av} = 0.3 + \left(\frac{A}{B}\right) C^{4/5} = 1.826 = \frac{q\,D}{\Delta T k}$$

Hence the heat flux, q, is

$$q = \frac{Nu_{D,\,av}\Delta T k}{D} = \frac{1.826 \times 2 \times 0.0257}{10 \times 10^{-6}} = 9385\,\text{W/m}^2$$

$$\dot{Q} = q \times (\text{area}) = q(\pi D L)$$

$$= 9385 \times \pi \times 10 \times 10^{-6} \times 5 \times 10^{-3} = 1.474\,\text{mW}$$

This is also the value of the electrical power dissipated, $\dot{Q} = I^2 R$, where $R = 140 \times 0.005 = 0.7$ ohms, and

$$I = \left(\frac{Q}{R}\right)^{1/2} = 45.9\,\text{mA}$$

5.8.3 Spheres

There are obvious similarities between the boundary layer development around a sphere and that around a cylinder in crossflow. There are many correlations, perhaps the most commonly cited is that due to Whitaker (1972), where the average Nusselt number is given by

$$Nu_{D,\,av} = 2 + (0.4 Re_D^{1/2} + 0.06 Re_D^{2/3}) Pr^{0.4} \left(\frac{\mu_\infty}{\mu_w}\right) \tag{5.92}$$

where all properties are evaluated at the freestream temperature T_∞, except μ_w which is evaluated at the surface temperature of the sphere. Equation (5.92) has been validated to within 30% by experimental measurements for the range $3.5 \leqslant Re_D \leqslant 7.6 \times 10^4$; $0.71 \leqslant Pr \leqslant 380$ and $1.0 \leqslant (\mu_\infty/\mu_w) \leqslant 3.2$.

5.9 ● Closing comments

Forced convection occurs as a result of superimposed motion, either by forcing the flow past a stationary surface or by moving a surface in a stationary fluid. By implication, the strength of the forced motion exceeds that due to natural buoyancy and we are able to neglect the buoyancy terms in the governing equations. As a consequence the Nusselt number depends mostly on the Reynolds and Prandtl numbers, and is independent of the surface-to-fluid temperature difference. The magnitude of the Reynolds number also governs transition from laminar to turbulent forced convection. The value of the transition Reynolds number depends on the presence of freestream turbulence, pressure gradient and so on, as well as the actual flow configuration; for flow in a pipe $Re_{trans} \approx 2300$; for external flow over a flat plate $Re_{trans} \approx 5 \times 10^5$.

Examples of forced convection reviewed in this chapter are parallel flow between two plates, laminar and turbulent pipe flow, laminar and turbulent flow over a flat plate, a rotating disc, wall jet, cylinder and sphere. Laminar flow and simple geometries such as flat plates and circular pipes make themselves amenable to analytical solutions. In this respect, the integral equations are particularly useful. For turbulent flow in circular pipes and flat plates, analytical solutions are possible using analogy methods, which link the shear stress and heat flux. Mathematical analysis of the more complex geometries, such as the flow over a sphere or a wall jet, becomes too difficult so empirical correlations derived from experimental data have been presented. The various correlations presented in this chapter are summarised in Table 5.1.

Table 5.1 Summary of forced convection correlations

Flow configuration	Equation	Comments
Laminar pipe flow	(5.35a)	$Re_D < 2300$, Constant heat flux
	(5.35b)	$Re_D < 2300$, Constant wall temperature
Turbulent pipe flow	(5.41)	$2300 \leqslant Re_D \leqslant 10^5$
	(5.42)	For heating/cooling of fluid
		$0.7 \leqslant Pr \leqslant 160$, $Re_D > 10^4$, $L/D \geqslant 10$
	(5.43)	Use where property variation is significant
	(5.44)	For rough surface
	(5.46)	Liquid metals $0.001 \leqslant Pr \leqslant 0.1$
Laminar flat plate	(5.67)	Local, $Re_x \leqslant Re_{x,\,trans}$
	(5.68)	Average, $Re_x \leqslant Re_{x,\,trans}$
	(5.69)	Unheated starting length
Turbulent flat plate	(5.83)	Local, $Re_x \geqslant Re_{x,\,trans}$
	(5.84)	Average, $Re_x \geqslant Re_{x,\,trans}$
	(5.85)	Unheated starting length
Rotating disc	(5.87)	Laminar $Re_\Omega < 3 \times 10^5$
	(5.88)	Turbulent $Re_\Omega > 3 \times 10^5$
Impinging jet	(5.89)	$2 \times 10^3 \leqslant Re \leqslant 4 \times 10^5$
		$2 \leqslant H/D \leqslant 12$ and $2.5 \leqslant r/D \leqslant 7.5$
Cylinder in crossflow	(5.90)	$Re_D Pr > 0.2$
Sphere	(5.92)	$3.5 \leqslant Re_D \leqslant 7.6 \times 10^4$;
		$0.71 \leqslant Pr \leqslant 380$ and $1.0 \leqslant (\mu_\infty/\mu_w) \leqslant 3.2$

5.10 ● References

Bejan, A (1993). *Heat Transfer*. New York: Wiley

Blasius, H. (1913). Das Ähnlichkeitsgesetz bei Reibungsvorgängen in Flüssigkeiten. *Forsch. Arb. IngWes*, No. 131

Childs, P.R.N. and Long, C.A. (1996). A review of forced convective heat transfer in stationary and rotating annuli. *Proc. Inst. Mech. Engrs*, **210**, 123–34

Churchill, S.W. and Bernstein, M. (1977). A correlating equation for forced convection from gases and liquids to a circular cylinder in crossflow, *ASME J. Heat Transfer* **99**, 300–6

Colebrook, C.F. (1939). Turbulent flow in pipes with particular reference to the transition region between the smooth and rough pipe laws. *J. Inst. Civil Engineers*, **11**, 133–56

Dittus, P.W. and Boelter, L.M.K. (1930). *Heat Transfer in Automobile Radiators of the Tubular Type* Vol. 2, pp. 443–61. Berkeley, CA: University of California

Dorfman, L.A. (1963). *Hydrodynamic Resistance and the Heat Loss from Rotating Solids*. Edinburgh: Oliver & Boyd

Gnielinski, V. (1976). New equations for heat and mass transfer in turbulent pipe and channel flow. *Int. Chem. Eng.*, **16**, 359–68

Kays, W.M. and Crawford, M.E. (1993). *Convective Heat and Mass Transfer* 3rd edn. New York: McGraw-Hill

Martin, H. (1977). *Advances in Heat Transfer*. Vol. 13: *Heat and Mass Transfer between Impinging Gas Jets and Solid Surfaces*. New York: Academic Press

Notter, R.H. and Sleicher, C.A. (1972). A solution to the turbulent Graetz problem III. Fully developed and entry region heat transfer rates. *Chem. Eng. Sci.*, **27**, 2073–93

Owen, J.M. and Rogers, R.H. (1989). *Flow and Heat Transfer in Rotating Disc Systems.*, Vol. 1: *Rotor–Stator Systems*. Taunton: Research Studies Press

Owen, J.M. and Rogers, R.H. (1995). *Flow and Heat Transfer in Rotating Disc Systems*. Vol. 2: *Rotating Cavities*. Taunton: Research Studies Press

Pohlhausen, E. (1921). Der Wärmeaustausch Zwischen festen Körpern und Flüssigkeiten mit kleiner Reibung und kleiner Wärmeleitung. *ZAMM*, **1**, 252–68

Sieder, E.N. and Tate, G.E. (1936). Heat Transfer and Pressure Drop of Liquids in Tubes. *Ind. Eng. Chem.*, **28**, 1429

Von Kármán, T. (1921). Uber Laminare und Turbulente Reibung, *ZAMM*, **1**,. 233–52

Whitaker, S. (1972). Forced convection and heat transfer correlations for flow in pipes, past flat plates, single cylinders, single spheres, and for flow in packed beds and tube bundles. *AIChE J.* **18**, 361–71

Zukauskas, A.A. (1987). Convective heat transfer in crossflow. *Handbook of Single Phase Convective Heat Transfer*, Ch. 6. New York: Wiley

5.11 ● End of chapter questions

5.1 An electric motor having a rotor diameter of 40 mm rotates at 50 000 rev/min. The annular (air) gap between the rotor and stationary outer casing is 0.2 mm. If both the rotor and casing have the same surface temperature, calculate the heat flux due to viscous dissipation (or windage). For air in the gap take $\mu = 1.8 \times 10^{-5}$ kg/m s and $k = 0.026$ W/mK

[±493 W/m²]

5.2 A vertical cylindrical container, 2 m in diameter and 5 m high, is filled with liquid nitrogen at 70 K. The top and bottom surfaces of the container may be considered to be perfectly insulated. The heat transfer coefficient on the inside (due to the liquid nitrogen) is given by the relation, $h_i = 195\Delta T^{1/3}$ (W/m²K), where ΔT is the appropriate surface-to-fluid temperature difference. Calculate the surface temperatures and the heat loss from the container when it is insulated with 0.15 m thickness of glass

wool ($k = 0.05$ W/m K), the ambient temperature is 20°C and a wind blows across the outer surface with a speed of 10 m/s. (Since the thickness of the insulation is small in relation to the diameter of the container, the curvature of the cylinder may be neglected for conduction effects.) For air take $\rho = 1.2$ kg/m³, $k = 0.026$ W/m K, $Pr = 0.71$ and $\mu = 1.8 \times 10^{-5}$ kg/m s.

[70.48 K, 289.45 K, 2289 W]

5.3 A pipeline heater, 3 m long and 40 mm dia., is used to heat a supply of carbon dioxide from 250 K to 500 K. The walls of the heater are heated to a constant temperature of 600 K. Using Equation (5.41), determine the mass flow that will be required.

For carbon dioxide at 375 K, and 1 bar, take $\rho = 1.41$ kg/m³, $k = 0.023$ W/m K, $Pr = 0.737$ and $\mu = 1.8 \times 10^{-5}$ kg/m s.

[0.0103 kg/s]

5.4 The supply of carbon dioxide in Question 5.3, is to be used to heat an adhesive which bonds a 3 mm thick, 40 mm dia. glass disc ($k = 1.4$ W/m K, $\rho = 2500$ kg/m^3 and $C = 750$ J/kg K) to a 40 mm dia. plastic disc of 20 mm thickness. The carbon dioxide is passed through a nozzle with a 6 mm exit dia. and located 30 mm above the surface of the glass disc; the ambient temperature is 300 K, and for carbon dioxide at 400 K and 1 bar take $\rho = 1.341$ kg/m^3, $k = 0.025$ W/m K, $Pr = 0.737$ and $\mu = 1.95 \times 10^{-5}$ kg/m s.

(a) Calculate the heat transfer coefficient

[996 W/m^{2}K]

(b) Assuming the plastic to be a perfect insulator, estimate the time for the temperature at the interface between the glass and plastic to reach 400 K. (Calculate the Biot number before starting this part and use the appropriate conduction model.)

[13 seconds]

5.5 Use Simpson's rule to evaluate the integral

$$\int_a^b \frac{dx}{x^{0.2}\left(1 - (4/x)^{0.9}\right)^{1/9}}$$

taking $a = 4.001$, $b = 6.001$, with a step length of 0.25 m.

[1.8883]

5.6 A single 2 m × 2 m flat plate solar collector is located in the middle of a flat roof which is 10 m × 10 m. Assuming turbulent flow, use the result from Question 5.5 to obtain an expression for the average Nusselt number when a wind of velocity U_∞ blows parallel to the surface of the roof.

$[Nu_{av,L} = 0.0389Re_L^{0.8}Pr^{1/3}]$

5.7 A heat exchanger for an air conditioning plant comprises a 10 mm i.d., 12 mm o.d. tube and a number of triangular fins of height $H = 0.15$ m. The fins are orientated so that air ($\rho = 1.2$ kg/m^3, $\mu = 1.8 \times 10^{-5}$ kg/m s, $k = 0.026$ W/m K and $Pr = 0.72$) from a cooling fan blows with a velocity $U = 2$ m/s across the base of each triangular fin with the apex at the trailing edge. The inside of the tube has a mean roughness of 0.1 mm and conveys 0.066 kg/s of refrigerant ($\rho = 1468$ kg/m^3, $\mu = 336 \times 10^{-6}$ kg/m s, $k = 0.0868$ W/m K and $Pr = 3.5$).

(a) Using a correlation of the general form for the local Nusselt number for a flat plate ($Nu_x = C \cdot Re_x^n$), show that the average Nusselt number for the triangular fin is

$$Nu_{av,H} = \frac{2C \cdot Re_H^n}{n(n+1)}$$

(b) Use an appropriate correlation to calculate the average heat transfer coefficient for the fin.

[19.4 W/m^{2}K]

(c) Calculate the average heat transfer coefficient for the inside of the tube.

[1678 W/m^{2}K]

(d) Calculate the average heat transfer coefficient for the outside (unfinned part) of the tube.

[44.3 W/m^{2}K]

(e) Is the use of a rough tube justified in this configuration?

5.8 Using Equations (5.87) and (5.88) show that the average Nusselt number for a mixed (laminar and turbulent) boundary layer on a disc of radius b, rotating in air ($Pr = 0.72$), having a quadratic surface temperature distribution ($n = 2$) is given by

$$Nu_{av,b} = 0.019Re_\Omega^{0.8}$$

$$+ \left[\left(\frac{r_{trans}}{b}\right)^3\left\{0.523Re_{\Omega,trans}^{0.5} - 0.019Re_{\Omega,trans}^{0.8}\right\}\right]$$

where $Re_{\Omega,trans}$ is the local Reynolds number, based upon the transition radius r_{trans}, at which transition from laminar to turbulent flow occurs.

5.9 Part of an industrial process plant requires a heated rotating disc as a heat source. The disc has a diameter of 0.6 m, its surface temperature varies with r^2, and rotates in air ($\rho = 1.2$ kg/m^3, $\mu = 1.8 \times 10^{-5}$ kg/m s, $k = 0.026$ W/m K and $Pr = 0.72$) at 800 rev/min. Assuming a transition Reynolds number of $Re_\Omega = 2 \times 10^5$, at what radius does transition occur? Using the result from Question 5.8, calculate the heat loss from one surface of the disc for a maximum surface-to-air temperature difference of 100°C.

$[r_{trans} = 0.189$ m, $\dot{Q} = 0.818$ kW$]$

5.10 A design to recover waste heat from hot water comprises a number of flat plate fins of length $L = 0.25$ m. Water ($\rho = 983$ kg/m^3, $\mu = 463 \times 10^{-6}$ kg/m s, $k = 0.653$ W/m K and $Pr = 2.97$) flows over the surface of the fins with a velocity of $U_\infty = 1$ m/s. A colleague has calculated the average heat transfer coefficient as having a value of 5140 W/m^{2}K. Why should you doubt this value? (Hint: Transition occurs at $Re_x = 3 \times 10^5$.) What is likely to be the correct average heat transfer coefficient?

[3244 W/m^{2}K]

5.11 Solve the integral momentum and energy equations (5.49) and (5.50) using linear velocity and temperature profiles, namely

$$\frac{(T - T_s)}{(T_\infty - T_s)} = \frac{u}{U_\infty} = \frac{y}{\delta}$$

Comment on the validity of these profiles.

[$\delta/x = 3.46Re_x^{-1/2}$ and $Nu_x = 0.289Re_x^{1/2}Pr^{1/3}$]

5.12 For turbulent forced convection from a horizontal flat surface, the Colburn analogy can be written as

$$Nu_x = 0.5\, C_{f,x} Re_x Pr^{1/3}$$

where the local skin friction coefficient is defined as $C_{f,x} = 2\tau_{s,x}/\rho U_\infty^2$.

(a) If $C_{f,x} = C_0(Re_x)^m$ (where C_0 and m are constants), show that the average Nusselt number for a surface of length L is given by

$$Nu_{av,L} = \frac{C_0 Re_L^{m+1} Pr^{1/3}}{2(m + 1)}$$

(b) Measurements of the shear stress from a prototype electronics circuit board show that the local skin friction coefficient and Reynolds number are related by

$$C_{f,x} = 0.12Re_x^{-0.2}$$

The circuit board measures 250 mm in the streamwise direction and 150 mm in the cross-stream direction and dissipates 60 W of heat from one side into the surrounding air ($\rho = 1$ kg/m^3, $\mu = 1.8 \times 10^{-5}$ kg/m s, $k = 0.026$ W/m K and $Pr = 0.7$) which is at 25°C. Neglect heat transfer by radiation and use the above expressions to estimate the free stream velocity of cooling air which is required to maintain a surface temperature of 40°C.

[12.4 m/s]

(c) Perform a simple calculation to show that radiative heat transfer from the board can be neglected.

5.13 An electrically heated thin foil of length $L = 25$ mm and width $W = 8$ mm is to be used as a wind speed meter. Wind with a temperature T_∞ and velocity U_∞ blows parallel to the longest side. The foil is internally heated by an electric heater dissipating $\dot{Q}$ (watts) from both sides, and is to be operated in air where $T_\infty = 20$°C, $C_p = 1.005$ kJ/kg K, $\nu = 1.522 \times 10^{-5}$ m^2/s, $\rho = 1.19$ kg/m^3 and $Pr = 0.72$. The surface temperature, T_s, of the foil is to be measured at the trailing edge, but can be assumed to be constant. Estimate the wind speed when $T_s = 32$°C and $\dot{Q} = 0.5$ W.

[18.3 m/s]

5.14 A 1 mm dia. spherical thermocouple bead ($C = 400$ J/kg K, $\rho = 7800$ kg/m^3) is required to respond to 99.5% change of the surrounding air ($\rho = 1.2$ kg/m^3, $\mu = 1.8 \times 10^{-5}$ kg/m s, $k = 0.0262$ W/m K and $Pr = 0.707$) temperature in 10 ms. What is the minimum air speed at which this will occur?

[4.5 m/s]

Chapter six

Free convection

6.1 ● Introduction

In free convection, fluid motion is self-induced by a buoyancy force. This arises from density variations in the fluid, generated from temperature differences between the fluid and neighbouring surface. A particularly good example of *external* free convection is the naturally induced flow in the vicinity of a heated vertical surface such as a central heating radiator (see Figure 6.1a). Air adjacent to the radiator is heated by it and as a result becomes less dense and rises. This motion leads to the formation of a boundary layer on the surface of the radiator. Fluid outside the boundary layer is entrained to balance the motion of rising fluid, creating a free convection 'loop'. The velocity is zero at the radiator surface and, if the space surrounding it is sufficiently large (that is, an order of magnitude or more larger than the boundary layer thickness), zero on the edge of the boundary layer.

If we reduce the space surrounding the radiator, by moving the surrounding walls closer and closer (and still maintain a temperature difference), there will come a point when the downward flow on the unheated wall is of comparable strength to the upward flow on the radiator. This is the limiting case of *internal* free convection. An example, both commonplace and of considerable significance, is the flow inside a cavity wall used to reduce heat loss from buildings; this is illustrated in Figure 6.1b.

Free convection is also referred to as natural convection. Both adjectives distinguish this mode of convective heat transfer from forced convection where, by implication, some external mechanism is responsible for forcing the flow. The buoyancy force is also present in forced convection, but by definition it is usually dominated by forces arising from the fluid inertia and/or the pressure gradient.

Further examples of free convection are the heat transfer from the fins of a motorcycle engine when at rest (as the motorcycle starts to move through the air, the buoyancy force becomes less significant, until at normal road speeds the heat transfer is due entirely to forced convection),

Fig. 6.1 ●
Free or natural
convection: (a)
external and
(b) internal

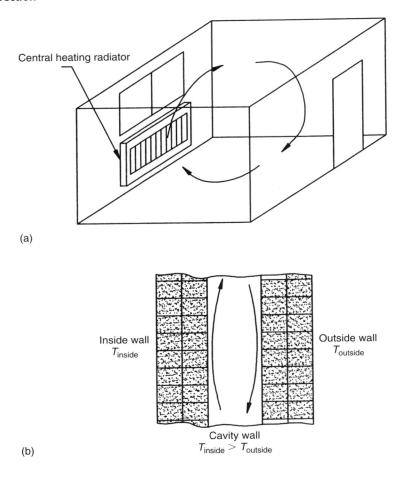

Central heating radiator

(a)

Inside wall
T_inside

Outside wall
T_outside

Cavity wall
$T_\text{inside} > T_\text{outside}$

(b)

a buoyant plume rising from a smokestack, the interior walls of buildings, the heat loss (or gain) from a tank of hot (or cold) fluid and, on a much larger scale, the heating of the earth by the sun that creates motion in the atmosphere that we experience as the wind. Following on in a similar way to the previous chapter, we shall review the simpler analytical solutions and correlations for certain simple idealised geometries, as this brings out important information and relationships that are of value both for design and for understanding the physical processes.

As noted in Section 4.4.2, the pertinent dimensionless groups used in free convection are the Grashof number, Gr, and the Prandtl number, Pr. The product $Gr \cdot Pr$ has particular significance in free convection and is referred to as the Rayleigh number, Ra. A generally accepted criterion to determine the significance of the buoyancy force is the parameter Gr_x/Re_x^2. For values of $Gr_x/Re_x^2 \gg 1$ the buoyancy force dominates and the flow may be considered as due entirely to free convection. Conversely, for $Gr_x/Re_x^2 \ll 1$ the buoyancy force is negligible and the flow should be considered as entirely forced convection. For $Gr_x/Re_x^2 \approx 1$ the flow is in a mixed convection regime where both free and forced convection effects are present.

In evaluating Gr_x/Re_x^2 the characteristic length for Re_x and Gr_x may or may not be the same. For example, natural convection from a vertical plate (height H, length L) takes as the characteristic length the height, H, of the plate. A forced flow parallel to the plate (at $90°$ to the gravity vector) has a characteristic length of the plate length, L.

Mathematically, free convection problems are more difficult to analyse than forced convection. In forced convection, the momentum equation can be solved in isolation from the energy equation. In free convection, the buoyancy force (which is temperature dependent) appears in the momentum equation. It is therefore not possible to solve the momentum and energy equations separately, but together and simultaneously. These complexities will restrict the mathematical analysis presented in this chapter. For details of the mathematical treatment of more complex free convection flows, the reader is urged to consult two excellent textbooks on the subject – Gebhart et al. (1988) and Jaluria (1980) – and the reviews by Ede (1967) for external free convection, and Ostrach (1972) for internal free convection.

6.2 ● Free convection from a vertical surface

Figure 6.2 depicts the velocity and temperature profiles in a free convection boundary layer forming on a vertical heated surface (the analysis will be identical for a cooled surface) for a fluid with Pr of order unity or greater (since $\delta > \delta_T$). At the wall, the no-slip condition applies and $(u)_{y=0} = 0$; also $(T)_{y=0} = T_s$. At the edge of the velocity boundary layer, $y = \delta$, $(u)_{y=\delta} = 0$ and $(T)_{y=\delta} = T_\infty$. The former condition is markedly different from that in forced convection where $(u)_{y=\delta} = U_\infty$.

Von Kármán's integral technique (applied to forced convection in Chapter 5) was also applied to natural convection from a vertical surface by Squire (1938). To apply this method to a surface of constant temperature we make the following assumptions:

1. The flow is laminar, steady and incompressible. The last point may appear to be a contradiction, since it is through variations in fluid density that the flow is driven. In this respect we mean that density variations are included only in the buoyancy force term; elsewhere in the equations the density is treated as constant.

2. Buoyancy effects are confined to the boundary layer region and $U_\infty = 0$.

3. The analysis applies for fluids with near unity Prandtl number, which implies that $\delta_T \approx \delta$.

4. Since the magnitude of the velocity is small, viscous dissipation is neglected.

Fig. 6.2 ● Free
convection from a
vertical heated
surface for a fluid
with $Pr \geqslant 1$

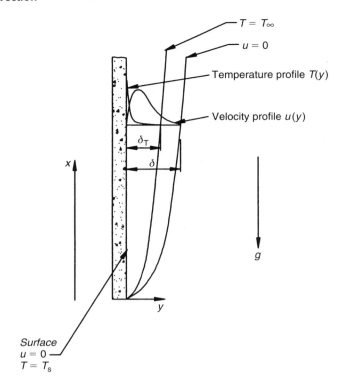

Incorporating these assumptions into the integral momentum and energy
equations, Equations (4.95) and (4.99):

$$\frac{d}{dx} \int_0^\delta u^2 dy = -\frac{\tau_s}{\rho} + \beta g \int_0^\delta (T - T_\infty) dy \qquad (6.1)$$

$$\frac{d}{dx} \int_0^\delta u(T - T_\infty) dy = \frac{q_s}{\rho C_p} \qquad (6.2)$$

Equation (6.1) implies that fluid inertia is balanced by a difference between
the buoyancy and the viscosity; the buoyance drives the flow and the
viscosity tends to retard it. Equation 6.2 indicates a balance between
convection and conduction.

Unlike the problem of forced convection from a horizontal surface,
discussed in Section 5.5, the variable T appears in both equations. The
equations are said to be coupled and they must be solved simultaneously.
Following the procedure used in Section 5.5, it is convenient to assume
polynomial distributions for the velocity component, u, and the
temperature in the boundary layer, T. Inspection of Figure 6.2 does,
however, reveal that for free convection, at $y = 0$ and at $y = \delta$, $u = 0$, so
the velocity must therefore have a maximum value inside the boundary

layer. A suitable form for the velocity profile in a laminar free convection boundary layer is given by

$$u = U_1\eta(1-\eta)^2 \tag{6.3}$$

where η is the usual (see Section 5.5) non-dimensional distance in the boundary layer, $\eta = y/\delta$ and U_1 is a constant. By differentiating Equation (6.3) we find that the maximum velocity in the boundary layer occurs at $\delta/3$ and has a value of $u_{max} = 4U_1/27$.

The boundary conditions for the velocity profile are

$$(u)_{\eta=0} = 0$$

$$(u)_{\eta=1} = 0$$

$$\left(\frac{\partial u}{\partial \eta}\right)_{\eta=0} = U_1$$

$$\left(\frac{\partial u}{\partial \eta}\right)_{\eta=1} = 0 \tag{6.4}$$

Similar to the way in which U_1 scales the velocity distribution, the distribution of temperature in the boundary layer is scaled by the surface-to-free-stream temperature difference $(T_s - T_\infty)$, hence

$$(T - T_\infty) = (T_s - T_\infty)(1 - \eta)^2 \tag{6.5}$$

Substituting $\theta = (T - T_\infty)/(T_s - T_\infty)$ gives $\theta = (1 - \eta)^2$. The boundary conditions are mostly self-evident:

$$(\theta)_{\eta=0} = 1 \quad \text{because } T = T_s \text{ at } y = 0$$

$$(\theta)_{\eta=1} = 0 \quad \text{because } T = T_\infty \text{ at } y = \delta$$

$$\left(\frac{\partial \theta}{\partial \eta}\right)_{\eta=1} = 0 \quad \text{because } q = 0 \text{ at } y = \delta \tag{6.6}$$

$$\left(\frac{\partial \theta}{\partial \eta}\right)_{\eta=0} = -2$$

The individual terms in the momentum integral equation (6.1) are then as follows:

$$\int_0^\delta u^2 \, dy = \delta \int_0^1 u^2 \, d\eta = \frac{\delta U_1^2}{105} \tag{6.7a}$$

$$\int_0^\delta (T - T_\infty) \, dy = \delta \int_0^1 (T - T_\infty) \, d\eta = \frac{\delta(T_s - T_\infty)}{3} \tag{6.7b}$$

and for the energy equation (6.2),

$$\int_0^\delta u(T - T_\infty) \, dy = \frac{\delta U_1(T_s - T_\infty)}{30} \tag{6.7c}$$

Substitution of (6.7a), (6.7b) and (6.7c) into (6.1) and (6.2) gives

$$\frac{1}{105}\frac{d}{dx}(U_1^2\delta) = \frac{1}{3}g\beta(T_s - T_\infty)\delta - \frac{\nu U_1}{\delta} \tag{6.8}$$

$$\frac{1}{30}(T_s - T_\infty)\frac{d}{dx}(U_1\delta) = \frac{2\alpha(T_s - T_\infty)}{\delta} \tag{6.9}$$

Equations (6.8) and (6.9) are coupled ordinary differential equations in U_1 and δ. To solve them, the assumption is made that $U_1 = C_1 x^{m_1}$ and $\delta = C_2 x^{m_2}$ where C_1, C_2, m_1 and m_2 are constants. Substitution into Equations (6.8) and (6.9) gives

$$\frac{C_1^2 C_2(2m_1 + m_2)x^{2m_1 + m_2 - 1}}{105} = \frac{1}{3}g\beta(T_s - T_\infty)C_2 x^{m_2} - \frac{\nu C_1 x^{m_1 - m_2}}{C_2} \tag{6.10}$$

and

$$\frac{C_1 C_2(m_1 + m_2)x^{m_1 + m_2 - 1}(T_s - T_\infty)}{30} = \frac{2\alpha(T_s - T_\infty)}{C_2 x^{m_2}} \tag{6.11}$$

The variables η and θ are independent of x (see definitions under Equations (6.3) and (6.5)) and are known as similarity variables. Similarity is maintained if both sides of Equations (6.10) and (6.11) are independent of x, which implies that the left and the right sides of these equations must have the same powers of x. From which, by equating powers of x, we get

$$m_1 = \frac{1}{2} \quad \text{and} \quad m_2 = \frac{1}{4} \tag{6.12}$$

Substitution back into Equations (6.10) and (6.11) gives

$$C_2 = 3.93\left(\frac{20}{21} + \frac{\nu}{\alpha}\right)^{1/4}\left\{\frac{g\beta(T_s - T_\infty)}{\nu^2}\right\}^{-1/4}\left(\frac{\nu}{\alpha}\right)^{-1/2}$$

$$C_1 = 5.17\nu\left(\frac{20}{21} + \frac{\nu}{\alpha}\right)^{-1/2}\left\{\frac{g\beta(T_s - T_\infty)}{\nu^2}\right\}^{1/2} \tag{6.13}$$

From the original definition, $\delta = C_2 x^{m_2}$, the boundary layer development is given by

$$\frac{\delta}{x} = 3.93(0.952 + Pr)^{1/4}Gr_x^{-1/4}Pr^{-1/2} \tag{6.14}$$

Also, for the heat flux at the surface of the plate,

$$q_s = -k\left(\frac{\partial T}{\partial y}\right)_{y=0} = -k(T_s - T_\infty)\left(\frac{\partial\theta}{\partial\eta}\right)_{\eta=0} = \frac{2k(T_s - T_\infty)}{\delta} \tag{6.15}$$

and from the definition of the local Nusselt number,

$$Nu_x = \frac{q_s x}{k(T_s - T_\infty)} = \frac{2x}{\delta} \tag{6.16}$$

So, for laminar free convection from a vertical plate, using Equation (6.14) for δ in Equation (6.16), the local Nusselt number is given by

$$Nu_x = 0.508 Pr^{1/2}(0.952 + Pr)^{-1/4} Gr_x^{1/4} \qquad (6.17a)$$

and for a plate of length L, the average Nusselt number is

$$Nu_{av} = 0.677 Pr^{1/2}(0.952 + Pr)^{-1/4} Gr_L^{1/4} \qquad (6.17b)$$

For air with $Pr = 0.72$, Equations (6.17a) and (6.17b) simplify to

$$Nu_x = 0.378 Gr_x^{1/4} \qquad (6.18a)$$

$$Nu_{av} = 0.504 Gr_L^{1/4} \qquad (6.18b)$$

The relations derived in Section 4.7.2 by applying scaling arguments to laminar-free convection are consistent with the above relations, found from a more detailed treatment. An exact solution of the boundary layer equations by Schmidt and Beckmann (1930) gives (for $Pr = 0.72$) $Nu_x = 0.36 Gr_x^{1/4}$. This serves to illustrate that despite the simplifications used, the integral technique yields answers that are surprisingly accurate.

Equation (6.17) was derived assuming $\delta_T \approx \delta$, i.e. $Pr \approx 1$. Table 6.1 gives values of the ratio $Nu_x/(Gr_x Pr)^{1/4}$ for fluids with Prandtl numbers in the range $0.01 \leqslant Pr \leqslant 1000$.

Until relatively recently it was thought that transition from laminar to turbulent-free convection occurred when $Gr \cdot Pr \approx 10^9$. More current work, and in particular that by Bejan and Large (1990) makes a convincing argument that the criterion for transition is $Gr_x \approx 10^9$. For laminar flow, as shown above and verified experimentally, Nu_{av} varies with $Gr_L^{1/4}$. For turbulent flow, experimental evidence is sparse and disputed, but it is generally accepted that in turbulent-free convection, Nu_{av} is approximately proportional to $Gr_L^{1/3}$. McAdams (1954) gives the following correlation for turbulent-free convection:

$$Nu_x = 0.1(Gr_x Pr)^{1/3} \qquad (6.19a)$$

which also implies that

$$Nu_{av} = 0.1(Gr_L Pr)^{1/3} \qquad (6.19b)$$

A solution of the integral momentum and energy equations for turbulent-free convection was obtained by Eckert and Jackson (1951). Assuming turbulent forms for the velocity and temperature profiles, together with a

Table 6.1 ● Values of $Nu_x/(Gr_x Pr)^{1/4}$ for fluids with non-unity Prandtl number

Pr	0.01	0.72	1	2	10	100	1000
$Nu_x/(Gr_x Pr)^{1/4}$	0.162	0.387	0.401	0.426	0.465	0.490	0.499

Source: Based on Bejan (1995)

modified form of the shear stress relationship given by Equation (5.78) and the Colburn analogy (4.90), they obtained the relation

$$Nu_{av} = 0.0246 Gr_L^{2/5} f(Pr) \tag{6.20}$$

where $f(Pr) = Pr^{7/15}(1 + 0.494 Pr^{2/3})^{-2/5}$. (The derivation of Equation (6.20) is set as a exercise at the end of this chapter.)

The following empirical correlations of experimental data proposed by Churchill and Chu (1975) cover a wider Rayleigh number range ($10^{-1} \leqslant Ra_L \leqslant 10^{12}$) than either Equations (6.18) or (6.19):

$$Nu_{av} = \{0.825 + A/B\}^2 \tag{6.21}$$

where: $A = 0.387 Ra_L^{1/6}$

$$B = [1 + (0.492/Pr)^{9/16}]^{8/27}$$

Improved accuracy for the range $Ra_L \leqslant 10^9$ can be obtained from

$$Nu_{av} = 0.68 + A/B \tag{6.22}$$

where $A = 0.67 Ra_L^{1/4}$

$$B = [1 + (0.492/Pr)^{9/16}]^{4/9}$$

For the case of a uniform heat flux at the surface (as opposed to a uniform surface temperature), the problem is to find $T_s - T_\infty$, since the flux, q, is known. Hence we define an alternative Rayleigh number, Ra_x^*, given by

$$Ra_x^* = \frac{g\beta x^4 q}{\alpha v k} \tag{6.23}$$

Recommended correlations for the local and average Nusselt numbers are given by Vliet and Liu (1969) as

$$Nu_x = 0.6 Ra_x^{*0.2} \tag{6.24a}$$

$$Nu_{av} = 0.75 Ra_x^{*0.2} \qquad \text{Laminar } 10^5 \leqslant Ra_x^* \leqslant 10^{13} \tag{6.24b}$$

$$Nu_x = 0.568 Ra_x^{*0.22} \tag{6.25a}$$

$$Nu_{av} = 0.645 Ra_x^{*0.22} \qquad \text{Turbulent } 10^{13} \leqslant Ra_x^* \leqslant 10^{16} \tag{6.25b}$$

All of the above expressions, for laminar and turbulent flow, may be applied to a vertical cylinder of height L if the thickness of the boundary layers, δ and δ_T, is much less than the diameter, D. For fluids with a Prandtl number of unity or greater, this condition is satisfied by the inequality due to Bejan (1995):

$$\frac{D}{L} \gg Ra_L^{-1/4} \tag{6.26}$$

Experimental evidence shows that the heat transfer is increased above the vertical plate value when the curvature of the cylinder is large enough (i.e. $D/L < Ra_L^{-1/4}$) to affect the development of the boundary layers.

Example 6.1

Estimate the cooling capacity, by natural convection in air, of the heat sink shown in Figure 6.3. The fins may be assumed to have a constant surface temperature of $T_s = 60°C$, the ambient air temperature $T_\infty = 20°C$ and the fin efficiency is 60%.

Solution   Evaluate properties at $T_m = \frac{1}{2}(T_s + T_\infty) = 40°C$, at which

$\rho = 1.1\,\text{kg/m}^3$

$\mu = 1.91 \times 10^{-5}\,\text{kg/m s}$

$k = 0.027\,\text{W/m K}$

$Pr = 0.70$

Evaluate β at T_∞ ($\beta = 1/293\,\text{K}^{-1}$):

$$Gr_L = \frac{\rho^2 g\beta\Delta T L^3}{\mu^2}$$

$$= \frac{1.1^2 \times 9.81 \times (1/293) \times (60 - 20) \times 0.02^3}{(1.91 \times 10^{-5})^2}$$

$$= 35\,536$$

From Table 6.1, $Nu_x = 0.387(Gr_x Pr)^{1/4}$, which implies that the local heat transfer coefficient, h, varies as $x^{-1/4}$. Hence, by integrating, the average Nusselt number is given by

$$Nu_{av} = \left(\frac{4}{3}\right)0.387\,(Gr_L Pr)^{1/4} = 0.516(Gr_L Pr)^{1/4}$$

$$= 0.516(35\,536 \times 0.7)^{1/4} = 6.48 = \frac{h_{av}L}{k}$$

$$h_{av} = \frac{Nu_{av}k}{L} = \frac{6.48 \times 0.027}{0.02} = 8.75\,\text{W/m}^2\text{K}$$

Fig. 6.3 ● Heat sink cooled by free convection (Example 6.1)

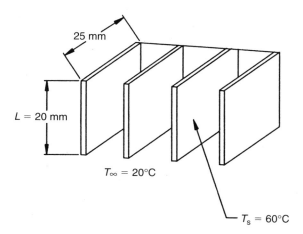

25 mm

$L = 20$ mm

$T_\infty = 20°C$

$T_s = 60°C$

The average heat flux, $q_{av} = h_{av}(T_s - T_\infty)$:

$$q_{av} = 8.75 \times (60 - 20) = 350 \, \text{W/m}^2$$

and the total heat dissipated by both surfaces of four fins, with an efficiency η, is

$$\dot{Q} = 4 \times 2 \times \eta \times q_{av} \times L \times W$$
$$= 8 \times 0.6 \times 360 \times 0.02 \times 0.025$$
$$= 0.84 \, \text{W}$$

Comment The heat loss by radiation, q_{rad}, will not be insignificant. Assuming black body behaviour to surroundings at T_∞,

$$q_{rad} = \sigma(T_s^4 - T_\infty^4)$$
$$= 56.7 \times 10^{-9} \times (333^4 - 293^4)$$
$$= 279 \, \text{W/m}^2$$

which means that it is contributing 44% of the total heat loss.

From Equation (6.14) the boundary layer thickness at $x = L = 0.02$ m is approximately 8 mm. If the fin spacing is significantly less than this, then fully developed channel flow will occur between the fins, and a relationship similar to Equation (6.40), but in Cartesian coordinates, will need to be used. It is left as an exercise to the reader to follow the methodology in Section 6.5 and show that the equivalent expression to Equation (6.40) for a vertical channel of width L and height H is $Nu_H = Ra_L/24$ (see also Questions 6.7 and 6.8 at the end of this chapter).

6.3 ● Inclined and horizontal surfaces

Figure 6.4 shows a schematic diagram of the flow induced by surfaces inclined at an angle θ to the vertical, or more correctly, to the gravitational vector. Under these circumstances, the buoyancy force can be resolved into a component acting parallel to the surface, or the x direction ($= g \cos\theta$), and a component acting in the y direction or normal to the surface. For a heated surface ($T_s > T_\infty$): the flow on the bottom side of the plate is kept in contact with the plate surface by the y-component of the buoyancy force; the flow on the top side tends to become three-dimensional under the action of the y component of the buoyancy force. The converse is seen to apply for the case of a cooled surface ($T_s < T_\infty$): the flow on the top side of the plate maintains contact with the surface; the flow on the bottom side tends to become three-dimensional.

It is generally accepted that for the top surface of a cool plate or the bottom surface of a hot plate, inclined at an angle θ, where $-60° \leqslant \theta \leqslant 60°$, the gravitational acceleration, g, in the Grashof number may be replaced by

Fig. 6.4 ● Natural convection from inclined surfaces

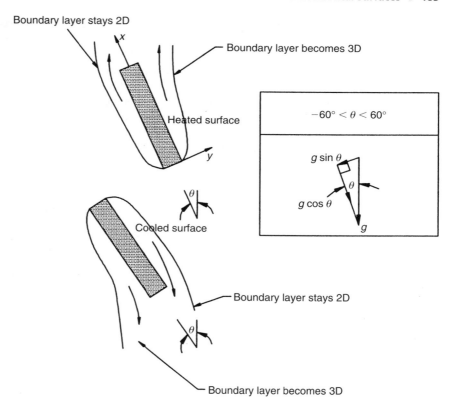

Fig. 6.5 ● Free convection from a horizontal surface

$g \cos \theta$. Equations (6.17), (6.19), (6.21) or the results shown in Table 6.1 can then be used to calculate the Nusselt number. For the bottom surface of a cool plate or the top surface of a hot plate, or for values of θ greater than 60°, no recommendations are given, but the interested reader is advised to consult recent published literature.

For a horizontal plate, as shown in Figure 6.5, there is only one component of the buoyancy force – that normal to the surface. The flow field is again

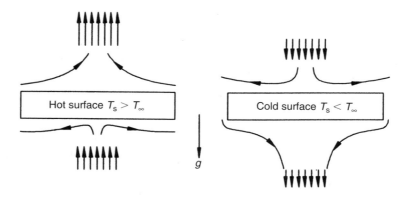

dependent on the heating configuration. For a hot surface ($T_s > T_\infty$) facing upwards, the flow is free to move upwards away from the plate. Conservation of mass requires that this is replenished by a flow of cooler fluid towards the plate. By contrast, for the downwards-facing side of a hot surface, the natural tendency of the flow to rise is impeded by the plate surface. Consequently the flow must move horizontally towards the edges of the plate before it can rise. As a result, compared with the hot side facing upwards, the heat transfer rate is reduced. Correlations for the different configurations for a horizontal plate (Incropera and DeWitt, 1996) are

Top side of a hot plate or bottom side of a cold plate

$$Nu_{\mathrm{av},L} = 0.54(Gr_L Pr)^{1/4} \qquad 10^4 \leqslant (Gr_L Pr) \leqslant 10^7 \tag{6.27}$$

$$Nu_{\mathrm{av},L} = 0.15(Gr_L Pr)^{1/3} \qquad 10^7 \leqslant (Gr_L Pr) \leqslant 10^{11} \tag{6.28}$$

Bottom side of a hot plate or top side of a cold plate

$$Nu_{\mathrm{av},L} = 0.27(Gr_L Pr)^{1/4} \qquad 10^4 \leqslant (Gr_L Pr) \leqslant 10^{10} \tag{6.29}$$

The *characteristic length, L*, used in Equations (6.27) to (6.29) *is the ratio of the plate surface area to the perimeter.*

Example 6.2

A helicopter platform for a hospital is 12 m square and covered with a non-slip coating with a solar absorptivity of $\alpha_s = 1$ and emissivity $\varepsilon = 1$. Estimate the surface temperature when subjected to an incident solar flux of 1 kW/m² and the surrounds are at 20°C.

Solution   Since we do not know the surface temperature, we evaluate the properties at $T_\infty = 20°C$. If better accuracy is required, we could make this part of an iterative cycle, where the first iteration gives an estimate of surface temperature, from which we can obtain T_m and so on. This procedure would probably not be worthwhile in this example, where the properties of air do not change dramatically over the particular range of temperatures.

For air at 20°C and 1 bar,

$$\rho = 1.19 \, \mathrm{kg/m^3}$$

$$\mu = 1.8 \times 10^{-5} \, \mathrm{kg/m \, s}$$

$$k = 0.026 \, \mathrm{W/m \, K}$$

$$Pr = 0.707$$

evaluate β at T_∞ ($\beta = 1/293 \, \mathrm{K^{-1}}$).

The convective loss from the helicopter platform is modelled as a horizontal plate with the hot side facing upwards. To establish if the flow is laminar or turbulent we calculate the Rayleigh number for an, as yet unknown, temperature difference $\Delta T = (T_s - T_\infty)$. The characteristic length, L, for a horizontal plate is the ratio of the area to the perimeter, i.e. $L = (12 \times 12)/(4 \times 12) = 3\,\text{m}$.

$$Gr_L = \frac{\rho^2 g \beta \Delta T L^3}{\mu^2}$$

$$= \frac{1.19^2 \times 9.81 \times (1/293) \times \Delta T \times 3^3}{(1.8 \times 10^{-5})^2}$$

$$= 3.95 \times 10^9 \Delta T$$

The flow is clearly turbulent for all likely values of ΔT, so we choose Equation (6.28),

$$Nu_{\text{av},L} = 0.15(Gr_L Pr)^{1/3}$$

from which

$$h_{\text{av}} = 0.15(k/L)(Gr_L Pr)^{1/3}$$

$$= 0.15 \times (0.026/3) \times (3.95 \times 10^9 \times 0.707)^{1/3} \Delta T^{1/3}$$

$$= 1.83 \Delta T^{1/3}$$

$$= 1.83(T_s - T_\infty)^{1/3}\,\text{W/m}^2\,\text{K}$$

From a heat balance on the surface of the helicopter platform, the solar flux in q_{solar} is equal to the convective and radiative loss:

$$q_{\text{solar}} = \sigma(T_s^4 - T_\infty^4) + h_{\text{av}}(T_s - T_\infty)\,\text{W/m}^2$$

$$1000 = 56.7 \times 10^{-9}(T_s^4 - T_\infty^4) + 1.83(T_s - T_\infty)^{4/3}$$

Since radiative effects are involved, the temperatures must be in kelvin, and for $T_\infty = 293\,\text{K}$ we can express the above equation in terms of the unknown temperature T_s as

$$56.7 \times 10^{-9} T_s^4 + 1.83(T_s - 293)^{4/3} - 1418 = 0$$

This may be solved by iteration. The Newton–Raphson method is recommended as a relatively simple iterative scheme – a simple exercise is given in Question 6.1 at the end of this chapter. For the $(n + 1)$th estimate of x, for the function $f_{(x)} = 0$, this is given by the sequence

$$x^{(n+1)} = x^{(n)} - \left\{ \frac{f_{(x)}}{df_{(x)}/dx} \right\}$$

In the context of finding the surface temperature, $f_{(x)}$ is given by the above equation involving T_s, and by differentiating this:

$$\frac{df_{(x)}}{dx} = 226.8 \times 10^{-9} T_s^3 + 2.44(T_s - 293)^{1/3}$$

Starting with $T_s^{(0)} = 293\,K$, we get

$T_s^{(1)} = 468.31\,K$

$T_s^{(2)} = 384.31\,K$

$T_s^{(3)} = 360.4\,K$

$T_s^{(4)} = 358.46\,K$

$T_s^{(5)} = 358.45\,K$

Hence, after five iterations, the surface temperature has converged to two decimal places and has a value of approximately 85°C.

Comment   This represents a maximum estimate of the surface temperature, since any wind will introduce a forced convection component acting to enhance the convective heat transfer and cool the surface. The calculated temperature of 85°C could pose a design problem, since the surface is probably made of a plastic material and at this temperature it could become soft or even melt.

6.4 ● Other external flows: cylinder, sphere and rotating disc

These important geometries are dealt with here fairly briefly. For further information the reader is advised to consult Gebhart *et al.* (1988).

6.4.1   Horizontal cylinder

The natural convection flow induced over the surface of a horizontal cylinder bears similarity to that over a vertical wall, the difference being that the surface of a cylinder is curved, whereas the wall is not. To account for this, the Nusselt number and Grashof number both use the diameter of the cylinder, D, as the characteristic length. A single correlation put forward by Churchill and Chu (1975), covering the range $Ra_D \leqslant 10^{12}$, is

$$Nu_{av,D} = \{0.6 + A/B\}^2 \tag{6.30}$$

where $A = 0.387\,Ra_D^{1/6}$

$$B = [1 + (0.559/Pr)^{9/16}]^{8/27}$$

6.4.2   Sphere

As with forced convection there are similarities between the flows around a cylinder and a sphere. Heat transfer measurements from spheres have been successfully correlated by Churchill (1983) for the range $Pr \geqslant 0.7$ and $Ra_D < 10^{11}$:

$$Nu_{av,D} = 2 + A/B \tag{6.31}$$

where $A = 0.589 Ra_D^{1/4}$

$$B = [1 + (0.469/Pr)^{9/16}]^{4/9}$$

Example 6.3

A copper heating coil is used to heat a large cylinder of water. The coil may be considered to be a horizontal cylinder of length 1.5 m, which has a uniform surface temperature of 80°C and an outer diameter of 0.025 m. Estimate the heat transfer to water at 10°C.

Solution For water at $T_m = \frac{1}{2}(80 + 10) = 45°C$,

$\rho = 990 \, \text{kg/m}^3$

$\mu = 605.1 \times 10^{-6} \, \text{kg/m s}$

$k = 0.64 \, \text{W/m K}$

$Pr = 3.9$

β at $T_\infty = 0.88 \times 10^{-4} \, \text{K}^{-1}$ (We cannot use $\beta = 1/T_\infty$ since water is not an ideal gas; it does not obey $\rho = p/RT$)

For a horizontal cylinder, Equation (6.30) gives

$$Nu_{av,D} = \{0.6 + A/B\}^2$$

where $A = 0.387 Ra_D^{1/6}$ and $B = [1 + (0.559/Pr)^{9/16}]^{8/27}$.
For $Pr = 3.9$,

$$Nu_{av,D} = \{0.6 + 0.445 Gr_D^{1/6}\}^2$$

$$Gr_D = \frac{\rho^2 g \beta \Delta T D^3}{\mu^2}$$

$$= \frac{990^2 \times 9.81 \times 0.88 \times 10^{-4} \times (80 - 10) \times 0.025^3}{(605.1 \times 10^{-6})^2}$$

$$= 2.52 \times 10^6$$

$Nu_{av,D} = \{0.6 + 0.445(2.52 \times 10^6)^{1/6}\} = 33.6 = h_{av}D/k$

$h_{av} = (k/D) \, Nu_{av,D} = (0.64/0.025) \times 33.6 = 860 \, \text{W/m}^2\text{K}$

$q_{av} = h_{av}\Delta T = 860 \times 70 = 60\,151 \, \text{W/m}^2$

$\dot{Q} = q_{av}A = q_{av}\pi DL = 60\,151 \times \pi \times 0.025 \times 1.5 = 7.1 \, \text{kW}$

6.4.3 Rotating disc

A rotating disc provides a good example of a flow that changes from pure natural convection when the disc is stationary to mixed convection and pure forced convection when the disc is rotating at a sufficiently fast speed. Richardson and Saunders (1963) correlated experimental data from a disc with a constant surface temperature rotating in air ($Pr = 0.72$) and suggested a mixed convection correlation of the form

$$Nu_{av} = 0.47(Re_\Omega^2 + Gr)^{1/4} \tag{6.32}$$

where

$$Gr = g\beta \Delta TL^3/v^2$$

$$Re_\Omega = \Omega b^2/v$$

$$Nu_{av} = h_{av}b/k$$

for a horizontal disc $L = b$ and for a vertical disc $L = \frac{1}{2}\pi b$ where b is the disc radius.

Example 6.4

A drive shaft inside a gas turbine engine is 0.1 m in diameter and has a solid end cap. The shaft rotates at a constant speed of 5000 rev/min in air at an absolute pressure of 4 bar and temperature of 400 K. The end cap has a surface temperature of 500 K. Is the heat loss from the end cap dominated by free or forced convection?

Solution The relative significance of free convection to the total convective heat loss can be determined from the ratio Re_Ω^2/Gr_L (6.32) where $L = \pi b/2 = \pi \times 0.05/2 = 0.0786$ m.

For air at 4 bar and $T_m = 450$ K,

$$\rho = \frac{p}{RT} = \frac{4 \times 10^5}{287 \times 450} = 3.1 \text{ kg/m}^3$$

$$\mu = 2.485 \times 10^{-5} \text{ kg/m s}$$

$$\beta = \frac{1}{T_\infty} = \frac{1}{400} \text{ K}^{-1}$$

$$\Omega = \frac{5000 \times 2\pi}{60} = 524 \text{ rad/s}$$

$$Re_\Omega^2 = \left(\frac{\rho \Omega b^2}{\mu}\right)^2$$

$$= \left(\frac{3.1 \times 524 \times 0.05^2}{2.485 \times 10^{-5}}\right)^2$$

$$= 2.67 \times 10^{10}$$

$$Gr_L = \frac{\rho^2 g \beta \Delta T L^3}{\mu^2}$$

$$= \frac{3.1^2 \times 9.81 \times (1/400) \times (500 - 400) \times 0.0786^3}{(2.485 \times 10^{-5})^2}$$

$$= 1.85 \times 10^7$$

$$Re_\Omega^2 / Gr_L = \frac{2.67 \times 10^{10}}{1.85 \times 10^7}$$

$$= 1441 \quad (\gg 1 \text{ so forced convection dominates})$$

Comment   Forced and free convection effects will be comparable at $\Omega = 13.8\,\mathrm{rad/s}$ (132 rev/min).

6.5 ● Vertical channel flow

The fins on the heat exchanger in Example 6.1 were assumed to be separated by a gap wide enough that independent boundary layers form on each fin surface. Equation (6.14) provides an estimate of the required distance for this to be the case, and as an order of magnitude we would require the ratio of fin spacing, D, to vertical height, H, to be $D/H > Ra_H^{-1/4}$. Imagine now that the fins are moved closer together, so that the separate boundary layers on each fin merge. This is a vertical channel, and is the free convection partner of parallel flow in forced convection (Section 5.2).

Figure 6.6 shows a vertical channel of circular cross-section with an outer radius r_o, a constant surface temperature T_s and height H. Flow enters the channel with a temperature T_∞, the component of velocity in the z direction is V_z, and the component in the radial direction is V_r.

The z direction momentum equation in cylindrical coordinates is (see the end of Section 4.2.2):

$$\rho \left\{ V_r \frac{\partial V_z}{\partial r} + V_z \frac{\partial V_z}{\partial z} \right\} = F_z - \frac{\partial p}{\partial z} + \mu \left\{ \frac{\partial^2 V_z}{\partial r^2} + \frac{1}{r} \frac{\partial V_z}{\partial r} + \frac{\partial^2 V_z}{\partial z^2} \right\} \tag{6.33}$$

For laminar flow of a fluid of Prandtl number unity ($\delta = \delta_T$) or greater ($\delta > \delta_T$), a fully developed velocity profile implies that the temperature profile is also fully developed and the radial velocity is zero. Equation (6.33) then reduces to

$$F_z - \frac{\partial p}{\partial z} + \mu \left\{ \frac{\mathrm{d}^2 V_z}{\mathrm{d} r^2} + \frac{1}{r} \frac{\mathrm{d} V_z}{\mathrm{d} r} \right\} = 0 \tag{6.34}$$

The *radial* momentum equation (which under these conditions simplifies to $\partial p / \partial r = 0$) tells us that pressure is a function of z only; $\partial p / \partial z = \mathrm{d}p/\mathrm{d}z$, which, as there is no superimposed pressure field, is equal to the hydrostatic

Fig. 6.6 ●
Naturally induced
flow in a vertical
circular channel:
(a) geometry;
(b) velocity;
(c) temperature

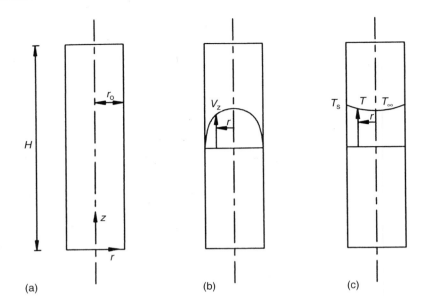

(a)   (b)   (c)

pressure gradient, dp_∞/dz or $-\rho_\infty g$. The body force per unit volume can be represented by $F_z = -\rho g$. Using the Boussinesq approximation (4.55) and (4.57), $g(\rho_\infty - \rho) = \rho g \beta (T - T_\infty)$ and Equation (6.34) becomes

$$\frac{1}{r}\frac{d}{dr}\left(r\frac{dV_z}{dr}\right) = -\rho g \beta \frac{(T - T_\infty)}{\mu} \tag{6.35}$$

Equation (6.35) cannot be solved without solving the energy equation, in which, V_z and T also appear together. This now requires a simultaneous solution of both energy and momentum equations. A great simplification can be made by recognising that for a fully developed thermal boundary layer, the temperature of the fluid is close to the surface temperature of the channel. Hence $T - T_\infty$ can be replaced by $T_s - T_\infty$.

Integrating Equation (6.35) twice and using the boundary conditions that $dV_z/dr = 0$ at $r = 0$ and $V_z = 0$ at $r = r_o$, we get the equation for the fully developed velocity profile:

$$V_z = \frac{\rho g \beta (T_s - T_\infty)r_o^2}{4\mu}\left\{1 - \left(\frac{r}{r_o}\right)^2\right\} \tag{6.36}$$

The heat flow from the walls, $\dot{Q}$, is balanced by the increase in enthalpy of the fluid, hence

$$\dot{Q} = \dot{m}C_p(T_s - T_\infty) \tag{6.37}$$

and is related to the heat flux, q, by

$$\dot{Q} = q(2\pi r_o H) \tag{6.38}$$

The mass flow, $\dot{m}$, in Equation (6.37) can be obtained by integrating the velocity profile (6.36):

$$\dot{m} = 2\pi\rho \int_0^{r_o} V_z r \, dr = \frac{\pi\rho^2 g\beta(T_s - T_\infty)r_o^4}{8\mu} \tag{6.39}$$

From the definition of the Nusselt number based on the channel height, H,

$$Nu_H = \frac{qH}{(T_s - T_\infty)k}$$

Using Equation (6.39) in (6.37) and relation (6.38) it is apparent that

$$Nu_H = \frac{Gr_D Pr}{128} \tag{6.40}$$

where the Grashof number Gr_D uses the diameter, $D = 2r_o$, of the channel for the characteristic length scale.

6.6 ● Free convection in enclosures

The foregoing discussion of free convection has so far been confined to external flows where the conditions away from the wall are not influenced significantly by the convection process. Figure 6.7 depicts a three-dimensional

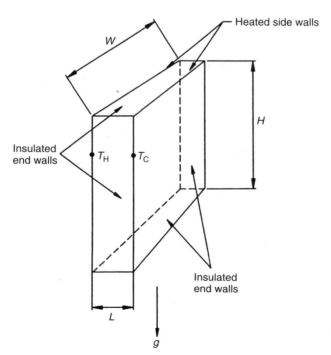

Fig. 6.7 ● Natural convection within a heated enclosure

view of an enclosure and the associated nomenclature to be used in this section. The enclosure is heated by the side walls where $T_H > T_C$, and the four end walls are assumed to be insulated. The aspect ratio, W/L, is assumed to be sufficiently large that the flow field inside the enclosure is in effect two-dimensional.

Natural convection within enclosures has obvious application to the heat loss from buildings where double-glazed windowpanes or cavity walls are used to reduce the heat loss. Other examples of enclosures are the circulation of air (and fire) within buildings, solar panels and, on a much larger scale, the atmospheric and oceanic circulation currents.

As is to be expected from the introductory remarks above, free convection (itself a complicated subject compared with forced convection) within enclosures is further complicated by the interactions between the two side wall boundary layers and a finite-sized fluid reservoir. Even a rudimentary analysis of an enclosure problem is beyond the scope of this textbook; the interested reader should consult Gebhart *et al.* (1988) or Bejan (1995) for such details, or the review by Ostrach (1972).

Enclosure flows are conveniently separated into two classes:

1. Enclosures heated from the side.
2. Enclosures heated from below.

Further refinements and subdivisons are possible (for example, triangular and circular enclosures, constant temperature and constant heat flux walls). For this level of detail, the interested reader should consult one or more of the above three texts.

6.6.1 Enclosure heated from the side

The possible regimes of free convection for an enclosure heated from the side are shown in Figure 6.8. The nature of the flow depends on the Rayleigh number, Ra_H, and the enclosure aspect ratio H/L. For regimes 1 and 2, conduction is the dominant mechanism of heat transport, and $Nu = 1$, whereas convection dominates in regimes 3 and 4.

For regime 3, the analysis given by Bejan (1995) gives the overall Nusselt number as

$$Nu_{av} = \frac{\dot{Q}L}{kH(T_H - T_C)} = f(H/L, Ra_H)\frac{L}{H}Ra_H^{1/4} \tag{6.41}$$

where $\dot{Q}$ is the overall heat flow (W/m) across the enclosure per unit length in the direction of W indicated on Figure 6.7. The values of $f(H/L, Ra_H)$ are correlated against the parameter $(H/L)^{4/7}Ra_H^{1/7}$ and are given in Table 6.2.

For a fixed value of Ra_H, as the aspect ratio H/L is made very small, then in the limit the enclosure is treated as shallow with heating at the end vertical walls. An analysis of natural convection in a shallow enclosure has been made by Bejan and Tien (1978): comparison with available experimental data is very good, not only for the shallow enclosure regime but also for

Fig. 6.8 ● Heat transfer regimes for natural convection within enclosures heated from the side: (1) conduction; (2) conduction, tall enclosure (3) boundary layer regime; and (4) shallow enclosure

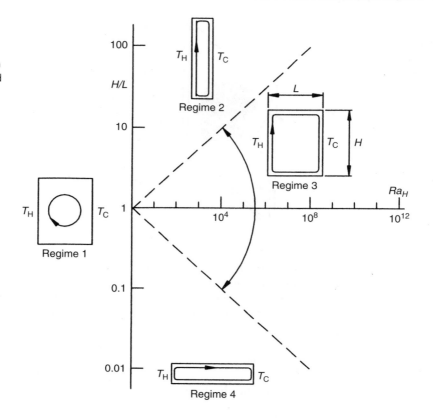

Table 6.2 ● Natural convection values of constants for Equation (6.41)

$(H/L)^{4/7}Ra_H^{1/7}$	1	2	3	4	5	10	50	100	1000
$f(H/L, Ra_H)$	0.238	0.277	0.292	0.3	0.315	0.338	0.354	0.361	0.364

regime 3. Bejan and Tien do not present a closed-form expression for the overall Nusselt number, and the results are shown graphically.

Example 6.5

A double-glazed window (Figure 6.9) unit has a height of 0.6 m and the two panes of glass are separated by a gap of 10 mm. Neglecting the thermal resistance of the glass, estimate the heat loss by convection through the window and the overall heat transfer coefficient due to the air gap when the surface temperatures on the inside and outside windowpanes are 10°C and 2°C, respectively.

Fig. 6.9 ●
Double-glazed
window for
Example 6.5

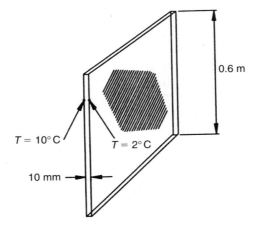

Solution The double-glazed unit clearly forms an enclosure heated from the side. Before proceeding further we need to establish the flow regime in accord with Figure 6.8. This involves evaluating two parameters: the aspect ratio H/L and the Rayleigh number Ra_H, based on the window height H.

$$\frac{H}{L} = \frac{0.6}{0.01} = 60$$

Evaluate properties in the air gap for $T_m = 6°C$:

$$\rho = 1.284 \, \text{kg/m}^3$$

$$\mu = 1.725 \times 10^{-5} \, \text{kg/m s}$$

$$k = 0.0244 \, \text{W/m K}$$

$$Pr = 0.71$$

Evaluate β at T_m:

$$\beta = \frac{1}{279} \, \text{K}^{-1}$$

$$Ra_H = Gr_H Pr$$

$$= \left[\frac{\rho^2 g \beta \Delta T H^3}{\mu^2} \right] Pr$$

$$= \left[\frac{1.284^2 \times 9.81 \times (1/279) \times (10 - 2) \times 0.6^3}{(1.725 \times 10^{-5})^2} \right] \times 0.71$$

$$= 2.4 \times 10^8$$

which, from Figure 6.8, establishes the flow as being borderline between regime 2 (conduction dominated, $Nu_{av} = 1$) and regime 3 (convection dominated). For the latter we may use Equation (6.41), with the aid of Table 6.2 to establish $f(H/L, Ra_H)$

$$(H/L)^{4/7} Ra_H^{1/7} = (60)^{4/7} \times (2.4 \times 10^8)^{1/7} = 163$$

and from Table 6.2, $f(H/L, Ra_H) \approx 0.361$. Hence from Equation (6.41),

$$Nu_{av} = 0.361(L/H) \, Ra_H^{1/4}$$

$$= 0.361 \times (1/60) \times (2.4 \times 10^8)^{1/4}$$

$$= 0.75$$

Since $Nu_{av} \approx 1$ this result implies that the heat transfer is governed by conduction; the heat flux is given by the usual relation:

$$q = \left(\frac{k}{L}\right)\Delta T = \frac{0.0244}{0.01}(10 - 2) = 19.52 \, \text{W/m}^2$$

and the overall heat transfer coefficient, U, is

$$U = \frac{k}{L} = \frac{0.0244}{0.01} = 2.44 \, \text{W/m}^2\text{K}$$

Comment   A wider gap will promote convection, leading to an increased heat loss. There is a trade-off between effective noise reduction (wide gap) and effective heat insulation (comparatively small gap).

6.6.2   Enclosure heated from below

With side heating, a small temperature difference will always give rise to a buoyancy force, and hence fluid motion will begin. For an enclosure heated from below, the temperature difference needs to exceed some critical value before the onset of buoyancy-induced fluid motion. This is usually characterised as the *critical Rayleigh number*, and a generally accepted criterion for the onset of convection is $Ra_L > Ra_{L,\text{crit}}$, where $Ra_{L,\text{crit}} = 1708$.

Figure 6.10 depicts the flow in a horizontal enclosure for $Ra_L < 1708$ and $Ra_L > 1708$. For values of Rayleigh number less than the critical value, there is no fluid motion, only layers of thermally stratified fluid; conduction dominates and $Nu_L = 1$. For Rayleigh numbers exceeding the critical value, the flow pattern consists of counterrotating cells known as *Bénard cells*. Experimental measurements in the range $3 \times 10^5 \leqslant Ra_L \leqslant 7 \times 10^9$ are well correlated by the expression (due to Globe and Dropkin, 1959):

$$Nu_{L,av} = 0.069 \, Ra_L^{1/3} Pr^{0.074} \tag{6.42}$$

A schematic diagram of an enclosure with the hot surface below the cold surface, and inclined at $0°$ (horizontal enclosure), θ and $180°$ is shown in

Fig. 6.10 ● The flow structure in a horizontal enclosure heated from below

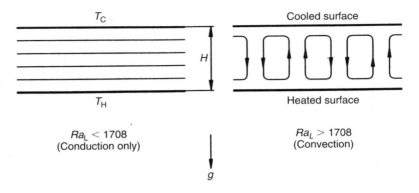

$Ra_L < 1708$
(Conduction only)

$Ra_L > 1708$
(Convection)

Figure 6.11. As depicted in the lower part of the figure, the angle of inclination θ has a marked effect on the heat transfer, as the flow mechanisms change from Bénard convection cells to a single-cell regime (enclosure heated from the side) at $\theta = 90°$. A local minimum in the Nusselt number appears at an angle $\theta*$ located somewhere between $0°$ and

Fig. 6.11 ● Natural convection within an inclined enclosure

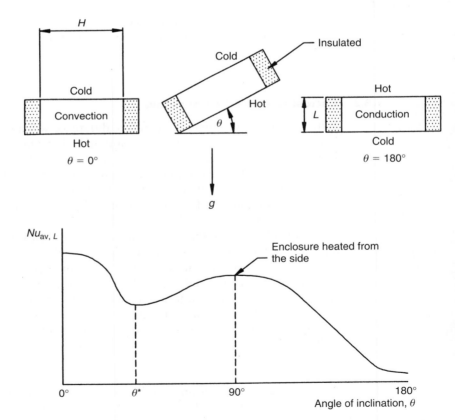

$90°$. Finally, convection is progressively attenuated by the hot wall being turned 'upside down', and at $\theta = 180°$ the fluid is stable and convection is absent.

The behaviour of the average Nusselt number with angle θ is given by the following relationships.

1. $\theta^* > \theta > 0°$ and $H/L > 10$ (Hollands *et al*, 1976):

$$Nu_{av,L} = 1 + [1.44\{1 - A\}^\dagger] [\{1 - A\sin(1.8\theta)^{1.6}\}^\dagger]$$
$$+ \left\{ \left(\frac{0.293}{A} \right)^{1/3} - 1 \right\}^\dagger \tag{6.43}$$

where $A = 1708/(Ra_L \cos\theta)$ and the terms denoted by † should be set to equal zero if they become negative.

2. $\theta^* > \theta > 0°$ and $H/L < 10$ (Catton, 1978):

$$Nu_{av,L} = \left[\left\{ \frac{Nu_{av,L}(90°)}{Nu_{av,L}(0°)} \right\} \sin(\theta^*)^{1/4} \right]^{\theta/\theta^*} \tag{6.44}$$

3. $90° > \theta > \theta^*$ (Ayyaswamy and Catton, 1973):

$$Nu_{av,L} = Nu_{av,L}(90°) \sin(\theta)^{1/4} \tag{6.45}$$

4. $180° > \theta > 90°$ (Arnold *et al.*, 1975):

$$Nu_{av,L} = 1 + [Nu_{av,L}(90°) - 1] \sin(\theta) \tag{6.46}$$

where the angle θ^* is related to the aspect ratio H/L as follows:

H/L	1	3	6	12	> 12
θ^*	25°	53°	60°	67°	70°

Example 6.6

The double-skin arrangement shown in Figure 6.12 is proposed for the horizontal roof (of dimensions 6 m × 3 m) of a large boiler. The convective heat transfer coefficient on the boiler side of the roof is $h_{in} = 20$ W/m²K and the air temperature inside the boiler is $T_{in} = 360°C$. The convective heat transfer coefficient on the outside of the boiler is given by the relation $h_{out} = 1.7(T_2 - T_{out})^{1/3}$, where T_2 is the surface temperature of the outside skin and T_{out} is the air temperature on the outside of the boiler, which is 40°C.

Estimate the heat loss by convection when the gap between the two surfaces is 0.3 m.

Fig. 6.12 ●
Double-skin boiler
roof (Example 6.6)

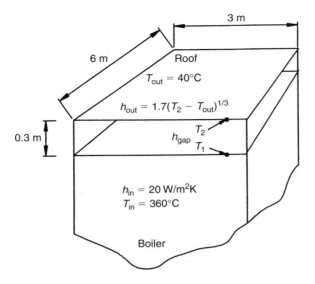

Tout = 40°C

hout = 1.7(T₂ − Tout)^(1/3)

$T_\text{out} = 40°C$

$h_\text{out} = 1.7(T_2 - T_\text{out})^{1/3}$

T_2

h_gap T_1

$h_\text{in} = 20\,\text{W/m}^2\text{K}$
$T_\text{in} = 360°C$

Boiler

Solution Recognising that the heat flow through the roof can be modelled as a horizontal enclosure, the heat transfer coefficient for the gap, h_gap, between the inside and outside skins will be given by Equation (6.42):

$$Nu_{L,\text{av}} = 0.069\,Ra_L^{1/3}\,Pr^{0.074}$$

Evaluate properties in the air gap for $T_\text{m} = \tfrac{1}{2}(360 + 40) = 200°C$

$\rho = 0.723\,\text{kg/m}^3$

$\mu = 2.57 \times 10^{-5}\,\text{kg/m s}$

$k = 0.037\,\text{W/m K}$

$Pr = 0.71$

Evaluate β at T_m

$\beta = 1/473\,\text{K}^{-1}$

From this we can calculate

$$Ra_L = Gr_L\,Pr$$

$$= [\rho^2 g\beta\Delta T L^3/\mu^2]\,Pr$$

$$= \left[\frac{0.723^2 \times 9.81 \times (1/473) \times \Delta T \times 0.3^3}{(2.57 \times 10^{-5})^2}\right] \times 0.71$$

$$= 3.15 \times 10^5\,\Delta T$$

$$Nu_{L,\text{av}} = 0.069\,Ra_L^{1/3}\,Pr^{0.074} = 0.069 \times (3.15 \times 10^5\,\Delta T)^{1/3} \times (0.71)^{0.074}$$

$$= 4.57\Delta T^{1/3} = h_\text{gap}L/k$$

$$h_\text{gap} = 4.57\Delta T^{1/3}(k/L) = 0.564\Delta T^{1/3} \qquad (\text{W/m}^2\text{K})$$

And in the context of the temperatures indicated on Figure 6.12:

$$h_{gap} = 0.564(T_1 - T_2)^{1/3} \qquad (W/m^2 K)$$

For one-directional heat flow through the roof,

$$q = h_{in}(T_{in} - T_1) = 20(360 - T_1) \qquad (W/m^2)$$

$$q = h_{gap}(T_1 - T_2) = 0.564(T_1 - T_2)^{4/3} \qquad (W/m^2)$$

$$q = h_{out}(T_2 - T_{out}) = 1.7(T_2 - 40)^{4/3} \qquad (W/m^2)$$

Combining the second and third equations gives

$$T_1 = 3.287T_2 - 91.49 \qquad (6.47)$$

Combining the first and third equations and using Equation (6.47) to substitute for T_1 gives

$$1.7(T_2 - 40)^{4/3} + 65.74T_2 - 9030 = 0 \qquad (6.48)$$

This may be solved by iteration using the Newton–Raphson method:

$$x^{(n+1)} = x^{(n)} - \left\{ \frac{f_{(x)}}{df_{(x)}/dx} \right\}$$

where $f(x)$ is given by (6.48), and by differentiation, $df_{(x)}/dx = 2.267(T_2 - 40)^{1/3} + 65.74$.

Starting at $T_2^{(0)} = 120°C$, we get

$$T_2^{(1)} = 127.37°C$$

$$T_2^{(2)} = 127.35°C \text{ (converged)}$$

Using Equation (6.47) $T_1 = 3.287T_2 - 91.49 = 327.1°C$ and

$$q = h_{gap}(T_1 - T_2) = 0.564(T_1 - T_2)^{4/3}$$

$$= 0.564(327.1 - 127.35)^{4/3}$$

$$= 658 \, W/m^2$$

So the total heat loss by convection from the boiler roof, $\dot{Q}$, is

$$\dot{Q} = qA$$

$$= 658 \times 6 \times 3$$

$$= 11.85 \, kW$$

Comment
As the gap is reduced, the conduction limit it attained when $Ra_L \leqslant 1708$. This would require a gap of around 9 mm, and the heat flux will be given by the one-dimensional conduction equation,

$$q = \frac{k}{L}(T_1 - T_2) = \frac{0.037}{0.009}(T_1 - T_2)$$

$$= 4.11(T_1 - T_2)$$

Proceeding in a similar way as before by writing out three equations for the heat flux and eliminating all but one of the unknown temperatures, we get the result $T_1 = 0.17T_2 + 298.6$ and $1.7(T_2 - 40)^{4/3} + 3.4T_2 - 1288 = 0$. From which

$$T_1 = 322°C$$

$$T_2 = 137.9°C$$

$$\dot{Q} = 13.63\,kW$$

So reducing the gap in order to suppress convection effects actually results in an increase in the heat transfer. This is because the thermal resistance of the narrow (conduction-dominated) gap is actually less than the wider (convection-dominated) gap. There is clearly some scope for further improvement, either by optimising or by making a roof of two or more narrow conduction-dominated gaps. This is left as an exercise for the interested student to pursue.

6.7 ● Closing comments

Free or natural convection is created by the buoyancy force which arises as a result of the temperature difference between the fluid and its surrounding surface. This leads to coupling of the momentum and energy equations, resulting in a more mathematically complex solution to the governing equations. Free convection effects are always present in forced convection; their significance can be estimated by evaluating the ratio Gr_L/Re_L^2. As with forced convection, it is helpful to categorise natural convection as either external or internal flows. For external flows (such as flows generated by heated vertical, horizontal or inclined surfaces), the Nusselt number can be obtained from a number of reliable correlations that involve the Grashof and Prandtl numbers. For a vertical surface, transition from laminar to turbulent flow is considered to occur when $Gr_x > 10^9$; and, depending on the heating configuration, for a horizontal surface, it is considered to occur when $Gr_x > 10^7$. For internal flows, the geometric aspect ratio of the enclosure is an additional influential parameter in determining the Nusselt number. Internal flows may exhibit a number of regimes, depending on the value of the Rayleigh number and aspect ratio of the enclosure.

Examples of free convection include the flow created on the outside of a central heating radiator; on the inside wall of a building; the flow generated by a heat sink (unventilated); atmospheric and oceanic circulation.

The correlations presented in this chapter are summarised in Table 6.3.

Table 6.3 Summary of free convection correlations

Flow configuration	Equation	Comments
Vertical plate – laminar	(6.17a)	$Pr \approx 1$, local
	(6.17b)	$Pr \approx 1$, average
	Table 6.1	$0.01 < Pr < 1000$, local
	(6.22)	Average, good accuracy for $Ra_L < 10^9$
	(6.24a)	Constant heat flux, local
	(6.24b)	Constant heat flux, average
Vertical plate – turbulent	(6.19)	Local and average ($Gr_x > 10^9$)
	(6.21)	Average, $0.1 \leqslant Ra_L \leqslant 10^{12}$
	(6.25a)	Constant heat flux, local
	(6.25b)	Constant heat flux, average
Inclined surface	Vertical plate	$-60° \leqslant \theta \leqslant 60°$ replace g with $g\cos\theta$
Horizontal surface L = area/perimeter	(6.27)	Top side of a hot plate or bottom side of a cold plate $10^7 \leqslant Gr_L Pr \leqslant 10^7$
	(6.28)	Top side of a hot plate or bottom side of a cold plate $10^7 \leqslant Gr_L Pr \leqslant 10^{11}$
	(6.29)	Bottom side of a hot plate or top side of a cold plate $10^4 \leqslant Gr_L Pr \leqslant 10^{10}$
Horizontal cylinder	(6.30)	$Ra_D \leqslant 10^{12}$
Sphere	(6.31)	$Pr \geqslant 0.7$, $Ra_D < 10^{11}$
Rotating disc	(6.32)	Air, $Pr \approx 0.7$
Vertical channel	(6.40)	Circular channel, fully developed laminar flow
Enclosure heated from the side	(6.41)	See also Figure 6.8 and Table 6.2
Enclosure heated from below	$Nu_L = 1$	$Ra_L < 1708$
	(6.42)	$3 \times 10^5 \leqslant Ra_L \leqslant 7 \times 10^9$
Inclined enclosure	(6.44)	$\theta^* > \theta > \theta°$ and $H/L < 10$
	(6.45)	$90° > \theta > \theta^*$
	(6.46)	$180° > \theta > 90°$

6.8 ● References

Arnold, J.N., Catton, I. and Edwards, D.K. (1975) Experimental Investigation of Natural Convection in Inclined Rectangular Regions of Differing Aspect Ratios. *ASME Paper 75-HT-62*

Ayyaswamy, P.S. and Catton, I. (1973). The boundary layer regime for natural convection in a differentially heated, tilted rectangular cavity. *J. Heat Transfer*, **95**, 543–5

Bejan, A. (1995). *Convection Heat Transfer* 2nd edn. New York: Wiley

Bejan, A. and Large, J.L. (1990). The Prandtl number effect on the transition in natural convection along a vertical surface. *J. Heat Transfer*, **112**, 787–90

Bejan, A. and Tien, C.L. (1978). Laminar natural convection heat transfer in a horizontal cavity with different end temperatures. *J. Heat Transfer*, **100**, 641–7

Catton, I. (1978) Natural Convection in Enclosures. Proceedings of the 6th International Heat Transfer Conference, Toronto, Canada. **6**, pp. 13–43

Churchill, S.W. (1983). Free convection around immersed bodies, in *Heat Exchanger Design Handbook* (Schlünder, E.U., ed.). New York: Hemisphere

Churchill, S.W. and Chu, H.H.S. (1975). Correlating equations for laminar and turbulent free convection from a vertical plate. *Int. J. Heat Mass Transfer*, **18**, 1323–9.

Eckert, E.R.G. and Jackson, T.W. (1951). Analysis of Turbulent Free Convection Boundary Layers on a Flat Plate. *NACA Report 1015*

Ede, A.J. (1967). *Advances in Heat Transfer*. Vol. 4: *Advances in Free Convection*, pp. 1–64. New York: Academic Press

Gebhart, B., Jaluria, Y., Mahajan, R.L. and Sammakia, B. (1988). *Buoyancy Induced Flows and Transport*. New York: Hemisphere

Globe, S. and Dropkin, D. (1959). Natural convection heat transfer in liquids confined by two horizontal plates and heated from below. *J. Heat Transfer*, **81**, 24–8.

Hollands, K.G.T., Unny, S.E., Raithby, G.D. and Konicek, L. (1976). Free convective heat transfer across inclined air layers. *J. Heat Transfer*, **98**, 189–93

Incropera, F.P. and DeWitt, D.P. (1996). *Fundamentals of Heat and Mass Transfer* 4th edn. New York: Wiley

Jaluria, Y. (1980). *Natural Convection Heat and Mass Transfer*. Oxford: Pergamon Press

McAdams, W.H. (1954). *Heat Transmission* 3rd edn. New York: McGraw-Hill

Ostrach, S. (1972). *Advances in Heat Transfer*. Vol. 8: *Natural Convection in Enclosures*, pp. 161–227. New York: Academic Press

Richardson, P.D. and Saunders, O.A. (1963). Studies of flow and heat transfer associated with a rotating disc. *J. Mech. Engng Sci.*, **5**(4), 336–42

Schmidt, E. and Beckmann, W. (1930). Das Temperatur und Geschwindigkeitsfeld von einer licher Wandtemperatur. *Forsch. Geb. IngWes.*, **1**, 391–406

Squire, H.B. (1938). In *Modern Developments in Fluid Dynamics* Vol. 2. (Goldstein, S., ed.). New York: Oxford University Press

Vliet, G.C. and Liu, C.K. (1969). An experimental study of turbulent natural convection boundary layers. *J. Heat Transfer*, **91**, 517–31

6.9 ● End of chapter questions

6.1 Solve the following equation using Newton–Raphson iteration:

$$1.2T^{4/3} - 0.76T = 146.2$$

[$T = 42.61$]

6.2 A 3 m high solid brick wall is 0.3 m thick and has a thermal conductivity of $k = 1.2$ W/m K. The wall separates a room at $T_{in} = 20°C$ from the outside at $T_{out} = -5°C$. Taking a value for the heat transfer coefficient on the outside surface as $h_{out} = 30$ W/m²K and that for the inside surface given by the correlation $Nu_x = 0.1(Gr_x Pr)^{1/3}$, calculate:

(a) The surface temperature on each side of the wall.

[6.3°C, −3.7°C]

(b) The value of the internal heat transfer coefficient.

[2.9 W/m²K]

(c) The heat loss through the wall.

[40 W/m²]

(d) The overall heat transfer coefficient

[1.6 W/m²K]

Assume $\rho = 1.19$ kg/m³, $\mu = 1.8 \times 10^{-5}$ kg/m s, $k = 0.026$ W/m K and $Pr = 0.707$.

6.3 Engine oil inside a thin steel sump of a longitudinal engine (0.6 m by 0.3 m), rear-wheel drive vehicle is maintained at 90°C. The underside of the sump is exposed to the airstream created by the vehicle as it moves forwards, with the largest dimension in the direction of the airstream. The average maximum outside temperature is 27°C. For design purposes it is necessary to maintain the surface temperature of the sump at below 80°C.

(a) Draw a sketch illustrating how the convective heat transfer can be modelled. List the assumptions used.

(b) Show that the heat transfer coefficient inside the oil sump (i.e. the oil-side) is given by

$$h_{inside} = 15.55(90 - T_s)^{1/4} \quad (\text{W/m}^2\text{K})$$

where T_s is the surface temperature of the sump.

(c) Show that the heat transfer coefficient on the outside (i.e. the air-side) is given by

$$h_{outside} = 6.5U_\infty^{0.8} \quad (\text{W/m}^2\text{K})$$

(d) What value of U_∞ is necessary to achieve the desired surface temperature?

[0.76 m/s]

(e) What implications does this have on the design of the cooling system?

	Oil	Air
ρ (kg/m³)	848	1.177
μ (kg/m s)	0.025	1.846×10^{-5}
k (W/m K)	0.137	0.026 24
Pr	395	0.707
β (K⁻¹)	0.0007	

For forced convection from a flat plate:

$$Nu_x = 0.0288\, Re_x^{0.8} Pr^{1/3}$$

For free convection from a horizontal plate (L = area/perimeter):

Upper surface heated or lower surface cooled

$$10^4 < (Gr_L Pr) < 10^7$$
$$Nu_{av,L} = 0.54(Gr_L Pr)^{1/4}$$

$$10^7 < (Gr_L Pr) < 10^{11}$$
$$Nu_{av,L} = 0.15(Gr_L Pr)^{1/3}$$

Lower surface heated or upper surface cooled

$$10^5 < (Gr_L Pr) < 10^{10}$$
$$Nu_{av,L} = 0.27(Gr_L Pr)^{1/4}$$

6.4 Compare the heat loss by free convection from a human body at a surface temperature of 35°C in water at 10°C with that in air at 10°C. Consider an idealised form for a human body as a vertical cylinder 0.3 m in diameter and 1.8 m high. Use the following property values, evaluated at $T_m = 22.5°\text{C}$ except for β which is evaluated at T_∞.

	Water	Air
ρ (kg/m³)	998	1.19
μ (kg/m s)	0.951×10^{-3}	1.8×10^{-5}
β (K⁻¹)	0.088×10^{-3}	$1/T_\infty$
k (W/m K)	0.607	0.026
Pr	6.56	0.707

$$[q_{water}/q_{air} = 108,\ q_{water} = 11.3\,\text{kW/m}^2]$$

6.5 The cargo hold of an oil tanker has a double-skin construction, where the vertical side of the ship is separated into a number of cells each 1 m high and 0.2 m wide and filled with air. The cargo hold itself has a vertical wall 5 m high and is filled to this level with oil which, to allow pumping of this viscous fluid, is maintained at 50°C.

Assuming a sea water temperature of 10°C, and that the average Nusselt number on the outside surface of the ship's hull, of length $L = 150$ m, is $Nu_{av,out} = 1.26 \times 10^6$, and neglecting the thermal resistance of the hull material, use the thermal properties tabulated below to show that the heat flux, through the hull is given by the following relations (where T_1 and T_2 are the respective temperatures on the inside and outside of the hull):

$$q = 15.03(50 - T_1)^{4/3}$$

$$q = 0.88(T_1 - T_2)^{5/4}$$

$$q = 5065T_2 - 50650$$

	Water 20°C	Air 30°C	Oil 40°C
ρ (kg/m³)	998	1.15	876
μ (kg/m s)	1.002×10^{-3}	1.86×10^{-5}	0.21
β (K⁻¹)	0.088×10^{-3}	$1/T_\infty$	7×10^{-4}
k (W/m K)	0.603	0.026	0.144
Pr	6.95	0.707	2870

Now estimate the heat loss by convection through the cargo hold.

[80 W/m²]

6.6 A spherical thermistor bead has a diameter of 3 mm. Consider the heat transfer by convection and radiation, and estimate the maximum power dissipation if the surface temperature of the thermistor is limited to 120°C when the air temperature is 25°C.

[0.118 W]

6.7 Perform an analysis for vertical channel flow, as in Section 6.5 except that the flow is between two vertical parallel plates of height H and separated by a gap L. Show that for each surface, the Nusselt number for fully developed flow is given by $Nu_H = Ra_L/24$.

6.8 A 100 mm high array of 100 vertical cooling channels, each separated by a gap of 1 mm, is used as an electronics heat sink. Calculate the convective heat transfer from the array (each fin is 100 mm high and 150 mm wide) when the surface temperature is 60°C and the air temperature is 20°C.

	Air 40°C
ρ (kg/m^3)	1.11
μ (kg/m s)	1.91×10^{-5}
β (K^{-1})	$1/T_\infty$
k (W/m K)	0.027
Pr	0.7

[4.3 W]

6.9 Repeat Question 6.8 for an array of ten channels, separated by a gap of 10 mm.

[74.7 W]

6.10 Consider the two cases mentioned in Questions 6.8 and 6.9, merged boundary layers and separate boundary layers, and show that if the lengths of the fin array is fixed, the optimum spacing D_{opt} (to achieve the maximum heat transfer rate) is given by the expression

$$\frac{D_{opt}}{H} = 2.314\,Ra_H^{-1/4}$$

6.11 A triangular fin of height H is to be used in a heat exchanger, with the base of the triangle at the leading edge of the flow, which relies on natural convection to provide the heat exchange. If the local Nusselt number on the surface of the fin is given by the expression $Nu_x = 0.387(Gr_xPr)^{1/4}$,

show that the average Nusselt number is given by the expression

$$Nu_{av,H} = \left(\frac{32}{21}\right)0.387(Gr_HPr)^{1/4}$$

6.12 Carry out an analysis similar to that in Section 6.2, except for turbulent-free convection. As a starting point, take Equations (6.1) and (6.2) using the following profiles for the velocity and temperature distribution in the boundary layer:
$$u = U_1\eta^{1/7}(1-\eta)^4$$

$$(T-T_\infty)/(T_s-T_\infty) = (1-\eta^{1/7})$$

together with the shear stress relationship

$$\tau_s/\rho = 0.0225U_1^2(v/U_1\delta)^{1/4}$$

and the Colburn analogy

$$Nu_x = \frac{1}{2}C_{f,x}\,Re_xPr^{1/3}$$

(a) Show that these substitutions result in the following coupled ordinary differential equations:

$$0.0523\frac{d}{dx}(U_1^2\delta) = 0.125g\beta(T_s-T_\infty)\delta$$

$$-0.0225\,U_1^2(v/U_1\delta)^{1/4}$$

$$0.0366\frac{d}{dx}(U_1\delta)$$

$$= 0.0225\,Pr^{-2/3}U_1(v/U_1\delta)^{1/4}$$

(b) By analogy with the laminar case outlined in Section 6.2, make the substitutions $U_1 = c_1x^{m_1}$ and $\delta = c_2x^{m_2}$ and show that $m_1 = 0.5$ and $m_2 = 0.7$.

(c) Continue the analysis to show that the local Nusselt number is given by

$$Nu_x = 0.0295\,Gr_x^{2/5}\,Pr^{7/15}\times$$
$$(1+0.494\,Pr^{2/3})^{-2/5}$$

Chapter seven

Condensation and boiling heat transfer

7.1 ● Introduction

The previous three chapters have discussed in some depth the physics and engineering applications of single-phase convective heat transfer. Fluids may also experience a change of phase, and evaporation and condensation are examples of the transition from a liquid to vapour and, the reverse process, vapour to liquid. A change of phase also occurs during melting of a solid to form a liquid, and the reverse process of solidification. The physical mechanisms of phase-change processes are extremely complex: the transition from one phase to another involves transfer of latent heat, and surface tension effects can play an important role. So, although for the purpose of analysis we may make use of the single-phase convection methodologies, we cannot adopt the single-phase correlations and relations so far discussed. Indeed, as noted in Table 1.2, the heat transfer co-efficients associated with boiling water and condensation of steam are orders of magnitude greater than their single-phase counterparts.

To cover all heat-transfer phase-change processes is clearly beyond the scope of this book. Only boiling and condensation are discussed here – note the restriction placed by the word 'boiling', which is used instead of the more general case of evaporation. The interested reader is referred to the excellent book by Lock (1996), which covers the thermodynamics of interface growth; droplet, spray and bubble dynamics; as well as separate chapters on melting, solidification, evaporation and condensation. Boiling and condensation are perhaps the most widely used phase-change processes in engineering. The generation of steam and its subsequent condensation is inherent to a significant portion of the world's power generation equipment, prime movers for large ships, refrigeration and air conditioning systems. A knowledge of heat transfer during these phase-change processes is therefore of importance to the practising engineer.

7.2 ● Condensation

Condensation of a vapour will occur on a surface which need be only slightly below the corresponding saturation temperature. On becoming a liquid, the latent heat of vaporisation is released by the vapour, so it becomes possible to transfer a large quantity of heat with a small temperature difference, and directly we see that the heat transfer coefficient ($h = q/\Delta T$) can, in principle, be virtually infinite. In practice, a continuous process of condensation can take place in the form of droplets or as a film. The condensate itself forms a layer on the condenser surface and imposes some extra thermal resistance. So heat transfer coefficients, although large in comparison with single-phase convective heat transfer, are not infinite.

7.2.1 Droplet condensation

If surface tension forces are relatively large, the condensate will coalesce into droplets which, after forming, slide off the condenser surface under the action of gravity. Droplets provide very little surface thermal resistance, and the associated heat transfer coefficients are, as a guide, approximately ten times greater than those for film condensation (Bejan, 1995). Consequently, there is considerable interest in hydrophobic coatings and non-wetting agents that can be introduced into the vapour. Coatings are of limited success, as they must be thin to offset the additional thermal resistance, so in practice they eventually wear off. The cost of non-wetting agents can outweigh the benefit from the increase in heat transfer coefficient. And as condensation rates increase, so does the number of drops and their tendency to form a continuous film. It is for these reasons that practical condenser design will assume film condensation. Figure 7.1 shows a schematic diagram of condensation on a vertical cold wall, on a horizontal cylinder and a vertical cylinder. The condensation process may be entirely laminar or, as shown in

Fig. 7.1 ● Film condensation: (a) vertical wall; (b) horizontal tube; and (c) vertical tube

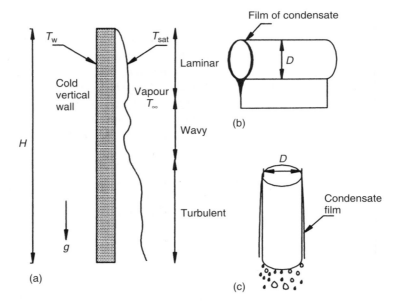

Figure 7.1a, the film may undergo transition to turbulence through a regime of 'wavy' flow.

7.2.2 Laminar film condensation on a vertical surface

Condensation of vapour into a laminar film on a vertical surface is probably the simplest case of phase-change convection, and renders itself amenable to analysis. In fact it was Nusselt (1916) who first derived the analytical relations for laminar film condensation. Figure 7.2 shows a schematic diagram of such a film, having a thickness δ, which varies with the vertical coordinate y. As with single-phase convection, the objective of this analysis is to find a relationship for $\delta(y)$ in terms of the relevant physical properties, dimensions and conditions. We can then relate this to the heat transfer coefficient from an energy balance.

An *a priori* assumption is that inertial forces are negligible in relation to those arising from gravity and viscous action. From inspection of Figure 7.2, the equation of motion applied to a control volume, of length dy, and unit depth into the paper is then

$$\tau\,dy = g(\rho_L - \rho_V)\,(\delta - x)\,dy \tag{7.1}$$

where ρ_L and ρ_V are the densities of the liquid and vapour phases respectively, and g the acceleration due to gravity. For laminar flow, the shear stress τ is related to the velocity gradient by

$$\tau = \mu_L \left(\frac{dv}{dx}\right) \tag{7.2}$$

Fig. 7.2 ● Laminar film condensation on a vertical wall

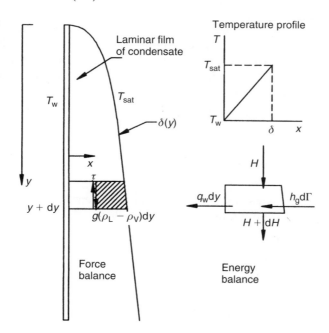

where, to maintain consistency, v is the velocity in the direction of y. Additionally, it is also assumed that $\rho_L \gg \rho_V$, so, $(\rho_L - \rho_V) \approx \rho_L$ and Equation (7.1) becomes, with the help of Equation (7.2),

$$\mu_L \left(\frac{dv}{dx} \right) = g\rho_L(\delta - x) \tag{7.3}$$

Integrating Equation (7.3) gives the downward velocity profile as

$$v = \frac{g\rho_L}{\mu_L} \left(\delta x - \frac{x^2}{2} \right) \tag{7.4}$$

The rate of mass flow, per unit length into the paper, through the control volume is given by (density × velocity × area),

$$\Gamma = \int_0^\delta \rho_L v \, dx$$

$$= \frac{g\rho_L^2 \delta^3}{3\mu_L} \tag{7.5}$$

Applying the first law of thermodynamics to the control volume, labelled 'energy balance' in Figure 7.2.

The vertical enthalpy flow, H, is made up from the specific enthalpy h_f of saturated liquid at T_{sat}, and since $T < T_{sat}$, an amount $C_p(T_{sat} - T)$ is subtracted to account for subcooling:

$$H = \int_0^\delta \rho_L v \left[h_f - C_p(T_{sat} - T) \right] dx \tag{7.6}$$

The net lateral enthalpy flow is made up from the specific enthalpy h_g of the vapour, at the liquid–vapour interface, multiplied by the mass of condensate produced in the control volume $d\Gamma$, and the heat transferred to the surface q_w multiplied by the area, per unit depth into the paper, dy. Hence, for the control volume energy balance,

$$H - (H + dH) + h_g \, d\Gamma - q_w \, dy = 0 \tag{7.7}$$

and from Fourier's law,

$$q_w = -k_L \left(\frac{dT}{dx} \right)_{x=0} \tag{7.8}$$

Following Nusselt's (1916) original argument, we assume a linear temperature distribution in the liquid, for which

$$\frac{(T_{sat} - T)}{(T_{sat} - T_w)} = 1 - \frac{x}{\delta} \tag{7.9}$$

hence

$$q_w = \frac{k_L}{\delta} (T_{sat} - T_w) \tag{7.10}$$

Using Equations (7.4) and (7.9) and integrating Equation (7.6),

$$H = \{h_f - \tfrac{3}{8} C_p (T_{sat} - T)\} \Gamma \tag{7.11}$$

Using Equations (7.8) to (7.11) in Equation (7.7), gives

$$\frac{(T_{sat} - T_w)k_L \, dy}{\delta} = \{h_{fg} + \tfrac{3}{8} C_p (T_{sat} - T_w)\} \, d\Gamma \tag{7.12}$$

where h_{fg} is the latent heat of vaporisation. For future reference, the term in the curly brackets will be denoted by h'_{fg}, as it represents an enhanced latent heat of vaporisation, including the latent heat and the sensible heat due to subcooling. This quantity can also be expressed as

$$h'_{fg} = h_{fg}\left(1 + \tfrac{3}{8} Ja\right) \tag{7.13}$$

where the ratio of the sensible heat to the latent heat is known as the *Jakob number, Ja*:

$$Ja = \frac{C_p(T_{sat} - T_w)}{h_{fg}} \tag{7.14}$$

Substituting the result for the rate of condensate flow given by Equation (7.5) into Equation (7.12) and noting that $\delta \, d\delta^3 = 3\delta^3 \, d\delta$,

$$\delta^3 \, d\delta = \frac{k_L \mu_L (T_{sat} - T_w) \, dy}{g\rho_L^2 h'_{fg}} \tag{7.15}$$

Integrating with the condition that $\delta = 0$ at $y = 0$

$$\delta = \left\{ \frac{4 k_L \mu_L (T_{sat} - T_w) y}{g\rho_L^2 h'_{fg}} \right\}^{1/4} \tag{7.16}$$

The local heat transfer coefficient, h, takes the usual definition, which is given by

$$h = \frac{q_w}{T_{sat} - T_w}$$

$$= \frac{k_L}{\delta} \qquad \text{(from Equation (7.10))} \tag{7.17}$$

$$= 0.707\left\{ \frac{g\rho_L^2 h'_{fg} k_L^3}{\mu_L(T_{sat} - T_w) y} \right\}^{1/4} \qquad \text{(from Equation (7.16))} \tag{7.18}$$

The average heat transfer coefficient, h_{av}, for a flat surface, of length H, and with a constant value of T_w is (see Section 4.4.3) simply

$$h_{av} = \tfrac{4}{3} h_{y=H} \tag{7.19a}$$

$$= 0.943\left\{ \frac{g\rho_L^2 h'_{fg} k_L^3}{\mu_L(T_{sat} - T_w)H} \right\}^{1/4} \tag{7.19b}$$

The result expressed by Equation (7.19b) was first derived by Nusselt, and one main assumption is that inertial effects are negligible. This has been tested by the analysis of Sparrow and Gregg (1959), who demonstrated that the error in using Equation (7.19b) is less than 3% for $Ja \leqslant 0.1$ and $1 \leqslant Pr_L \leqslant 100$.

The total heat transfer, $\dot{Q}$, to a surface of area A is given by the usual relation:

$$\dot{Q} = h_{av} A (T_{sat} - T_w) \tag{7.20}$$

and the total mass flow rate of condensate, $\dot{m}$, is given by

$$\dot{m} h_{fg} = \dot{Q}$$

and from Equation (7.20)

$$\dot{m} = \frac{h_{av} A (T_{sat} - T_w)}{h'_{fg}} \tag{7.21}$$

Equations (7.20) and (7.21) hold true for any geometrical configuration and any flow regime.

7.2.3 Turbulent film condensation on a vertical surface

As in all flows, the stable laminar regime will eventually break down into turbulent flow. As shown in Figure 7.1a, for a condensing film, the transition from laminar to turbulent flow is accompanied by a transitional 'wavy' regime. The criterion which delimits these regimes is the *film Reynolds number Re_y*. For a condensate, the limit of the laminar flow regime is marked by $Re_y \sim 30$. For $30 \leqslant Re_y \leqslant 1800$ the film undergoes a 'wavy' transition regime and for $Re_y > 1800$ the film is fully turbulent.

A definition for Re_y can be obtained by following the practice for pipe flow. A Reynolds number, Re, may be defined in the usual way as

$$Re = \frac{\rho_L V_b D_h}{\mu_L} \tag{7.22}$$

where ρ and μ are evaluated as properties of the liquid and V_b is a characteristic bulk average velocity in the downward y direction. For a section of dimensions δ in the x direction and depth L into the paper, the equivalent, or hydraulic, diameter D_h is defined as

$$D_h = \frac{4 \times \text{Cross-sectional area}}{\text{Wetted perimeter}}$$

$$= \frac{4\delta L}{\delta + L}$$

$$\approx 4\delta \qquad \text{since } \delta \ll L \tag{7.23}$$

From continuity, $\rho_L V_b \delta = \Gamma$, so

$$Re_y = \frac{4\Gamma}{\mu_L} \tag{7.24}$$

The subscript y is used not to denote a characteristic length, but as a reminder that the Reynolds number is evaluated with the *mass flow* at this value of y.

Returning to the definition of the heat transfer coefficient expressed by Equation (7.17), $h = k_L/\delta$, and using Equation (7.5) to express the film thickness in terms of the local mass flow rate, we obtain

$$h = \left(\frac{g\rho_L^2 k_L^3}{3\mu_L \Gamma}\right)^{1/3} \tag{7.25}$$

From the definition of the film Reynolds number given by Equation (7.24), and also Equation (7.19a), the laminar film average heat transfer coefficient given by Equation (7.19b) may be alternatively expressed as

$$\frac{h_{av}}{k_L}\left(\frac{\mu_L^2}{\rho_L^2 g}\right)^{1/3} = 1.47 Re_H^{-1/3} \qquad (Re_H < 30) \tag{7.26}$$

It is easily verified that the quantity $(\mu_L^2/\rho_L^2 g)^{1/3}$ has dimensions of length, and the group on the left side of Equation (7.26) is in effect a Nusselt number.

Chen *et al.* (1987) carried out an extensive review of data and existing correlations and proposed a universal correlation for the wavy and turbulent regimes:

$$\frac{h_{av}}{k_L}\left(\frac{\mu_L^2}{\rho_L^2 g}\right)^{1/3} = (Re_H^{-0.44} + 5.82 \times 10^{-6} Re_H^{0.8} Pr_L^{1.3})^{1/2} \qquad (Re_H \geqslant 30) \tag{7.27}$$

Example 7.1

Saturated steam at 2 bar condenses on a vertical tube 60 mm dia. and 1.5 m long, which is maintained at a surface temperature of 80°C by a flow of cooling water through the inside. Calculate:

(a) The heat transfer coefficient, h_{av}.

(b) The rate of heat transfer, $\dot{Q}$.

(c) The rate of condensate production, $\dot{m}$.

Solution   For saturated steam at 2 bar,

$$T_{sat} = 120.2°C$$

$$h_{fg} = 2202 \text{ kJ/kg K}$$

$$\rho_v = 1.129 \text{ kg/m}^3$$

For the liquid phase, we evaluate properties at the mean film temperature $T_m = \frac{1}{2}(T_{sat} + T_w) = 100°C$, for which

$$\rho_L = 958 \, kg/m^3$$

$$C_p = 4.219 \, kJ/kg \, K$$

$$\mu_L = 0.279 \times 10^{-3} \, kg/m \, s$$

$$Pr = 1.73$$

First we assume that $\delta \ll D$ and that the condensation process is entirely laminar (we can check these assumptions after a first cycle through the calculations). Also since $\rho_L \gg \rho_V$, we are justified in replacing $(\rho_L - \rho_V)$ with ρ_L.

For laminar film condensation on a vertical surface (7.19b)

$$h_{av} = 0.943 \left\{ \frac{g\rho_L^2 h'_{fg} k_L^3}{\mu_L(T_{sat} - T_w)H} \right\}^{1/4}$$

where, from Equations (7.13) and (7.14)

$$h'_{fg} = h_{fg}\left[1 + \tfrac{3}{8}Ja\right]$$

$$Ja = \frac{C_p(T_{sat} - T_w)}{h_{fg}}$$

hence

$$Ja = \frac{4219 \times (120.2 - 80)}{2202 \times 10^3} = 0.077$$

$$h'_{fg} = 2202 \times 103(1 + 0.625 \times 0.077)$$

$$= 2266 \times 10^3 \, J/kg$$

For laminar film condensation

$$h_{av} = 0.943 \times \left\{ \frac{9.81 \times 958^2 \times 2266 \times 10^3 \times 0.681^3}{0.279 \times 10^{-3} \times (120.2 - 80) \times 1.5} \right\}^{1/4}$$

$$h_{av} = 4172 \, W/m^2K$$

$$\dot{Q} = h_{av}A(T_{sat} - T_w)$$

$$= 4172 \times \pi \times 0.08 \times 1.5 \times (120.2 - 80)$$

$$= 63.2 \, kW$$

$$\dot{m} = \frac{\dot{Q}}{h'_{fg}} = \frac{63.2 \times 10^3}{2266 \times 10^3} = 0.0279 \, kg/s$$

We will now check the assumptions. Firstly, from Equation (7.16)

$$\delta = \left\{ \frac{4 k_L \mu_L (T_{sat} - T_w) y}{g \rho_L^2 h'_{fg}} \right\}^{1/4}$$

$$= \left\{ \frac{4 \times 0.681 \times 0.279 \times 10^{-3} \times (120.2 - 80) \times 1.5}{9.81 \times 958^2 \times 2266 \times 10^3} \right\}^{1/4}$$

$$= 0.218 \, \text{mm}$$

which satisfies the condition that $\delta \ll D$.

Secondly, we evaluate the film Reynolds number to see if the flow really is laminar. From Equation (7.24)

$$Re_y = \frac{4\Gamma}{\mu_L}$$

where $\Gamma = \dot{m}/\pi D = 0.0279/\pi \times 0.08 = 0.111 \, \text{kg/m s}$. For a tube of length H,

$$Re_H = \frac{4 \times 0.111}{0.279 \times 10^{-3}} = 1591$$

This value of Re_H signifies that the flow is not entirely laminar and we need to use Equation (7.27), whence for $Re_H \geqslant 30$

$$\frac{h_{av}}{k_L} \left(\frac{\mu_L^2}{\rho_L^2 g} \right)^{1/3} = \left(Re_H^{-0.44} + 5.82 \times 10^{-6} \, Re_H^{0.8} \, Pr_L^{1.3} \right)^{1/2}$$

Clearly it is not possible to solve the above relation directly, since we do not know h_{av} or Re_H (our previous estimate of Re_H was obtained from a mass flow rate derived from the assumption of laminar flow). An iterative solution is therefore necessary. However, the mechanics of this can be simplified by noting from Equations (7.21) and (7.24) that

$$h_{av} = \frac{h'_{fg} \mu Re_H}{4H(T_{sat} - T_w)}$$

from which, after substituting this result and values for the known physical quantities into Equation (7.27), we obtain

$$Re_H = 12\,658 \left(Re_H^{-0.44} + 11.87 \times 10^{-6} Re_H^{0.8} \right)^{1/2}$$

which is satisfied for $Re_H = 2480$, and since

$$h_{av} = \frac{h'_{fg} \mu Re_H}{4H(T_{sat} - T_w)}$$

then

(a) $h_{av} = \dfrac{2266 \times 10^3 \times 0.279 \times 10^{-3} \times 2480}{4 \times 1.5 \times (120.2 - 80)} = 6500 \, \text{W/m}^2\text{K}$

(b) $\dot{Q} = h_{av}A(T_{sat} - T_w) = 6500 \times \pi \times 1.5 \times 0.08 \times (120.2 - 80) = 98.5\,\text{kW}$

(c) $\dot{m} = \dfrac{\dot{Q}}{h'_{fg}} = \dfrac{98.5 \times 10^3}{2266 \times 10^3} = 0.0435\,\text{kg/s}$

Comment The assumptions that the film thickness is less than the tube diameter and that the density of the vapour is less than that of the liquid phase are certainly justified. The initial assumption of laminar flow is incorrect, so a more complicated correlation was used for the heat transfer coefficient in the wavy and turbulent regimes. This requires an iterative solution, and when converged the heat transfer and condensate rates are 55% greater than predicted by the laminar flow correlation. Note also the large value of heat transfer coefficient ($h = 6500\,\text{W/m}^2\text{K}$), associated with the process of condensation.

7.2.4 Other surface configurations

The discussion so far has, for the sake of simplicity, been limited to a vertical plane. Dhir and Lienhard (1971) have shown that for a surface inclined at ϕ to the vertical it is acceptable to replace g with $g\cos\phi$. For the inside or outside of a vertical pipe (Figure 7.1c) of diameter D, any of the vertical wall results may be used, providing that $D \gg \delta$. For laminar film condensation on spheres and horizontal tubes (Figure 7.1b), the leading constant in Equation (7.19b) takes a different value and

$$h_D = 0.826\left\{\frac{g\rho_L^2\,h'_{fg}\,k_L^3}{\mu_L(T_{sat} - T_w)D}\right\}^{1/4} \qquad \text{(spheres)} \qquad (7.28)$$

$$h_D = 0.729\left\{\frac{g\rho_L^2\,h'_{fg}\,k_L^3}{\mu_L(T_{sat} - T_w)D}\right\}^{1/4} \qquad \text{(horizontal tubes)} \qquad (7.29)$$

where the diameter, D, is used in place of the wall length, L. Of further, obvious interest is the averaged heat transfer coefficient for N horizontal tubes, stacked one above the other. This is obtained by replacing D with ND in Equation (7.29). However, this tends to provide a conservative estimate of the heat transfer since it assumes that a continuous sheet flows from one tube to the next. In practice the condensate will tend to leave one surface as droplets which impinge on the next tube give rise to an enhancement in the heat transfer:

$$h_D = 0.729\left\{\frac{g\rho_L^2\,h'_{fg}\,k_L^3}{N\mu_L(T_{sat} - T_w)D}\right\}^{1/4} \qquad (N \text{ horizontal tubes}) \qquad (7.30)$$

Example 7.2

Figure 7.3a shows a cross-section through a bundle of horizontal condenser tubes and Figure 7.3b shows the same bundle rotated through 90°. Assuming laminar film condensation on the external surface of the tubes, determine the ratio of the rate of condensate production of geometry (b) to that of geometry (a).

Solution From Equation (7.30), for each column of N tubes the average heat transfer coefficient is given by

$$h_D = 0.729 \left\{ \frac{g \rho_L^2 h'_{fg} k_L^3}{N \mu_L (T_{sat} - T_w) D} \right\}^{1/4}$$

and from (7.20), for each column of N tubes

$$\dot{Q} = h_D N \pi D L (T_{sat} - T_w)$$

For the same dimensions and fluid properties, this may be written as

$$\dot{Q} = C N^{3/4}$$

where C is a constant.
    For geometry (a):

1st column,	$N = 2$	$\dot{Q}_1 = C \cdot 2^{3/4}$
2nd column,	$N = 3$	$\dot{Q}_2 = C \cdot 3^{3/4}$
3rd column,	$N = 4$	$\dot{Q}_3 = C \cdot 4^{3/4}$
4th column,	$N = 3$	$\dot{Q}_4 = C \cdot 3^{3/4}$
5th column,	$N = 2$	$\dot{Q}_5 = C \cdot 2^{3/4}$

$$\text{Total heat transfer} = \dot{Q}_1 + \dot{Q}_2 + \dot{Q}_3 + \dot{Q}_4 + \dot{Q}_5$$
$$= C \left\{ 2^{3/4} + 3^{3/4} + 4^{3/4} + 3^{3/4} + 2^{3/4} \right\}$$
$$= 10.75C$$

Fig. 7.3 ● Bundle of horizontal condenser tubes (Example 7.2)

Tube, diameter D, length L

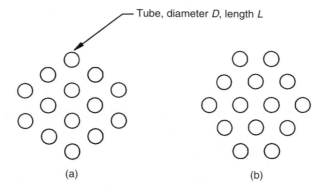

(a)

(b)

For geometry (b):

1st column, $N = 1$ $\dot{Q}_1 = C$

2nd column, $N = 2$ $\dot{Q}_2 = C \cdot 2^{3/4}$

3rd column, $N = 3$ $\dot{Q}_3 = C \cdot 3^{3/4}$

4th column, $N = 2$ $\dot{Q}_4 = C \cdot 2^{3/4}$

5th column, $N = 3$ $\dot{Q}_5 = C \cdot 3^{3/4}$

6th column, $N = 2$ $\dot{Q}_6 = C \cdot 2^{3/4}$

7th column, $N = 1$ $\dot{Q}_7 = C$

$$\text{Total heat transfer} = \dot{Q}_1 + \dot{Q}_2 + \dot{Q}_3 + \dot{Q}_4 + \dot{Q}_5 + \dot{Q}_6 + \dot{Q}_7$$
$$= C\left\{1 + 2^{3/4} + 3^{3/4} + 2^{3/4} + 3^{3/4} + 2^{3/4} + 1\right\}$$
$$= 11.6C$$

From Equation (7.21) the rate of production of condensate is directly proportional to the heat transfer rate, so

$$\frac{\text{Geometry (b)}}{\text{Geometry (a)}} = \frac{11.6}{10.75} = 1.08$$

Comment   Clearly an improvement in the overall heat transfer for a tube bundle can be achieved if the stacking of individual tubes is restricted to as small a number as possible for each column.

7.2.5   Forced flow condensation

The process of condensation discussed so far assumes that the vapour outside the condensate film is quiescent. These are, then, examples of condensation by free convection. Condenser design often requires the use of tubes, through or over which the condensing vapour is forcibly pumped.

For a horizontal cylinder of diameter D exposed to a crossflow of vapour, with freestream velocity U_∞ (Shekriladze and Gomelauri, 1966):

$$h_{av}\frac{D}{k_L} = 0.64 Re_D^{1/2} \left[1 + \left\{1 + \frac{1.69gh'_{fg}\mu_L D}{U_\infty^2 k_L(T_{sat} - T_w)}\right\}^{1/2}\right]^{1/2} \tag{7.31}$$

where $Re_D = \rho_L U_\infty D/\mu_L$. Equation (7.31) is valid for $Re_D < 10^6$. As the freestream velocity of vapour becomes small, Equation (7.31) approaches the free convection limit given by Equation (7.29). Conversely, when gravitational effects are small, then $h_{av}D/k_L = 0.64 Re_D^{1/2}$, which bears obvious similarity to single-phase laminar forced convection from a plate.

For forced condensation inside horizontal tubes, at values of the vapour flow Reynolds number $\rho_V U_V D / \mu_V < 3.5 \times 10^4$, Chato (1962) found that

$$
h_D = 0.555 \left\{ \frac{g \rho_L^2 h_{fg}' k_L^3}{\mu_L (T_{sat} - T_w) D} \right\}^{1/4}
\tag{7.32}
$$

which is similar to Equation (7.29) for film condensation.

7.3 ● Boiling

Boiling takes place at a solid–liquid boundary; this distinguishes it from the process of evaporation which occurs at a liquid–vapour boundary. A liquid will boil when the temperature of a nearby surface exceeds the saturation temperature, T_{sat}, of the liquid. Boiling of a liquid without any externally imposed flow or agitation is referred to as pool boiling, since the conceptual model is one of a large quiescent 'pool' of liquid. Boiling of a liquid forced over a heated surface, as in the tubes of a boiler, is referred to as forced convection, or flow, boiling.

7.3.1 Pool boiling

Pool boiling corresponds to the everyday experience of seeing water boil in a saucepan placed on the cooker. Anyone who has watched this happen will be able to recall that as the temperature of the saucepan is increased there is some initial evidence of motion, followed by bubbles; the bubbles get larger and more vigorous and eventually break through the free surface of the water. If our cooker had enough power, we would observe that these individual bubbles lose their identity and coalesce with neighbours to form larger, irregular bubbles of vapour.

These casual observations can be formalised and distinct regimes of pool boiling marked out in a boiling curve. A boiling curve for water at atmospheric pressure is shown in Figure 7.4. This curve shows the relation between heat flux and temperature difference? The gradient represents the heat transfer coefficient.

Boiling begins when the wall temperature, T_w, exceeds the saturation temperature of the liquid, T_{sat}. When the excess temperature, taken as the quantity $\Delta T = T_w - T_{sat}$, is small ($\Delta T \leqslant 4°C$) then the liquid in the vicinity of the heated surface becomes superheated and rises. Although some bubbles are produced, these are mostly of air (air is less soluble in water at higher temperatures). Fluid motion is sustained by a process of free convection and the heat transfer relations discussed in Chapter 6 may be used. As ΔT is increased, individual bubbles of vapour will form at preferential sites known as *nucleation sites*. Any small imperfection on a surface, such as a crack, will constitute a possible nucleation site. Fluid 'sitting' in a crack will be heated by a larger surface area, and therefore

Fig. 7.4 ● Pool boiling curve for water at 1 atm

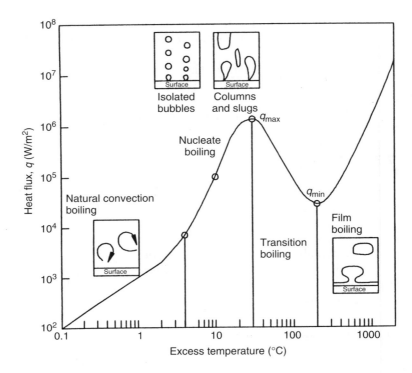

have a greater tendency to change phase into a vapour bubble. If the bulk of the liquid is below the saturation temperature, these vapour bubbles rise and then condense in the liquid – this is referred to as subcooled boiling. If, as is assumed here, the bulk temperature of the liquid is above the saturation point, then vapour bubbles rise through the liquid and eventually break free from the surface – this is referred to as saturated nucleate boiling and is indicated in Figure 7.4. Further increases in ΔT bring about an increase in the production rate of vapour bubbles, which eventually combine to form the larger slugs and columns of vapour. The motion of these bubbles causes agitation of the liquid and, as shown in Figure 7.4, compared with free convection boiling, the heat transfer coefficients in nucleate boiling are orders of magnitude larger. The point of inflection on the curve, at $\Delta T \approx 10°C$, sees a change in the behaviour of the increase of heat transfer coefficient. Beyond this point there is a reduction in the rate of increase of heat transfer coefficient with ΔT. Physically this corresponds to a transition from the generation of individual bubbles to the generation of columns and slugs of vapour, where the larger volume of vapour causes an increase in the thermal resistance from the wall to the bulk of the fluid. At atmospheric pressure, the maximum or peak heat flux, q_{max}, occurs at $\Delta T \approx 35°C$, and the path of the boiling curve beyond this point depends on the nature of our 'experiment'. There are two choices:

1. We can control ΔT and measure the heat flux q, as if we had immersed a heating pipe conveying hot liquid in the pool of water.

2. We can control q and measure the temperature difference ΔT, as if we had immersed an electric resistance heater in the water.

For a *temperature-controlled* experiment, further increases in ΔT, beyond the temperature associated with the peak heat flux, cause a reduction in q (as shown in Figure 7.4). A distinct regime, known as the transition boiling regime, is present where progressively larger areas of the heater surface are covered with an unstable vapour film. Being unstable, this is not a permanent feature, and the flow structure during the transition boiling regime vacillates between vapour film and the individual columns and slugs of nucleate boiling. The end of this regime, and the beginning of the film boiling regime, is marked by the minimum heat flux, q_{min}, which occurs at the so-called *Leidenfrost* temperature.

In film boiling, a stable film of vapour covers the heater surface and the heat flux will again increase with ΔT. Radiation from the wall to the liquid makes an increasingly significant contribution to the total heat transfer. Film boiling will persist as the temperature difference ΔT is increased, and if the heater wall has a sufficiently high melting point, it is possible for the heat flux in film boiling to exceed the peak heat flux at the juncture of the nucleate boiling regime and the transition boiling regime.

For an experiment where the *heat flux* is controlled, as the heat flux is increased beyond the peak heat flux, there is a sudden increase in ΔT to the value corresponding to the beginning of film boiling. The transition boiling regime is therefore absent.

As should be apparent by now, the physical mechanisms involved in the boiling of a liquid are extremely complex. It is beyond the scope of this book to examine the dynamics of bubble formation and growth, the effect of surface characteristics, the role of surface tension, density differences and so on.

For nucleate boiling, an early correlation of experimental data produced by Rohsenow (1952) still provides a useful working relationship to establish the link between heat flux, q, and excess temperature difference $\Delta T = T_w - T_{sat}$:

$$q = \mu_L h_{fg} \left(\frac{g(\rho_L - \rho_V)}{\sigma} \right)^{1/2} \left(\frac{C_{p,L}(T_w - T_{sat})}{C_{s,f} h_{fg} Pr_L^n} \right)^3 \tag{7.33}$$

The subscript 'L' is used to denote a property of the liquid phase, and 'V' for the vapour phase. Equation (7.33) involves two empirical constants: the role of different surface-to-fluid combinations and the surface finish are accounted for by $C_{s,f}$; the exponent, n, of the Prandtl number takes a value of 1 for water or 1.7 for all other fluids. Values of $C_{s,f}$ are given in Table 7.1; values of surface tension, σ, and latent heat, h_{fg}, are given in Table 7.2 for water and Table A.5a for some other fluids.

The value of the peak heat flux in nucleate pool boiling can be obtained from the following equation (Lienhard and Dhir, 1973):

$$q_{max} = C_o h_{fg} \rho_V \left(\frac{\sigma g (\rho_L - \rho_V)}{\rho_V^2} \right)^{1/4} \tag{7.34}$$

where $C_o = 0.149$ for a large horizontal surface

$\qquad C_o = 0.116$ for a large horizontal cylinder

Table 7.1

Values of the constant $C_{s,f}$, in the Rohsenow equation (7.33) for various surface–fluid combinations

Fluid and surface	$C_{s,f}$	Reference
Water–copper		
Scored surface	0.0068	Vachon et al. (1968)
Polished surface	0.0130	Vachon et al. (1968)
Water–stainless steel		
Teflon-coated surface	0.0058	Vachon et al. (1968)
Ground and polished surface	0.0068	Vachon et al. (1968)
Mechanically polished surface	0.0130	Vachon et al. (1968)
Chemically etched surface	0.0130	Vachon et al. (1968)
Water–brass	0.0060	Cryder and Finalbargo (1937)
Water–nickel	0.0060	Vachon et al. (1968)
Water–platinum	0.0130	Addoms (1948)
n-Pentane–copper		
Lapped surface	0.0049	Vachon et al. (1968)
Polished surface	0.0154	Vachon et al. (1968)
n-Pentane–chromium	0.0150	Cichelli and Bonilla (1945)
Ethyl alcohol–chromium	0.0027	Cichelli and Bonilla (1945)

Table 7.2

Surface tension and latent heat values for water

Saturation temperature, T_{sat} (°C)	Surface tension σ (N/m)	Latent heat h_{fg} (kJ/kg)
0	0.0755	2500.8
20	0.0729	2453.7
40	0.0695	2406.2
60	0.0661	2357.9
80	0.0627	2308.3
100	0.0589	2256.7
150	0.0487	2113.4
200	0.0378	1939.3
250	0.0261	1714.7
300	0.0143	1406.2
350	0.0036	916.1

Example 7.3

Calculate the heat flux and heat transfer coefficient associated with the pool boiling of water at 100°C and 1 bar, with an excess temperature difference of 10°C, in a stainless steel container with a ground and polished surface. Take the surface tension and latent heat as 0.058 N/m and 2257 kJ/kg, respectively. Compare this with the maximum heat flux. Repeat the calculation for a mechanically polished surface.

Solution For an excess temperature of 10°C, nucleate boiling occurs and we are able to use Equation (7.33) to predict the heat transfer coefficient. From Equation (7.33) with $n = 1$ for water:

$$q = \mu_L h_{fg} \left[\frac{g(\rho_L - \rho_V)}{\sigma} \right]^{1/2} \left[\frac{C_{p,L}(T_w - T_{sat})}{C_{s,f} h_{fg} Pr_L} \right]^3$$

For ground and polished stainless steel, $C_{s,f} = 0.0068$.
For water at 100°C,

$$\mu_L = 0.278 \times 10^{-3} \, \text{kg/m s}$$

$$C_{p,L} = 4216 \, \text{J/kg K}$$

$$k_L = 0.681 \, \text{W/m K}$$

$$\rho_L = 958 \, \text{kg/m}^3$$

and for vapour at 100°C,

$$\rho_V = 0.598 \, \text{kg/m}^3$$

Hence $Pr_L = \mu_L C_{p,L}/k = 0.278 \times 10^{-3} \times 4216/0.681 = 1.72$, so

$$q = 0.278 \times 10^{-3} \times 2257 \times 10^3 \times \left(\frac{9.81 \times (958 - 0.598)}{0.058} \right)^{1/2}$$

$$\times \left(\frac{4216 \times (110 - 100)}{0.0068 \times 2257 \times 10^3 \times 1.72} \right)^3$$

$$= 1.03 \times 10^6 \, \text{W/m}^2$$

$$h = \frac{q}{\Delta T} = \frac{1.03 \times 10^6}{10} = 103 \times 10^3 \, \text{W/m}^2\text{K}$$

The maximum heat flux for a horizontal surface is given by Equation (7.34):

$$q_{max} = 0.149 \, h_{fg} \, \rho_V \left(\frac{\sigma g(\rho_L - \rho_V)}{\rho_V^2} \right)^{1/4}$$

$$= 0.149 \times 2257 \times 10^3 \times 0.598 \times \left(\frac{0.058 \times 9.81 \times (958 - 0.598)}{0.598^2} \right)^{1/4}$$

$$= 1.25 \times 10^6 \, \text{W/m}^2$$

For a mechanically polished surface $C_{s,f} = 0.0130$ (Table 7.1), and since $q \sim C_{s,f}^{-3}$,

$$q = 1.03 \times 10^6 \left(\frac{0.013}{0.0068}\right)^{-3} = 0.1474 \times 10^6 \, \text{W/m}^2$$

$$h = \frac{q}{\Delta T} = \frac{0.1474 \times 10^6}{10} = 14.7 \times 10^3 \, \text{W/m}^2\text{K}$$

Comment The magnitude of the heat transfer coefficient is much greater than in single-phase convection, and for the ground and polished surface the heat flux is near the value of the maximum heat flux. Also, the value of heat flux (and heat transfer coefficient) is highly sensitive to the value of the surface finish coefficient $C_{s,f}$. In this example, doubling $C_{s,f}$ leads to a sevenfold reduction in the heat flux and heat transfer coefficient. If the temperature difference is maintained constant, the heat transfer coefficient increases with increasing pressure (i.e. saturation temperature). Primarily, this is due to the reduction with increasing temperature of dynamic viscosity, surface tension and latent heat.

The value of the minimum heat flux, $q_{\min}$, for a large horizontal plate was derived analytically by Zuber (1958); the leading coefficient in Equation (7.35) was obtained from subsequent experimental data

$$q_{\min} = 0.09 h_{fg} \, \rho_V \left(\frac{\sigma g(\rho_L - \rho_V)}{(\rho_L + \rho_V)^2}\right)^{1/4} \tag{7.35}$$

The film boiling regime has a number of similarities to condensation. Consequently, many of the established correlations for the average heat transfer coefficient, $h_{av} = q_{av}/(T_w - T_{sat})$, are in a similar form to those in Section 7.2:

$$h_{av} = C_o \left\{\frac{g \rho_V (\rho_L - \rho_V) \, h'_{fg} \, k_V^3}{\mu_V (T_w - T_{sat}) L}\right\}^{1/4} \tag{7.36}$$

where $h'_{fg} = h_{fg} + 0.4 C_{p,V}(T_w - T_{sat})$, and for a large diameter tube (diameter D) or a horizontal surface,

$$C_o = \left(0.59 + \frac{0.69\lambda}{D}\right), \quad L = \lambda$$

and $\lambda = 2\pi \left\{\dfrac{\sigma}{g(\rho_L - \rho_V)}\right\}^{1/2}$ (Westwater and Breen, 1962)

For a sphere $C_o = 0.67$ and $L = D$

(Lienhard and Dhir, 1973)

For a horizontal cylinder $C_o = 0.62$ and $L = D$

(Bromley, 1950)

As previously noted, at high temperatures in the film boiling regime, thermal radiation becomes significant and Bromley (1950) has suggested using an overall heat transfer coefficient

$$h = h_{av} + 0.75 h_{rad} \qquad (7.37)$$

where h_{av} is found from Equation (7.36) and for $h_{av} \gg h_{rad}$, h_{rad} from

$$h_{rad} = \sigma \varepsilon \frac{(T_w^4 - T_{sat}^4)}{(T_w - T_{sat})} \qquad (7.38)$$

where σ (to be consistent with the use of this symbol elsewhere in this book) is the Stefan–Boltzmann constant ($\sigma = 56.7 \times 10^{-9}$ W/m²K⁴) and ε is the emissivity of the heated surface.

Since the temperature differences in film boiling are large, it is recommended that the vapour properties in Equation (7.36) are evaluated at the mean film temperature.

Example 7.4

The surface temperature of the horizontal surface in Example 7.3 is increased to 300°C. Calculate the heat transfer coefficient for a Teflon-coated stainless steel surface ($\varepsilon = 0.9$).

Solution   First we need to establish the mode of boiling. This is easily done by calculating the heat flux assuming nucleate boiling (7.33) and comparing with the maximum heat flux (7.34). If the calculated heat flux exceeds the peak heat flux, we can assume that boiling occurs in the film boiling regime.

From Equation (7.33) with $C_{s,f} = 0.058$ (Teflon-coated stainless steel) and $n = 1$ (water),

$$q = 0.278 \times 10^{-3} \times 2257 \times 10^3 \times \left[\frac{9.81 \times (958 - 0.598)}{0.058} \right]^{1/2}$$

$$\times \left[\frac{4216 \times (300 - 100)}{0.0058 \times 2257 \times 10^3 \times 1.72} \right]^3$$

$$= 13.26 \times 10^9 \text{ W/m}^2$$

Since this greatly exceeds the peak heat flux, we assume that film boiling occurs and use Equation (7.36) to evaluate the heat transfer coefficient:

$$h_{av} = C_o \left\{ \frac{g\rho_V(\rho_L - \rho_V)h'_{fg}k_V^3}{\mu_V(T_w - T_{sat})\lambda} \right\}^{1/4}$$

Note that for film boiling we need to use the properties of the vapour phase. For water vapour at the mean film temperature of $\frac{1}{2}(300 + 100) = 200°C$,

$$k_V = 0.0375 \text{ W/m K}$$

$$\rho_V = 7.85 \text{ kg/m}^3$$

$$\mu_V = 15.7 \times 10^{-6} \text{ kg/m s}$$

$$C_{p,V} = 2910 \text{ J/kg K}$$

$$\begin{aligned} h'_{fg} &= h_{fg} + 0.4\, C_{p,V}(T_w - T_{sat}) \\ &= 2257 \times 10^3 + \{0.4 \times 2910 \times (300 - 100)\} \\ &= 2490 \times 10^3 \text{ J/kg} \end{aligned}$$

and for a tube of large diameter or a horizontal surface,

$$\begin{aligned} \lambda &= 2\pi \left\{ \frac{\sigma}{g\,(\rho_L - \rho_V)} \right\}^{1/2} \\ &= 2\pi \times \left\{ \frac{0.058}{9.81 \times (958 - 7.85)} \right\}^{1/2} \\ &= 0.0157 \text{ m} \end{aligned}$$

$$\begin{aligned} C_o &= \left(0.59 + \frac{0.69\lambda}{D} \right) \\ &= 0.59 \quad \text{since for a horizontal surface } D \to \infty. \end{aligned}$$

Hence the film boiling heat transfer coefficient is

$$h_{av} = 0.59 \times \left\{ \frac{9.81 \times 7.85 \times (958 - 7.85) \times 2490 \times 10^3 \times 0.0375^3}{15.7 \times 10^{-6}(300 - 100) \times 0.0157} \right\}^{1/4}$$

$$= 392 \text{ W/m}^2\text{K}$$

In addition we must use Equations (7.37) and (7.38) to account for radiation. The 'radiative' heat transfer coefficient is given by the former,

$$h_{rad} = \frac{\sigma\varepsilon(T_w^4 - T_{sat}^4)}{T_w - T_{sat}}$$

where σ is the Stefan–Boltzmann constant $\sigma = 56.7 \times 10^{-9} \text{ W/m}^2\text{K}^4$, so

$$h_{rad} = \frac{56.7 \times 10^{-9} \times 0.9\{(273 + 300)^4 - (273 + 100)^4\}}{(300 - 100)}$$

$$= 22.6 \text{ W/m}^2\text{K}$$

And from Equation (7.37), the 'total' heat transfer coefficient is given by

$$h = h_{av} + 0.75h_{rad}$$
$$= 392 + (0.75 \times 22.6)$$
$$= 409 \text{ W/m}^2\text{K}$$
$$q = h\Delta T$$
$$= 409 \times (300 - 100)$$
$$= 8.18 \times 10^4 \text{W/m}^2$$

Comment The heat flux and heat transfer coefficient in this example of film boiling are less than in nucleate boiling and less than the peak heat flux. Teflon has a relatively high emissivity, and the contribution of the radiative heat transfer, although small, has been included. The contribution of radiation to the total heat transfer will become more significant at higher surface temperatures. For a polished surface, the emissivity would be small ($\varepsilon \approx 0.05$ for polished stainless steel) and radiative effects could be safely ignored. Note the use of absolute temperature (kelvin) in the evaluation of h_{rad}.

7.3.2 Forced convection boiling

The forced flow of a liquid, heated by a surface, leading to boiling and generation of vapour is considerably more complicated than pool boiling. The process is also referred to as flow boiling and, like forced convection in single-phase flow, we can distinguish between internal and external forced convection boiling.

Figure 7.5 shows a schematic diagram of the flow boiling regimes in a heated vertical pipe. Beneath the figure is a graph showing the qualitative behaviour of the local heat transfer coefficient. Pure liquid enters the pipe, and in this regime the heat transfer may be estimated using the single-phase pipe flow correlations discussed in Chapter 5. Further along the pipe, vapour bubbles start to form; the transport of latent heat and the motion of the bubbles leads to an increase in the local heat transfer coefficient. As in pool boiling, further heating leads to an increase in the production rate of vapour bubbles, which coalesce forming larger slugs of vapour, with a further rise in the local heat transfer coefficient. The bubbly flow and slug flow regimes are associated with a vapour quality of approximately 1%, so their practical significance in the design of hardware intended to produce saturated and/or superheated vapour is limited. Rohsenow (1973) recommends that in this regime of nucleate flow boiling, the total heat flux, q_{tot}, should be obtained by adding the separate contributions from single-phase forced convection, q_c, and nucleate pool boiling, q_b:

$$q_{tot} = q_c + q_b \tag{7.39}$$

where q_b can be obtained from Equation (7.33).

Fig. 7.5 ●
Forced convection
(flow) boiling in a
vertical tube

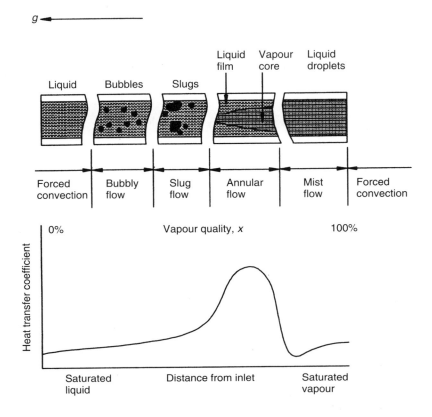

Increased heating leads to further production of vapour which, since it has a lower density than the liquid phase, speeds up and stratifies into a separate region in the centre of the pipe. A thin annular liquid film covers the pipe surface, and the relatively high velocity of the vapour core tends to suppress the nucleation of bubbles. Heat is therefore transferred across the thin liquid film, and vaporisation takes place at the interface with the vapour core. The range of vapour quality associated with this annular flow regime depends on the fluid properties and the specific geometry being considered. At vapour qualities of 25% or more, the flow is characterised by a mist flow regime. The walls of the pipe are now no longer coated by a film of relatively high thermal conductivity liquid, so the heat transfer coefficient decreases. This point is also referred to as *dryout*. When the vapour quality reaches 100%, single-phase forced convection pipe flow correlations may again be used with the appropriate fluid (vapour) properties.

A relatively straightforward correlation of internal forced convection boiling heat transfer has been made by Klimenko (1988). This applies to a boiling liquid at a pressure p, in a pipe, with thermal conductivity k_w and cross-sectional area A_c. The correlation is valid for both nucleate boiling and annular film boiling regimes up to the point of dryout.

First we determine if nucleate boiling or annular film boiling dominates by evaluating the parameter Φ:

$$\Phi = \frac{\dot{m}h_{fg}}{qA_c}\left\{1 + x\left(\frac{\rho_L}{\rho_V} - 1\right)\right\}\left(\frac{\rho_V}{\rho_L}\right)^{1/3} \qquad (7.40)$$

The quality of a volume of vapour comprising m_V kg of vapour and m_L kg of liquid is

$$x = \frac{m_V}{(m_L + m_V)} \qquad (7.41)$$

Nucleate boiling, $\Phi < 1.6 \times 10^4$:

$$Nu = 7.4 \times 10^{-3}q^{*0.6}P^{*0.5}Pr_L^{-1/3}\left(\frac{k_w}{k_L}\right)^{0.15} \qquad (7.42)$$

where $\quad Nu = \dfrac{h_b L_c}{k_L}$

$$L_c = \left(\frac{\sigma}{g(\rho_L - \rho_V)}\right)^{1/2}$$

$$q^* = \frac{qL_c}{h_{fg}\rho_V \alpha_L}$$

$$P^* = \frac{PL_c}{\sigma}$$

$$\alpha_L = \left(\frac{k}{\rho C_p}\right)_L$$

Annular film boiling, $\Phi > 1.6 \times 10^4$:

$$Nu = 8.7 \times 10^{-2}Re^{0.6}Pr_L^{1/6}\left(\frac{\rho_V}{\rho_L}\right)^{0.2}\left(\frac{k_w}{k_L}\right)^{0.09} \qquad (7.43)$$

where $\quad Re = \dfrac{\rho_L V L_c}{\mu_L}$

$$V = \frac{\dot{m}}{A_c \rho_L}\left\{1 + x\left(\frac{\rho_L}{\rho_V} - 1\right)\right\}$$

all properties being evaluated at T_{sat}.

The actual, effective, heat transfer coefficient due to boiling and single-phase forced convection is obtained from a similar relation to Equation (7.39):

$$h = \left(h_b^3 + h_c^3\right)^{1/3} \qquad (7.44)$$

7.4 ● Closing comments

Heat transfer during a process of phase change is a complex subject, since many more factors are influential compared with single-phase heat transfer. A change of phase brings about a transfer of *latent* heat which can take place with a very small temperature difference, so phase-change heat transfer results in very high heat transfer coefficients. This chapter has provided a brief description of the processes of condensation and boiling, together with working formulae for quantifying the heat transfer rate.

For condensation, it is advisable to assume that the condensate forms a continuous film. This film will be laminar for film Reynolds numbers $Re_y < 30$, wavy for $30 < Re_y < 1800$ and turbulent for $Re_y > 1800$. An analytically derived expression is available for the heat transfer coefficient in the laminar regime – the leading constant takes different values for a vertical surface, horizontal tube or a sphere. For condensation in the wavy and turbulent regimes, a correlation is available based on experimental data. Condensation will also take place when the condensate is forced past the condenser surface. Some correlations for the flow over and through tubes have been presented for forced flow condensation.

Boiling may be classified as either a free convection type flow known as pool boiling or a forced convection flow known as flow boiling. The pool boiling curve is useful in both a descriptive sense and in terms of quantitative information. The pool boiling of a liquid undergoes a number of flow regimes which depend on the magnitude of the excess temperature defined as $\Delta T = T_w - T_{sat}$. For small values of ΔT, motion is generated by free convection effects and so it is acceptable to use the single-phase correlations presented in Chapter 6. Larger values of ΔT lead to the production of vapour bubbles and an increase in the heat transfer coefficient. These coalesce to form slugs and columns of vapour with further increases in ΔT. The maximum heat flux occurs at the end of this nucleate boiling regime; for a temperature-controlled experiment, further heating leads to an unstable transitional boiling regime with a resulting decline in the heat transfer coefficient (a regime to be avoided in practical design). At the Leidenfrost temperature, a stable vapour film covers the surface and the heat transfer coefficient then rises with increasing values of ΔT. In this film boiling regime, heat transfer by radiation becomes increasingly significant. Heat transfer in the nucleate pool boiling regime is strongly dependent on surface characteristics (reflected by the constant $C_{s,f}$) and surface tension. Film boiling bears a number of physical similarities to film condensation, and established correlations are similar in form to those for condensation.

Flow boiling is more complex; qualitatively the process exhibits similar regimes to pool boiling: beginning of nucleation, coalescence of bubbles to form larger slugs, the formation of a film, and eventually formation of saturated and superheated vapour. Attempts have been made to predict the heat transfer rate using superposition of boiling and forced convection heat fluxes. A slightly more complicated procedure allows prediction of heat transfer coefficients up to the point of dryout in flow boiling.

The correlations cited in this chapter are summarised in Table 7.3.

Table 7.3 Summary of correlations for condensation and boiling

Flow configuration	Equation	Comments
Film condensation		
Vertical wall	(7.18)	Local for laminar, $Re_y < 30$
	(7.19) (7.26)	Average for laminar
	(7.27)	Average for $Re_y > 30$
Sphere	(7.28)	
Horizontal tube	(7.29)	
N horizontal tubes	(7.30)	
Forced flow condensation		
Cylinder in crossflow	(7.31)	Valid for $Re_{D,\,liquid} < 10^6$
Inside a horizontal tube	(7.32)	Valid for $Re_{D,\,vapour} < 3.5 \times 10^4$
Pool boiling		
Nucleate boiling heat flux	(7.33)	Valid for $q < q_{max}$, Table 7.1 for $C_{s,f}$
Nucleate boiling peak heat flux	(7.34)	Value of q_{max}
Minimum heat flux	(7.35)	Value of q_{min}
Film boiling heat transfer coeff.	(7.36) (7.37)	Evaluate vapour properties at mean film
	(7.38)	temperature
Forced convection boiling		
Nucleate boiling	(7.42)	Equation (7.40) delineates nucleate and
Annular film boiling	(7.44)	annular film boiling regimes. Properties evaluated at T_{sat}

7.5 ● References

Addoms, J.N. (1948). Heat transfer at high rates to water boiling outside cylinders. *DSc Thesis*, Massachusetts Institute of Technology

Bejan, A. (1995). *Convection Heat Transfer*. New York: Wiley

Bromley, A.L. (1950). Heat transfer in stable film boiling. *Chem. Eng. Prog.*, **46**, 221–7

Chato, J.C. (1962). Laminar condensation inside horizontal and inclined tubes. *J. ASHRAE*, **4**,. 52–60

Chen, S.L., Gerner, F.M. and Tien, C.L. (1987). General film condensation correlations. *Experimental Heat Transfer*, **1**, 93–107

Cichelli, M.T. and Bonilla, C.F. (1945). Heat transfer to liquids boiling under pressure. *Trans. AIChE*, **41**, 755–87

Cryder, D.S. and Finalbargo, A.C. (1937). Heat transmission from metal surfaces to boiling liquids: effect of the temperature of the liquid on the film coefficient. *Trans. AIChE*, **33**, 346–62

Dhir, V.K. and Lienhard, J.H. (1971). Laminar film condensation on plane and axisymmetric bodies in non-uniform gravity. *Journal of Heat Transfer*, **93**, 97–100

Klimenko, V.V. (1988). A generalised correlation for two-phase forced heat transfer. *Int. J. Heat Mass Transfer*, **31**, 541–52.

Lienhard, J.H. and Dhir, V.K. (1973). Extended Hydrodynamic Theory of the Peak and Minimum Pool Boiling Heat Fluxes, *NASA CR-2270*

Lock, G.S.H. (1996). *Latent Heat Transfer: An Introduction to Fundamentals*. Oxford: Oxford University Press

Nusselt, W. (1916). Die Oberflächenkondensation der Wasserdampfes, *Z. Vereines deutscher Ingenieure*, **60**, 541–69

Rohsenow, W.M. (1952). A method for correlating heat transfer data for surface boiling of liquids. *Trans. ASME*, **74**, 969–76

Rohsenow, W.M. (1973). Boiling. In *Handbook of Heat Transfer* (Rohsenow, W.M. and Hartnett, J.P. eds). New York: McGraw-Hill

Shekriladze, I.G. and Gomelauri, V.I. (1966). Theoretical study of laminar film condensation of flowing vapour. *Int. J. Heat Mass Transfer*, **9**, 581–91

Sparrow, E.M. and Gregg, J.L. (1959). A boundary layer treatment of laminar film condensation. *Journal of Heat Transfer*, **81**, 13–18

Vachon, R.I., Nix, G.H. and Tanger, G.E. (1968). Evaluation of constants for the Rohsenow pool-boiling correlation, *Trans. ASME Journal of Heat Transfer*, **90**, 239–47

Westwater, J.W., and Breen B.P. (1962). Effect of diameter of horizontal tubes on film boiling heat transfer. *Chem. Eng. Prog.*, **52**, 67–72

Zuber, N. (1958). On the stability of boiling heat transfer. *Trans. ASME*, **80**, 711–20

7.6 ● End of chapter questions

7.1 Assuming laminar film condensation, calculate the ratio of the heat transfer to a vertical tube to that for a horizontal tube of the same diameter, D, and length, H. Briefly comment on the implications of this to condenser design.

$$[1.294(D/H)^{1/4}]$$

7.2 Saturated steam at 1 bar condenses on the outside of a thin tube of 6 mm dia., and 1 m long. The tube is cooled by a flow of 0.053 53 kg/s of water at 15°C through the inside. Calculate:

(a) The heat transfer coefficient on the tube inside using the Dittus–Boelter relation (5.42).

$$[8305 \text{ W/m}^2\text{K}]$$

(b) The tube wall temperature.

$$[67.7°C]$$

(c) The heat transfer coefficient on the outside of the tube.

$$[13\,531 \text{ W/m}^2\text{K}]$$

(d) The total heat transfer.

$$[8238 \text{ W}]$$

(e) The rate of production of condensate.

$$[3.65 \times 10^{-3} \text{ kg/s}]$$

(f) The film Reynolds number.

$$[52]$$

For water on the tube inside take

$Pr = 8$; $k = 0.595$ W/m K;

$\mu = 1.136 \times 10^{-3}$ kg/m s

For steam at 1 bar take

$h_{fg} = 2257 \times 10^3$ J/kg K

For the condensate take

$\rho = 958$ kg/m^3; $k = 0.681$ W/m K;

$C_p = 4219$ J/kg K; $\mu = 0.279 \times 10^{-3}$ kg/m s

7.3 A design for a refrigerator, where condensate is pumped through tubes and condenses on the inside of them, has a thermal load of 400 watts. Calculate the length of 5 mm dia. tube required for two different refrigerants. In each case the saturation temperature of the liquid is 30°C and the tube wall temperature is 26°C.

(a) Dichlorodifluoromethane (Freon 12)

$\rho_L = 1295$ kg/m^3; $C_p = 984$ J/kg K;

$\mu = 2.47 \times 10^{-4}$ kg/m s; $k = 0.071$ W/m K;

$h_{fg} = 135 \times 10^3$ J/Kg K

$$[3.2 \text{ m}]$$

(b) Ammonia

$\rho_L = 602$ kg/m^3; $C_p = 4900$ J/kg K;

$\mu = 1.31 \times 10^{-4}$ kg/m s; $k = 0.51$ W/m K;

$h_{fg} = 1.146 \times 10^6$ J/kg K

$$[0.53 \text{ m}]$$

Comment on the above results.

7.4 A 10 mm dia., 1 m long copper tube with a scored surface is to be used to boil water adjacent to the external surface at atmospheric pressure. Calculate the surface temperature of the tube so that it operates at half the maximum heat flux. Find also the heat dissipation rate and the evaporation rate of water.

$$[107.8°C, 38.4 \text{ W}, 0.06 \text{ kg/h}]$$

7.5 A 10 mm dia., ground and polished stainless steel tube ($\varepsilon = 0.05$) is maintained at a surface temperature of 300°C while boiling water at atmospheric pressure. Identify the regime of pool boiling and calculate the heat transfer coefficient and heat flux.

[Film boiling, 461 W/m^{2}K, 92.2 × 10^3 W/m^2]

7.6 A 50 mm dia. vertical evaporator tube ($k_w = 20$ W/m K) conveys 1 kg/s of steam at 15.55 bar of $x = 0.2$. The tube is subjected to an incident heat flux of 106 W/m^2. Identify the regime of flow boiling and calculate the convective heat transfer coefficient and surface temperature of the tube.

[Nucleate flow boiling, 66 × 10^3 W/m^{2}K, 215.2°C]

Chapter eight

Radiative heat transfer

8.1 ● Introduction

Radiative heat transfer or thermal radiation is a distinctly separate mechanism from conduction and convection for the transport of heat. Thermal radiation is an electromagnetic phenomenon which occurs as a manifestation of the absolute temperature of matter; therefore all states of matter (solids, liquids and gases) emit thermal radiation. Unlike convection and conduction, heat can be transferred by thermal radiation through a vacuum; for example, the heat energy from the sun is transported in this way. Figure 8.1 shows that part ($10^{-5} \leqslant \lambda \leqslant 10^{3} \, \mu m$, $1 \, \mu m = 10^{-6} \, m$) of the electromagnetic spectrum which embraces gamma radiation, X-rays, ultraviolet, the visible spectrum, infrared and microwaves. To be more precise, thermal radiation is the name given to electromagnetic radiation in the range of wavelengths $0.1 < \lambda < 100 \, \mu m$, which spans part of the ultraviolet spectrum, the entire visible spectrum and that known as the infrared.

Fig. 8.1 ● The electromagnetic spectrum for $10^{-5} < \lambda < 10^{3}$ micrometres

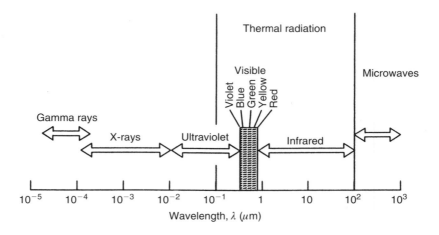

In the limited space available within this chapter, it has not been possible to discuss all aspects of radiative heat transfer and the narrative is restricted to those most relevant to engineering undergraduates. In particular, I have chosen not to include any material on radiation in absorbing and emitting media such as gases, luminous flames and liquids. Instead, I have chosen to concentrate on the useful, yet simpler case of radiative behaviour of solid surfaces. The reader is encouraged to consult the comprehensive text by Siegel and Howell (1992) for information on these subjects and others omitted from this text.

For solids, thermal radiation is absorbed and emitted within about $1\,\mu m$ of the surface. It is therefore a surface effect and it is mostly the surface finish rather than the material itself that governs the various radiative properties. In general, the energy in thermal radiation has a spectral distribution (variation with wavelength) and a directional distribution (variation with angle); this can complicate any analysis. The terms are illustrated graphically in Figure 8.2. For the purpose of most engineering analysis, it is possible to invoke the following simplifications:

1. A diffuse emitter (one that emits uniformly in all directions).

2. Total emission (the spectral emission integrated over all wavelengths).

In addition, it is particularly useful to introduce the concept of a *black body*; an ideal radiative surface which has the following properties:

1. It will absorb all incident radiation.

2. For a given wavelength and temperature, no surface can emit more energy as thermal radiation than a black body.

3. A black body is a diffuse emitter.

The black body is therefore a perfect absorber and emitter. Although no 'real' surface is a perfect black body, black body radiation is a key concept in the understanding of radiative heat transfer, and is used in the analysis

Fig. 8.2 ●
Spectral and directional distributions of emissive power

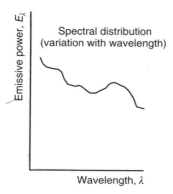

Spectral distribution (variation with wavelength)

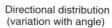

Directional distribution (variation with angle)

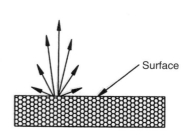

of radiation from engineering surfaces. In some respects, it is helpful to think of this concept in a similar way to the Carnot cycle used in engineering thermodynamics – a benchmark against which to compare real (non-ideal) processes. It should also be noted that the adjective 'black' does not refer to the visible colour; we perceive this owing to the characteristics of human eyesight, which is attuned to the visible part of the electromagnetic spectrum. For example, at 300 K white paint approximates black body behaviour almost as well as black paint.

Figure. 8.3 shows the variation of spectral thermal radiative power, $E_{\lambda,b}$, emitted by a black body (denoted by the subscript 'b') with wavelength (denoted by the subscript λ) at a number of different absolute temperatures ($T = 500, 1000, 2000$ and 5800 K). These curves are obtained by evaluating Equation (8.9). As will shortly be demonstrated, this is in turn related to the equation for the spectral radiative intensity, $I_{\lambda,b}$, derived by Planck from considerations of quantum statistical thermodynamics; a derivation is given in most undergraduate physics texts, for example (Beiser, 1969).

$$I_{\lambda,b} = \frac{2hc^2}{\lambda^5 [\exp(hc/\lambda kT) - 1]} \tag{8.1}$$

Fig. 8.3 ● The Planck distribution for a black body at four different temperatures

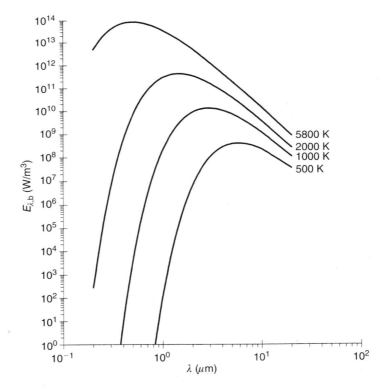

where T is the absolute temperature, h is Planck's constant ($6.6256 \times 10^{-34}\,\mathrm{J\,s}$), c is the velocity of electromagnetic radiation ($2.998 \times 10^{8}\,\mathrm{m/s}$) and k is the Boltzmann constant ($1.3806 \times 10^{-23}\,\mathrm{J/K}$).

The intensity of thermal radiation is therefore a function of just temperature and wavelength, because the other parameters appearing in Equation (8.1) are physical constants. The spectral intensity is defined as the energy per unit area normal to the direction of propagation, per unit solid angle about the direction, per metre of wavelength (in watts per cubic metre). A steradian is a dimensionless measurement of solid (three-dimensional) angle, as the radian is a dimensionless measurement of plane (two-dimensional) angle.

An important relationship between intensity and radiative power, $\dot{Q}$, can be derived from consideration of Figure 8.4. The radiation from the elementary area dA is completely intercepted by the hemisphere of radius r, at the base of which lies dA. The area dA_1 is an elemental area on the normal from dA to the hemisphere, the solid angle subtended at dA is $d\omega$ and by definition $d\omega = dA_1/r^2$ (hence the solid angle subtended by a complete hemisphere is $2\pi r^2/r^2 = 2\pi$). If the radiative energy from dA through dA_1 is $d\dot{Q}_1$ then the intensity of radiation, I, is related to the energy flow by

$$d\dot{Q}_1 = I_n\, d\omega_1\, dA \tag{8.2}$$

where the subscript 'n' is used to denote that the receiving area is normal to the radiation emitted from dA.

Fig. 8.4 ●
Radiation from an elemental area dA to a hypothetical hemisphere of radius r

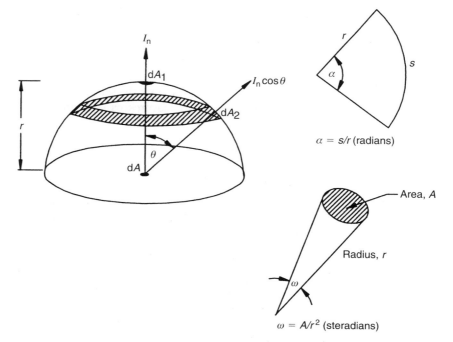

A further elemental ring area, located at an angle θ with the normal from dA, will receive radiation from dA given by

$$d\dot{Q}_2 = I_\theta \, d\omega_2 \, dA \qquad (8.3)$$

where $d\omega_2 = dA_2/r^2$ and I_θ is given by Lambert's cosine law

$$I_\theta = I_n \cos\theta \qquad (8.4)$$

The general form for the solid angle for each elemental ring area on the surface of the hemisphere, at angle θ from the normal to dA is given by

$$d\omega = \frac{2\pi r \sin\theta \, r \, d\theta}{r^2} = 2\pi \sin\theta \, d\theta \qquad (8.5)$$

Combining Equations (8.3), (8.4) and (8.5) allows us to find the thermal radiation passing through each elemental ring area:

$$d\dot{Q}_\theta = I_n \cos\theta \, 2\pi \sin\theta \, d\theta \, dA \qquad (8.6)$$

The total radiation is obtained by integrating over the entire hemisphere

$$\dot{Q} = I_n 2\pi \, dA \int_0^{\pi/2} \cos\theta \sin\theta \, d\theta$$

$$= I_n \pi \, dA \qquad (8.7)$$

From which

$$E = (\dot{Q}/dA) = I_n \pi \qquad (8.8)$$

Using this relation in Equation (8.1) gives the following form for the distribution of the *spectral emissive power*, $E_{\lambda,b}$:

$$E_{\lambda,b} = \frac{2\pi hc^2}{\lambda^5[\exp(hc/\lambda kT) - 1]} \qquad (8.9)$$

which is the equation that describes the set of curves depicted in Figure 8.3.

The following important features can be immediately recognised from this figure:

1. The spectral emissive power varies continuously with wavelength.
2. For any wavelength, the magnitude of the emitted radiation increases with increasing temperature.
3. The wavelength corresponding to the peak in spectral emissive power depends on temperature, and as the temperature is increased, then more radiation occurs at shorter wavelengths.
4. For objects at a high temperature – the value of 5800 K used here refers to the black body surface temperature of the sun – a significant proportion of the emitted radiation occurs in the visible part of the spectrum. For objects at a lower temperature, say 500 K, the emission is mostly in the infrared region and is barely visible to the human eye. As a simple demonstration of this, watch an electric cooker ring as it heats up.

Equation (8.1) can be differentiated to give a relation between wavelength and maximum emissive power. This is the so-called Wien displacement law, which has the form

$$\lambda_{max} T = \text{constant} = 2.8978 \text{ mm K} \tag{8.10}$$

(for example, at $T = 1000°C$, $\lambda_{max} = 2.3 \mu m$).

Example 8.1

At what wavelength is the maximum radiative power emitted from an engine exhaust pipe which is at 630°C?

Solution

$$\lambda_{max} T = 2.8978 \text{ mm K}$$

$$\lambda_{max} = \frac{2.8978}{(630 + 273)}$$

$$= 3.2 \mu m$$

Comment   Most of the radiation is in the infrared part of the spectrum and not visible to the human eye. At higher temperatures, the peak emission occurs at shorter wavelengths. Note the use of absolute temperature in Kelvin.

A further simplification can be made which is of great importance to engineering calculations. As previously noted, the distribution of energy depends on the wavelength, and this is denoted by the symbol $E_{\lambda,b}$. Equation (8.9) for the spectral emissive power can be integrated over all wavelengths, the dependence on λ is removed and the result is the *total* emissive power emitted by a black body:

$$E_b = \int_0^\infty \frac{2\pi h c^2}{\lambda^5 [\exp(hc/\lambda kT) - 1]} \, d\lambda \tag{8.11}$$

which, once the integral is carried out, can be written as

$$E_b = (\text{constant}) \times T^4$$

or more usually

$$E_b = \sigma T^4 \tag{8.12}$$

The constant σ is known as the the *Stefan–Boltzmann* constant and can be expressed in terms of the various fundamental constants as

$$\sigma = \frac{8\pi^5 k^4}{15 c^3 h^3}$$

giving it the value, in the SI system of units, $\sigma = 56.7 \times 10^{-9}\,\text{W/m}^2\text{K}^4$. Equation (8.12) and its variants will be used extensively throughout this chapter. The equation was first proposed by Joseph Stefan in 1879 after examination of experimental results. Boltzmann actually derived the equation using analytical considerations (a thermodynamic cycle analysis in which the radiation pressure was assumed to be the pressure of the working fluid).

Although thermal radiation is treated in this chapter more or less in isolation from the other modes of heat transfer (conduction and convection), it is always present whenever there is a temperature difference. Sometimes the contribution of radiation to the total heat transfer may be insignificant, sometimes it may not. In general, radiation will always be significant in free convection flows. In forced convection, radiation effects will become significant only at high temperatures, for example in combustion systems, furnaces, boilers.

Example 8.2

A central heating 'radiator' has a surface temperature of 70°C and heats a room maintained at 20°C. Using the following correlation for free convection ($Nu_L = 0.118 Gr_L^{1/3} Pr^{1/3}$), determine the contributions of convection and radiation to the total heat transfer from the radiator ($\rho = 1.2\,\text{kg/m}^3$, $\mu = 1.8 \times 10^{-5}\,\text{kg/m\,s}$, $k = 0.026\,\text{W/m\,K}$ and $Pr = 0.71$).

Solution Assuming black body behaviour, the net *radiative* heat transfer is

$$E_{\text{rad}} = \sigma T_{\text{rad}}^4$$

$$E_{\text{room}} = \sigma T_{\text{room}}^4$$

$$\begin{aligned} E_{\text{rad}-\text{room}} &= \sigma(T_{\text{rad}}^4 - T_{\text{room}}^4) \\ &= 56.7 \times 10^{-9}((273 + 70)^4 - (273 + 20)^4) \\ &= 367\,\text{W/m}^2 \end{aligned}$$

$$q_{\text{radiation}} = 367\,\text{W/m}^2$$

The net *convective* heat transfer is obtained from

$$Nu_L = 0.118(Gr_L Pr)^{1/3}$$

from which the convective heat transfer coefficient, h, is:

$$h = 0.118\,k\left(\frac{\rho^2 g\beta(T_{\text{rad}} - T_{\text{room}})}{\mu^2}\right)^{1/3} Pr^{1/3}$$

$$= 0.118 \times 0.026 \times \left\{\frac{1.2^2 \times 9.81 \times (1/293) \times (70-20)}{(1.8 \times 10^{-5})^2}\right\}^{1/3} 0.71^{1/3}$$

$$= 5.34\,\text{W/m}^2\text{K}$$

$$q_{\text{convection}} = h(T_{\text{rad}} - T_{\text{room}})$$

$$= 5.34(70 - 20) = 267 \, \text{W/m}^2$$

$$q_{\text{total}} = q_{\text{radiation}} + q_{\text{convection}}$$

$$= 367 + 267 = 634 \, \text{W/m}^2$$

Comment   Radiation accounts for 58% of the total heat transfer and convection for 42%. This supports the comments made earlier about the significance of radiation in relation to free convection.

8.2 ● Radiative properties

8.2.1   Emission, irradiation and radiosity (E, G and J)

A surface will emit radiation as a direct result of its absolute temperature; this is termed the *emission*, E. For a black body at temperature T, its emission is quantified by Equation 8.12. Also, radiation emitted from other objects, known as *irradiation* and denoted by G, can fall onto a surface and some fraction of this will be reflected from it. Figure 8.5 illustrates these effects. The total radiation leaving a surface (the sum of the reflected and emitted components) is termed the *radiosity* and given the symbol J.

8.2.2   Absorptivity, reflectivity and transmissivity

As mentioned above, some fraction of the irradiation may be reflected, the rest is either absorbed by the material or transmitted through it. These fractions are described by the properties known as reflectivity, absorptivity and transmissivity, which are given the respective symbols ρ, α and τ. In general, they have spectral dependence (values vary with wavelength), and at any particular wavelength their monochromatic value is indicated by the subscript λ (for example, ρ_λ, α_λ and τ_λ). However, for ease of understanding, we will refer to total values obtained by integrating the appropriate fraction of the irradiation, G, over the total spectrum; this

Fig. 8.5 ● Surface irradiation, emission, reflection and radiosity

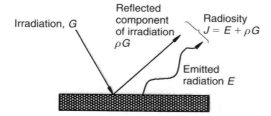

Irradiation, G

Reflected component of irradiation ρG

Radiosity $J = E + \rho G$

Emitted radiation E

means we can omit the subscript. Values of absorptivity, reflectivity and transmissivity for most materials and various conditions of surface finish can be found in the extensive tables in Touloukian and DeWitt (1972).

From the above discussion, it is apparent that the sum of the reflected, absorbed and transmitted components must equal the irradiation, and so

$$\rho G + \alpha G + \tau G = G$$

or

$$\rho + \alpha + \tau = 1 \qquad (8.13)$$

Most gases have a high value of τ (they transmit thermal radiation), and a low value of ρ and α (they reflect and absorb very little). For example, air at room temperature and pressure is transparent to thermal radiation; $\tau = 1$ and $\rho = \alpha = 0$. Carbon dioxide and water vapour are two notable exceptions to this. At terrestrial temperatures they both have high values of α, whereas at the temperature of the source of solar irradiation (5800 K) they have relatively low values of absorptivity. This property is in part responsible for creating the 'greenhouse effect'.

By contrast, most solids (except glass) transmit little thermal radiation, but reflect and absorb significant amounts. Since in this chapter most of the discussion concentrates on solids, Equation (8.13) for $\tau = 0$ becomes

$$\rho + \alpha = 1 \qquad (8.14)$$

Because of its common use, glass deserves a special mention. Glass is a highly selective transmitter of thermal radiation. It is difficult to make general statements referring to all glasses since the radiative properties depend on the composition of the glass. At short wavelengths ($\lambda < 1\,\mu m$) $\tau \rightarrow 1$; at longer wavelengths (corresponding to emission at 300 to 400 K), glass becomes virtually opaque to thermal radiation ($\tau \rightarrow 0$). Germanium and sapphire glasses are used when it is necessary to transmit thermal radiation at these relatively long wavelengths.

Referring to Figure 8.5, the radiosity J from a surface is given by

$$J = E + \rho G$$

and so, for $\tau = 0$

$$J = E + (1 - \alpha)G \qquad (8.15)$$

Irradiation and radiosity both have the same units as emission, W/m^2, which identifies them as fluxes.

8.2.3 Emissivity

The black body behaviour, given by the Planck distribution (8.9) is a theoretical upper bound for the thermal radiation emitted by a body. As shown in Figure 8.6, real surfaces (non-black or grey) emit less radiative power and can also be non-isotropic (their radiative properties are dependent on direction).

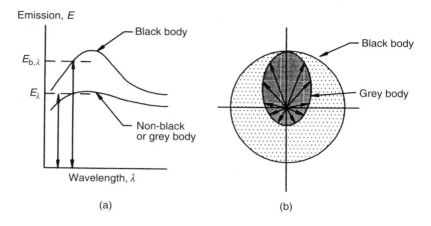

Fig. 8.6 ● Black body and grey body behaviour: (a) spectral and (b) directional dependence

The ratio of radiation emitted by a surface to that of a black body *at the same temperature* is known as the emissivity, ε. As with ρ, α and τ, emissivity has spectral dependence denoted by ε_λ ($\varepsilon_\lambda = E_\lambda/E_{b,\lambda}$); but here we will consider only the total emissivity. From the above, the definition of emissivity may be expressed formally as

$$\varepsilon = \frac{E}{E_b} \tag{8.16}$$

A comprehensive set of tables of emissivity can be found in Touloukian and DeWitt (1972). Some values are given in Table 8.1 as examples but the reader is urged to consult the extensive lists in the appropriate references. A range of values for some selected engineering materials is presented in Table A.9. It is important to note that ε depends on surface finish and temperature, and most tables give values of emissivity for various stated surface conditions and temperatures.

In most engineering problems, where radiation interchange occurs between surfaces at *comparable* absolute temperatures, it is usually adequate to assume that the emissivity is independent of temperature and therefore wavelength. Notable exceptions to this occur when we consider

Table 8.1 ● Typical values of total emissivity for some common materials

Surface	ε (range)[a]
Polished metals	0.02–0.2
Metals	0.1–0.7
Glass	0.75–0.95
Ceramics	0.6–0.9
Water	0.9–0.95
Wood	0.8–0.9
'Matt' paints	0.9–0.98

[a] Actual value depends on composition and temperature

thermal radiation from the sun, or from high-temperature sources such as gas flames, to surfaces at or around room temperature. For example, as shown in Figure 8.3, the peak emission in solar thermal radiation occurs at 5800 K at $\lambda \approx 0.5\,\mu m$, whereas peak emission at temperatures in the region 300–500 K are in the range 5–10 μm. Surface properties cannot be considered to remain constant over this comparatively wide range of wavelengths.

8.2.4 Kirchhoff's law of radiation

Consider, as illustrated in Figure 8.7, a number of small bodies in a large enclosure, in a thermal steady state and in thermal equilibrium with each other and the enclosure. Since the bodies are small, they have negligible effect on the radiative properties of the enclosure. The enclosure therefore acts as a black body, absorbing all incident radiation and emitting radiation diffusely and in accord with Equation (8.12). An energy balance may be written for the enclosure: the incident radiation is equal to the emitted radiation,

$$G = E_b \qquad (8.17)$$

For small body 1, with absorptivity α_1 and emissivity ε_1, in a steady state all the thermal radiation absorbed must be emitted:

$$\alpha_1 G = E_1 \qquad (8.18)$$

Using Equations (8.16) and (8.17) in Equation (8.18) gives

$$\alpha_1 = \frac{E_1}{G} = \frac{E_1}{E_b} = \varepsilon_1 \qquad (8.19)$$

Hence, for any small body in an enclosure, the total absorptivity and total emissivity are equal, $\alpha = \varepsilon$. The above arguments also apply to the mono-chromatic values, so $\alpha_\lambda = \varepsilon_\lambda$.

Fig. 8.7 ●
Radiative exchange between a number of small bodies in an enclosure

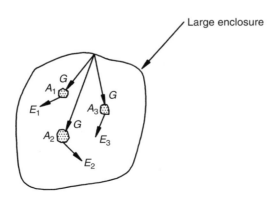

8.3 ● View factors

The view factor, F (alternatively referred to as the radiation configuration factor, or shape factor), expresses the fraction of radiation emitted by one object which is incident on another. It is therefore of use when analysing radiation exchange between bodies. Usually, F is given two subscripts, e.g. F_{ij}, F_{12}, F_{ab}, etc. The first denotes the surface that *emits* the radiation, the second denotes the surface that *receives* the radiation.

Consider the general geometry shown in Figure 8.8 and in particular the elemental areas dA_i and dA_j located on surfaces i and j, respectively. Let $d\dot{Q}_{ij}$ be the rate of heat transfer, in watts, from surface dA_i to dA_j. The components normal to these surfaces are also indicated. From Equations (8.3) and (8.4), the radiation leaving dA_i and falling on area dA_j is

$$d\dot{Q}_{ij} = (I_{n,i} \cos \theta_i)\, d\omega_j\, dA_i$$

Since $d\omega_j = (dA_j \cos \theta_j)/r^2$,

$$d\dot{Q}_{ij} = \frac{I_{n,i} \cos \theta_i \cos \theta_j\, dA_i\, dA_j}{r^2} \tag{8.20}$$

where $I_{n,i}$ is the total intensity of radiation (over all λ) leaving surface i. For a diffuse surface, where $I_{n,i}$ is independent of θ_i, Equation (8.20) may be integrated over the entire areas of the emitting and receiving surfaces A_i and A_j to give

$$\dot{Q}_{ij} = I_{n,i} \int_{A_j} \int_{A_j} \frac{\cos \theta_i \cos \theta_j\, dA_i\, dA_j}{r^2} \tag{8.21}$$

From Equation (8.7) the total radiation emitted from area A_i can be written as

$$\dot{Q}_i = \pi I_{n,i}\, dA_i \tag{8.22}$$

Fig. 8.8 ● View factor geometry

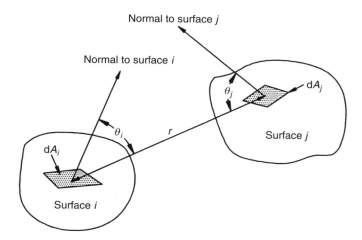

Normal to surface j

Normal to surface i

θ_j

dA_j

Surface j

θ_i

r

dA_i

Surface i

then F_{ij} (the ratio of radiation received by surface j to that emitted from surface i) is simply the ratio of Equations (8.21) and (8.22):

$$F_{ij} = \frac{1}{\pi A_i} \int_{A_j} \int_{A_i} \frac{\cos \theta_i \cos \theta_j \, dA_i \, dA_j}{r^2} \qquad (8.23a)$$

A similar expression can be derived for F_{ji}, the fraction received by area i which is emitted from surface j:

$$F_{ji} = \frac{1}{\pi A_j} \int_{A_i} \int_{A_j} \frac{\cos \theta_i \cos \theta_j \, dA_j \, dA_i}{r^2} \qquad (8.23b)$$

In practice, Equations (8.23) present a formidable task to evaluate analytically. Howell (1982) provides an extensive source of view factors for hundreds of geometries. Some of them are presented as closed-form analytical expressions, others are evaluated numerically and the results are depicted in graphical form. It is also worth noting that the integral in Equation (8.23) is carried out as part of a finite element analysis, and this could be used to calculate view factors for geometries where no tabulated information is available.

A selection, taken from Howell (1982), of view factor expressions for common two-dimensional geometries is presented in Table 8.2 and for some three-dimensional geometries in Table 8.3. The factors in Table 8.3 are also shown graphically in Figures 8.9, 8.10 and 8.11.

A useful relation can be derived from combining Equations (8.23a) and (8.23b):

$$A_i F_{ij} = A_j F_{ji} \qquad (8.24)$$

This is often referred to as the *reciprocity relation*. Another useful relationship is applied to an enclosure with N surfaces, and is referred to as the *summation rule*:

$$\sum_{j=1}^{N} F_{ij} = 1 \qquad (8.25)$$

Two other points are useful in calculating view factors:

1. For a flat or a convex surface, $F_{ii} = 0$; it does not see any part of itself.
2. For a concave surface, $F_{ii} > 0$, since it sees parts of itself.

The above relations constitute the rules for *view factor algebra* which allows us to manipulate geometric relations, such as area, and view factors to calculate all the pertinent view factors for a geometry.

Example 8.3

Obtain all the view factors for the three geometries shown in Figure 8.12:

(a) Two spheres or infinite concentric cylinders of surface areas A_1 and A_2 (Figure 8.12a).

Table 8.2 ● View factor expressions for some common two-dimensional geometries

Geometry		Expression

Parallel plates with centrelines connected by perpendicular

$$F_{ij} = \frac{\{(B+C)^2+4\}^{1/2} - \{(C-B)^2+4\}^{1/2}}{2B}$$

$$B = b/a$$
$$C = c/a$$

Two plates with a common edge at 90° to one another

$$F_{ij} = \frac{1+H-(1+H^2)^{1/2}}{2}$$

$$H = h/w$$

Two inclined plates of equal width with a common edge

$$F_{ij} = F_{ji} = 1 - \sin(\alpha/2)$$

Three-sided enclosure

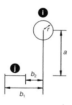

$$F_{ij} = \frac{a+b-c}{2a}$$

Infinite plane to a row of pipes

$$F_{ij} = 1 - (1-D^2)^{1/2} + D\tan^{-1}\{(1/D^2)-1\}^{1/2}$$
$$D = a/b$$

Cylinder to parallel rectangle

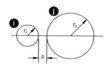

$$F_{ij} = \frac{1}{2\pi}(\tan^{-1}B_1 - \tan^{-1}B_2)$$
$$B_1 = b_1/a$$
$$B_2 = b_2/a$$

Parallel cylinders of differing radii

$$F_{ij} = \frac{1}{2\pi}[\pi + \{C^2-(R+1)^2\}^{1/2}$$
$$- \{C^2-(R-1)^2\}^{1/2}$$
$$+ (R-1)\cos^{-1}\{(R/C)-(1/C)\}$$
$$- (R+1)\cos^{-1}\{(R/C)+(1/C)\}]$$

$$R = r_2/r_1$$
$$S = s/r_1$$
$$C = 1+R+S$$

Table 8.3 ● View factor expressions for some common three-dimensional geometries

Geometry		Expression
Identical, parallel directly opposed rectangles		$F_{ij} = 2/(\pi XY)$ $\times \{\ln[(1+X^2)(1+Y^2)/(1+X^2+Y^2)^{1/2}$ $+ X(1+Y^2)^{1/2}\tan^{-1}[X/(1+Y^2)^{1/2}]$ $+ Y(1+X^2)^{1/2}\tan^{-1}[Y/(1+X^2)^{1/2}]$ $- X\tan^{-1}X - Y\tan^{-1}Y\}$ $X = a/c$ $Y = b/c$
Two finite rectangles having the same length, one common edge and at 90° to one another	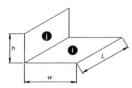	$F_{ij} = (1/\pi W)\{W\tan^{-1}(1/W)$ $+ H\tan^{-1}(1/H) - R\tan^{-1}(1/R)$ $+ 0.25\ln[\{(1+W^2)(1+H^2)/(1+R^2)\}$ $\times \{W^2(1+R^2)/R^2(1+W^2)\}^{W^2}$ $\times \{H^2(1+R^2)/R^2(1+H^2)\}^{H^2}]\}$ $H = h/L$ $W = w/L$ $R^2 = W^2 + H^2$
Two coaxial parallel discs	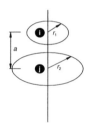	$F_{ij} = \frac{1}{2}[X - \{X^2 - 4(R_2/R_1)^2\}^{1/2}]$ $R_1 = r_1/a$ $R_2 = r_2/a$ $X = 1 + \{(1+R_2^2)/R_1^2\}$

(b) Small surface to a coaxial disc of diameter D, separated by an axial distance L (Figure 8.12b).

(c) Given that $F_{12} = 0.11$ (Figure 8.11), evaluate all the view factors for the conical geometry shown in Figure 8.12c.

Solution **(a)** $F_{11} + F_{12} = 1$ (summation rule)

$F_{11} = 0$ (convex surface)

and so

$F_{12} = 1$

$A_1 F_{12} = A_2 F_{21}$ (reciprocity rule)

$F_{21} = A_1/A_2$ (since $F_{12} = 1$)

$F_{22} = 1 - (A_1/A_2)$ (summation rule)

Fig. 8.9 ● View factors for identical, parallel directly opposed rectangles

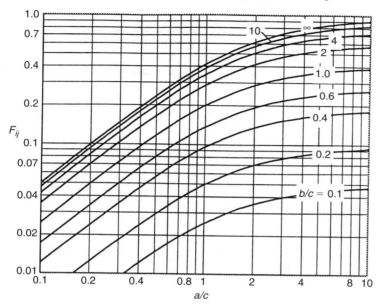

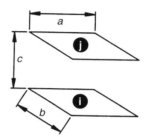

Fig. 8.10 ● View factors for two coaxial parallel discs

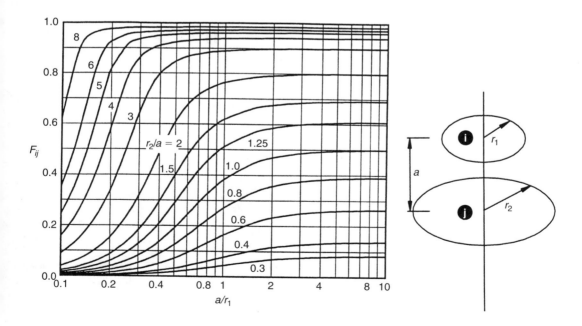

Fig. 8.11 ● View factors for two finite rectangles having the same length, one common edge and at 90° to one another

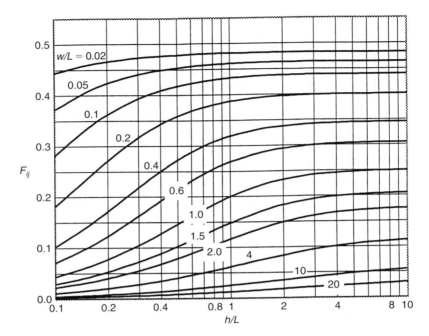

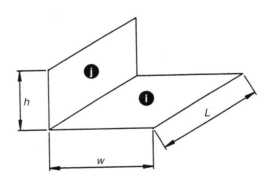

Fig. 8.12 ● Geometries for Example 8.3: (a) two concentric spheres, cylinders; (b) small surface to a coaxial disc; (c) view factor for a cone

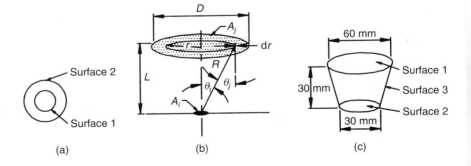

(b) From Equation (8.23a),

$$F_{ij} = \frac{1}{\pi A_j} \int_{A_j} \int_{A_i} \frac{\cos \theta_i \cos \theta_j \, dA_i \, dA_j}{R^2}$$

Now, since A_i is small, $dA_i = A_i$. It also follows that θ_i, θ_j and r can be considered independent of their position on A_i. This implies that we can let $\theta_i = \theta_j = \theta$, from which

$$F_{ij} = \frac{1}{\pi} \int_{A_j} \frac{\cos^2 \theta}{R^2} \, dA_j$$

The above equation may be transformed into radial coordinates, $R^2 = L^2 + r^2$, $\cos \theta = L/R$ and $dA_j = 2\pi r \, dr$, from which

$$F_{ij} = 2L^2 \int_0^{D/2} \frac{r \, dr}{(L^2 + r^2)^2}$$

$$= \frac{D^2}{D^2 + 4L^2}$$

(c) $A_1 = 900 \, \pi \, \text{mm}^2$; $A_2 = 225 \, \pi \, \text{mm}^2$; $A_3 = 1350 \, \pi \, \text{mm}^2$

$F_{12} = 0.11$	(given)
$F_{11} = 0$	(flat surface)
$F_{13} = 1 - 0.11 = 0.89$	(summation rule)
$F_{21} = F_{12} A_1 / A_2 = 0.44$	(reciprocity)
$F_{22} = 0$	(flat surface)
$F_{23} = 1 - 0.44 = 0.56$	(summation rule)
$F_{31} = F_{13} A_1 / A_3 = 0.593$	(reciprocity)
$F_{32} = F_{23} A_2 / A_3 = 0.0933$	(reciprocity)
$F_{13} = 1 - 0.593 - 0.0933 = 0.3133$	(summation rule)

Comment Case (a) shows the application of simple view factor algebra to obtain the view factors. In case (b) the complex integral is simplified as a consequence of one of the surfaces being very small in relation to the other. Case (c) illustrates a practical application, where all the view factors for a geometry are obtained using one view factor (either previously calculated or obtained from tables) and applying view factor algebra.

8.4 ● Radiation exchange between black bodies

In general, thermal radiation may leave a surface by both emission and reflection. When it reaches another surface the radiation may be absorbed or reflected, or both. This more complex case is treated in Section 8.5. A simplification which is sometimes adequate is to assume black body behaviour, in which case the only energy to leave a surface is by emission, since for a black body $\rho = 0$.

For radiation between two black bodies i and j and using Equation (8.12),

$$E_{b,i} = \sigma T_i^4 \quad (W/m^2)$$

The fraction of this incident on body j is simply, $F_{ij}E_{b,i}$. So the total radiative heat transfer transmitted from body i to body j is

$$\dot{Q}_{i-j} = \sigma F_{ij} A_i T_i^4$$

It follows that the total radiative transfer from j to i is

$$\dot{Q}_{j-i} = \sigma F_{ji} A_j T_j^4$$

If $T_i > T_j$ then the *net* radiative heat transfer from i to j is as follows (recall from the reciprocity rule (8.24) that $F_{ij}A_i = F_{ji}A_j$):

$$\dot{Q}_{ij} = \dot{Q}_{i-j} - \dot{Q}_{j-i} = \sigma F_{ij} A_i (T_i^4 - T_j^4) \tag{8.26}$$

Equation (8.26) demonstrates that the net rate at which radiation leaves surface i is equal to the net rate at which surface j receives radiation due to interaction with i. This may be generalised to analyse the net radiative heat transfer between N black surfaces at differing temperatures of an enclosure. (It is worthwhile to note here that in circumstances where there is not a complete enclosure, such as two opposing surfaces, an enclosure may be created by surrounding them with a third hypothetical surface.) The net radiation from surface i due to interactions with all the other surfaces is

$$\dot{Q}_i = \sum_{j=1}^{N} \sigma F_{ij} A_j (T_i^4 - T_j^4) \tag{8.27}$$

Example 8.4

In a boiler, heat is radiated from the burning fuel (surface 1) to the side walls (surface 3) and the boiler tubes (surface 2) at the top.

(a) Assuming black body behaviour and assuming the side walls are perfectly insulated, derive an expression for the temperature of the side walls T_3, as a function of the temperatures for the fuel bed T_1 and the boiler tubes T_2, and the corresponding areas A_1, A_2.

(b) If $A_1 = A_2 = 0.25\,m^2$, $F_{12} = F_{23} = F_{13} = 0.5$ and $T_1 = 1700°C$, $T_2 = 300°C$, what is the net radiative heat transfer to the boiler tubes?

Solution (a) For black body behaviour and insulated side walls, $\dot{Q}_3 = 0$ so, using Equation (8.27) for $i = 3$, $j = 1,2,3$, results in

$$A_3\sigma F_{31}(T_3^4 - T_1^4) + A_3\sigma F_{32}(T_3^4 - T_2^4) + A_3\sigma F_{33}(T_3^4 - \overset{0}{\cancel{T_3^4}}) = 0$$

From Equation (8.24), $A_3 F_{31} = A_1 F_{13}$ and $A_3 F_{32} = A_2 F_{23}$, so

$$T_3 = \left\{ \frac{A_1 F_{13} T_1^4 + A_2 F_{23} T_2^4}{A_1 F_{13} + A_2 F_{23}} \right\}^{1/4}$$

(b) The net radiative heat transfer to the boiler tubes, $\dot{Q}_2$, is obtained from Equation (8.27):

$$\dot{Q}_2 = A_2 \sigma F_{21}(T_2^4 - T_1^4) + A_2 \sigma F_{23}(T_2^4 - T_3^4)$$

Substituting the result derived above for T_3 and rearranging gives

$$\dot{Q}_2 = \sigma(T_2^4 - T_1^4) \left\{ A_2 F_{21} + \frac{(A_2 F_{23} A_1 F_{13})}{(A_1 F_{13} + A_2 F_{23})} \right\}$$

Inserting the appropriate values gives

$$\dot{Q}_2 = 56.7 \times 10^{-9}(573^4 - 1973^4)$$

$$\times \left\{ 0.25 \times 0.5 + \frac{(0.25 \times 0.5 \times 0.25 \times 0.5)}{(0.25 \times 0.5 + 0.25 \times 0.5)} \right\}$$

$$= -160 \, \text{kW}$$

Comment

The negative sign indicates a net heat flow into the boiler tubes. Without side walls, the heat flow to the tubes would be $\sigma A_2 F_{21}(T_2^4 - T_1^4)$. The presence of adiabatic side walls therefore enhances the total heat flow by the amount $[A_2 F_{23} A_1 F_{13}/(A_1 F_{13} + A_2 F_{23})]$. Such a calculation would be used in conjunction with the appropriate convection correlations for the heat transfer in the tubes to calculate the water (or steam) exit conditions. This would most probably form part of an iterative sequence.

8.5 ● Radiation exchange between grey bodies

The simplifying assumption of black body behaviour, although convenient, cannot be used in many radiation problems of engineering concern. In general, surfaces have an emissivity significantly less than unity. Reflections occur at the surface and the reflected components are, in turn, incident upon another surface, from that surface to another and so on.

8.5.1 Grey body enclosures

To analyse the general problem of radiative interchange between grey surfaces in an enclosure, we will make the following assumptions:

1. Each surface of the enclosure is at a constant temperature.
2. Each surface has uniform radiosity and irradiation; transmissivity $\tau = 0$, and the surface characteristics are diffuse.

The question posed is either to predict the radiant heat fluxes from knowing the surface properties and temperatures, or to calculate temperatures from (implicitly or explicitly) knowing the radiant and convective heat fluxes.

Recall (see Figure 8.5) that the net rate at which radiation leaves a surface, say surface i, denoted by $\dot{Q}_i$, is equal to the difference between the surface radiosity and irradiation. If this is zero then energy needs to be transferred to the surface by other means to maintain it at a constant temperature.

$$\dot{Q}_i = A_i(J_i - G_i) \tag{8.28}$$

For a solid, the transmissivity $\tau = 0$, and from Equation (8.15)

$$J_i = E_i + \rho_i G_i \tag{8.29}$$

Substitution of Equation (8.29) into (8.28) gives

$$\dot{Q}_i = A_i(E_i - \alpha_i G_i) \tag{8.30}$$

Using the definition of emissivity (8.16) and Kirchhoff's law (8.19), Equation (8.29) becomes

$$J_i = \varepsilon_i E_{b,i} + (1 - \varepsilon_i)G_i \tag{8.31}$$

Substituting for G_i in Equation (8.28) and arranging into a more convenient form gives the result

$$q_i = \frac{\dot{Q}_i}{A_i} = \frac{E_{b,i} - J_i}{(1 - \varepsilon_i)/\varepsilon_i} \tag{8.32}$$

To use this equation we must determine the radiosity, J. This can be done by considering exchange between the surfaces that form the enclosure. Sometimes the geometry will not be a full enclosure. There may be open ends which allow radiation to the ambient surroundings, for example a box with an open top. It is possible to form a hypothetical enclosure by considering these openings as extra surfaces – acting as a black body at the temperature of the surrounds.

From the definition of the view factor, the total rate at which radiation reaches surface i from all other surfaces, including i, is

$$A_i G_i = \sum_{j=1}^{N} F_{ji} A_j J_j$$

and using the reciprocity relationship (8.24)

$$A_i G_i = \sum_{j=1}^{N} F_{ij} A_i J_j \tag{8.33}$$

Substitution of this into Equation (8.28) for G_i gives

$$\dot{Q}_i = A_i \left\{ \sum_{j=1}^{N} F_{ij}(J_i - J_j) \right\} \tag{8.34}$$

and combining this with Equation (8.32) gives the following result:

$$\sum_{j=1}^{N} A_i F_{ij}(J_i - J_j) = \frac{A_i(E_{b,i} - J_i)}{(1 - \varepsilon_i)/\varepsilon_i} = \dot{Q}_i \tag{8.35}$$

To use Equation (8.35) we write this equation for each surface in the enclosure. The result is a set of N simultaneous equations, with $J_1, J_2, \ldots,$ J_N the unknowns, which for a simple system (say $N = 3$) may be solved by hand; for larger systems computer solutions using matrix manipulation are recommended. This procedure is best demonstrated by use of an example.

Example 8.5

High-temperature gas flows through the inside of a pipe of outer radius $r_2 = 30\,\text{mm}$. To reduce the thermal radiation emitted to an electrical control panel mounted nearby, a semicircular radiation shield of radius $r_1 = 100\,\text{mm}$ is placed concentrically around the pipe. Thermal radiation emitted from the pipe is radiated to both the shield and the surroundings which are at 310 K. The radiation view factor for the shield to itself is $F_{11} = 0.3345$.

(a) Deduce the remaining view factors.

(b) Consider a radiation balance on the system. The pipe temperature is $T_2 = 900\,\text{K}$ and the shield temperature is T_1; the emissivity of the inner surface of the shield is $\varepsilon_1 = 0.8$ and the emissivity of the outer surface of the pipe is $\varepsilon_2 = 0.5$. Assuming the surroundings can be approximated by a black body at 310 K, show that for surfaces 1 and 2:

$$J_1 = 4.861 \times 10^{-8} T_1^4 + 0.032\,15 J_2 + 57.9 \quad (\text{W/m}^2)$$
$$J_2 = 18\,731 + 0.25 J_1 \quad (\text{W/m}^2)$$

(c) Neglecting any heat loss by convection, estimate the surface temperature of the shield when the emissivity of the outer surface of the radiation shield is $\varepsilon_o = 0.1$.

(d) Recalculate part (c) but with $\varepsilon_0 = 0.8$ and $\varepsilon_1 = 0.1$.

(e) Recalculate part (c) but assuming an additional loss by convection with $h = 10\,\text{W/m}^2\text{K}$.

Solution Figure 8.13 shows a cross-section through the pipe and shield. A third, imaginary surface, surface 3, is also drawn to form the enclosure. The pipe and shield are assumed to be long in relation to their diameters, hence it is reasonable to neglect the radiation from the open ends.

Fig. 8.13 ●
Radiation shield
(Example 8.5)

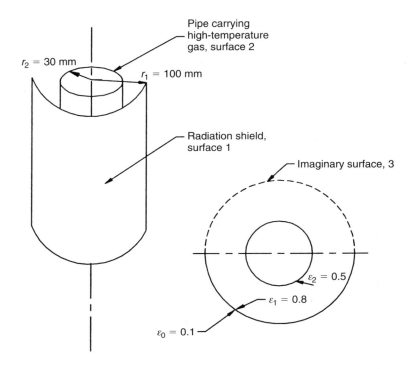

$r_2 = 30$ mm

Pipe carrying
high-temperature
gas, surface 2

$r_1 = 100$ mm

Radiation shield,
surface 1

Imaginary surface, 3

$\varepsilon_2 = 0.5$

$\varepsilon_1 = 0.8$

$\varepsilon_0 = 0.1$

(a) So $F_{21} + F_{22} + F_{23} = 1$ but $F_{22} = 0$ by symmetry $F_{21} = F_{23} = 0.5$, so

$$F_{11} = 0.3345 \text{ (given)}$$
$$F_{12} = F_{21} A_2 / A_1 = 0.15$$
$$F_{13} = 1 - F_{12} - F_{11} = 0.5155$$

$$F_{32} = F_{23} A_2 / A_3 = 0.15$$
$$F_{33} = F_{11} = 0.3345 \text{ (symmetry)}$$
$$F_{31} = 1 - F_{32} - F_{33} = 0.5155$$

(b) Using Equation (8.35),

$$\sum_{j=1}^{N} F_{ij}(J_i - J_j) = \frac{E_{b,i} - J_i}{(1 - \varepsilon_i)/\varepsilon_i}$$

Applying this to surface 1 (that is, with $i = 1$ and $j = 1,2,3$),

$$\frac{E_{b,1} - J_1}{(1 - \varepsilon_1)/\varepsilon_1} = (J_1 - J_1)^0 F_{11} + (J_1 - J_2)F_{12} + (J_1 - J_3)F_{13}$$

Note that $(1 - \varepsilon_1)/\varepsilon_1 = 0.25$, also $E_{b,1} = \sigma T_1^4$ and, if the surroundings can be approximated as a black body, then $J_3 = \sigma T_3^4$, from which

$$J_1 = \frac{\sigma T_1^4 + 0.25\,F_{12}\,J_2 + 0.25 F_{13}\,\sigma T_3^4}{1 + 0.25 F_{12} + 0.25 F_{13}}$$

$$= \{(56.7 \times 10^{-9}\,T_1^4) + (0.25 \times 0.15\,J_2)$$

$$(0.25 \times 0.5155 \times 56.7 \times 10^{-9} \times 310^4)\}/1.1664$$

$$= (4.861 \times 10^{-8}\,T_1^4) + 0.03215\,J_2 + 57.9 \qquad (\mathrm{W/m^2})$$

Similarly, application of Equation (8.35) to surface 2 (that is, with $i = 2$ and $j = 1,2,3$) gives

$$\frac{E_{b,2} - J_2}{(1 - \varepsilon_2)/\varepsilon_2} = (J_2 - J_1)F_{21} + (J_2 - J_3)F_{23}$$

In which

$$(1 - \varepsilon_2)/\varepsilon_2 = 1$$

$$E_{b,2} = \sigma T_2^4$$

$$J_3 = \sigma T_3^4$$

and so

$$J_2 = \frac{\sigma T_2^4 + J_1 F_{21} + \sigma T_3^4 F_{23}}{1 + F_{21} + F_{23}}$$

and with $T_2 = 900\,\mathrm{K}$

$$J_2 = \frac{56.7 \times 10^{-9} \times 900^4 + 0.5\,J_1 + 0.5 \times 56.7 \times 10^{-9} \times 310^4}{2}$$

$$= 18\,731 + 0.25\,J_1 \qquad (\mathrm{W/m^2})$$

(c) Combining the above equations to eliminate J_2, gives

$$J_1 = 4.9 \times 10^{-8}\,T_1^4 + 665 \qquad (\mathrm{W/m^2})$$

For the inside of the radiation shield ($i = 1$),

$$q_1 = \frac{E_{b,1} - J_1}{(1 - \varepsilon_1)/\varepsilon_1}$$

and for the outside, the external radiative heat flux q_{ext} is related to that on the inside q_1 by

$$q_{ext} = -q_1 = \varepsilon_o \sigma (T_1^4 - T_3^4)$$

Hence combining the two expressions for q_1 gives

$$\frac{E_{b,1} - J_1}{(1 - \varepsilon_1)/\varepsilon_1} = -\varepsilon_0 \sigma (T_1^4 - T_3^4)$$

And substituting $J_1 = 4.9 \times 10^{-8} T_1^4 + 665 \text{ W/m}^2$,

$$T_1^4 = \frac{0.025 \sigma T_3^4 + 665}{\sigma - 49 \times 10^{-9} + 0.025 \sigma}$$

$$T_1 = 522 \text{ K}$$

(d) With $\varepsilon_0 = 0.8$ and $\varepsilon_1 = 0.1$, applying Equation (8.35) with $\varepsilon_1 = 0.1$ gives the following result for surface 1:

$$J_1 = 8.112 \times 10^{-9} T_1^4 + 0.193 J_2 + 347.6 \quad (\text{W/m}^2)$$

and for surface 2 (the same result as before since the value of ε_1 does not feature in the equation):

$$J_2 = 18\,731 + 0.25 J_1 \quad (\text{W/m}^2)$$

Continuing as before, combining the above two equations, we get

$$J_1 = 8.534 \times 10^{-9} T_1^4 + 4163.5 \quad (\text{W/m}^2)$$

and from an energy balance on the outside surface

$$\frac{E_{b,1} - J_1}{(1 - \varepsilon_1)/\varepsilon_1} = -\varepsilon_0 \sigma (T_1^4 - T_3^4)$$

Then substituting $J_1 = 8.534 \times 10^{-9} T_1^4 + 4163.5 \quad \text{W/m}^2$ and $E_{b,1} = \sigma T_1^4$,

$$T_1 = 465 \text{ K}$$

(e) With a convective loss on the outside surface,

$$q_{\text{ext}} = -q_1 = \varepsilon_0 \sigma (T_1^4 - T_3^4) + h(T_1 - T_3)$$

and so, continuing in a similar way to above,

$$\frac{E_{b,1} - J_1}{(1 - \varepsilon_1)/\varepsilon_1} = -\varepsilon_0 \sigma (T_1^4 - T_3^4) - h(T_1 - T_3)$$

which, with $h = 10 \text{ W/m}^2 \text{ K}$, $T_3 = 310 \text{ K}$, $\varepsilon_1 = 0.8$, $\varepsilon_0 = 0.1$, gives

$$9.1175 \times 10^{-9} T_1^4 + 2.5 T_1 - 1453 = 0$$

which is satisfied for

$$T_1 = 442 \text{ K}$$

Comment   The answer does not depend on the geometric configuration of the hypothetical surface, since the view factors F_{31}, F_{32} and F_{33} do not appear in the above equations. Clearly, reversing the radiation shield so that the side with the low emissivity faces the hot surface and that with the larger emissivity faces the outside, results in a cooler surface temperature. The physical explanation for this is that a small emissivity (and hence absorptivity) reduces the amount of radiation absorbed, the large emissivity on the outside increases the radiation emitted. Convection is also seen to make a substantial difference to the surface temperature.

8.5.2   Radiative exchange between two grey bodies

The radiative heat transfer between *two* grey bodies is best considered as a special case. As with the analysis of radiative exchange in enclosures, we will assume that each body (or surface) is at a constant temperature, has uniform radiosity and irradiation, $\tau = 0$, and behaves as a diffuse reflector. There are three separate cases that are of practical significance.

Case 1: Interchange between two small grey bodies
For two small bodies (small in relation to their distance apart) with emissivities ε_1 and ε_2, a negligible proportion of the reflected radiation will return to the emitting body.
The energy emitted by body 1 is

$$A_1 \varepsilon_1 \sigma T_1^4$$

The fraction of this incident on body 2 is $F_{12}(A_1 \varepsilon_1 \sigma T_1^4)$. The amount absorbed by body 2 is

$$\dot{Q}_{1-2} = \alpha_2 (F_{12} A_1 \varepsilon_1 \sigma T_1^4)$$
$$= \varepsilon_1 \varepsilon_2 F_{12} A_1 \sigma T_1^4 \quad (\text{since } \alpha = \varepsilon)$$

It therefore follows that

$$\dot{Q}_{2-1} = \varepsilon_2 \varepsilon_1 F_{21} A_2 \sigma T_2^4$$

And since $A_1 F_{12} = A_2 F_{21}$, the net radiative heat transfer, $\dot{Q}$, is given by

$$\dot{Q} = \dot{Q}_{1-2} - \dot{Q}_{2-1}$$
$$= \varepsilon_1 \varepsilon_2 F_{12} A_1 \sigma (T_1^4 - T_2^4) \tag{8.36}$$

Case 2: Interchange between two large parallel grey plates
In this case the plates are large in relation to the distance between them, so the view factor can be assumed to be unity. We consider radiation leaving plate 1 and its subsequent attenuation by absorption and reflection (since the areas are equal, we need consider only the fluxes).

The radiative heat flux emitted by plate 1 is

$$q_1 = \varepsilon_1 \sigma T_1^4$$

The amount absorbed by surface 2 on the first incidence is

$$\alpha_2 \left(\varepsilon_1 \sigma T_1^4\right) = \varepsilon_2 \varepsilon_1 \sigma T_1^4$$

The remaining flux reflected from surface 2 is $\rho_2(\varepsilon_1 \sigma T_1^4)$. The amount of this absorbed by surface 1 is

$$\alpha_1 \left(\rho_2 \varepsilon_1 \sigma T_1^4\right) = \rho_2 \varepsilon_1^2 \sigma T_1^4$$

The remaining heat flux reflected from surface 1 is $\rho_1(\rho_2\varepsilon_1\sigma T_1^4)$. The amount absorbed by surface 2 on the second incidence is

$$\alpha_2 \left(\rho_1 \rho_2 \varepsilon_1 \sigma T_1^4\right) = \rho_1 \rho_2 \varepsilon_1 \varepsilon_2 \sigma T_1^4$$

Continuing this argument we can see that on the $(n + 1)$th incidence, the radiative heat flux absorbed by surface 2 is $(\rho_1\rho_2)^n \varepsilon_1 \varepsilon_2 \sigma T_1^4$. Similar reasoning can be applied to the radiative heat flux from surface 2. Hence the net exchange between these two surfaces is

$$\dot{Q} = \dot{Q}_{1-2} - \dot{Q}_{2-1}$$
$$= \varepsilon_1 \varepsilon_2 A \sigma (T_1^4 - T_2^4)\{1 + \rho_1\rho_2 + (\rho_1\rho_2)^2 + \ldots\} \tag{8.37}$$

The reflectivities ρ_1 and ρ_2 are each less than unity and, using the binomial theorem, the series in the curly brackets is equivalent to the expansion of $(1 - \rho_1\rho_2)^{-1}$. Equation (8.37) can then be written as

$$\dot{Q} = \frac{\varepsilon_1 \varepsilon_2 A \sigma (T_1^4 - T_2^4)}{1 - \rho_1\rho_2}$$
$$= \frac{\varepsilon_1 \varepsilon_2 A \sigma (T_1^4 - T_2^4)}{1 - (1 - \varepsilon_1)(1 - \varepsilon_2)} \tag{8.38}$$
$$= \frac{A\sigma (T_1^4 - T_2^4)}{(1/\varepsilon_1) + (1/\varepsilon_2) - 1}$$

Case 3: Interchange between a small grey body and a grey enclosure
Although the enclosure has an emissivity of less than unity, it acts as a black body (see Section 8.2.4).

The radiative energy emitted from the small body, surface 1, is

$$\dot{Q}_1 = \sigma A_1 \varepsilon_1 T_1^4$$

And since it acts as a black body, this will be entirely absorbed by the enclosure, so

$$\dot{Q}_{1-2} = \dot{Q}_1 = \sigma A_1 \varepsilon_1 T_1^4$$

The emission from the enclosure, surface 2, is

$$E_2 = \sigma A_2 \varepsilon_2 T_2^4$$

The fraction of this incident on surface 1 is

$$\dot{Q}_2 = F_{21}(\sigma A_2 \varepsilon_2 T_2^4)$$

Of which the amount absorbed by surface 1 is

$$\dot{Q}_{2-1} = \alpha_1(F_{21}\sigma A_2 \varepsilon_2 T_2^4)$$
$$= \sigma F_{21} A_2 \varepsilon_1 \varepsilon_2 T_2^4 \qquad (\text{since } \alpha_1 = \varepsilon_1)$$

The net radiative transfer between the enclosure and the small body, $\dot{Q}$, is the difference $\dot{Q}_{1-2} - \dot{Q}_{2-1}$, Now, as

$$\dot{Q} = \sigma A_1 \varepsilon_1 T_1^4 - \sigma F_{21} A_2 \varepsilon_1 \varepsilon_2 T_2^4$$

then in order for there to be zero net transfer when the temperatures T_1 and T_2 are equal implies that $A_1 = F_{21}A_2\varepsilon_2$, and hence

$$\dot{Q} = \sigma A_1 \varepsilon_1 (T_1^4 - T_2^4) \tag{8.39}$$

8.6 • Environmental radiation

8.6.1 General principles

Thermal radiation from the sun and other high-temperature sources such as hydrocarbon flames, requires special attention. A surface subjected to heating from solar energy or a gas flame 'sees' the emission at this elevated temperature, but emits at a much lower temperature, so Kirchoff's law ($\alpha = \varepsilon$) does not apply.

Radiation of thermal energy from the sun is necessary for life on earth. The sun provides energy for photosynthesis, and we obtain our food and most fuels either directly or indirectly from plants. Even in the United Kingdom, solar radiation (from 7 kWh/m² for December to 172 kWh/m² for June) can provide significant amounts of energy for space heating and domestic water. In countries with greater average solar radiation, solar power is used for desalination, where for 1 m² of absorber surface, subjected to 500 W/m² of solar irradiation it is possible to obtain 0.25 litres/hour of freshwater. Solar energy is also used as a source of energy for space vehicles, either to produce electricity from photovoltaic cells or as a source of heat.

The sun is approximately 1.39×10^9 m in diameter and 1.5×10^{11} m from the earth. As shown in Figure 8.14, absorption and scattering phenomena attenuate the extraterrestrial radiation incident upon the upper edge of the atmosphere. The extraterrestrial solar irradiation $G_{s,o}$ depends on the angle θ between the incident rays from the sun and a normal to the surface of the earth (the complement of the angle of the sun above the horizon), so

$$G_{s,o} = fS_c \cos\theta \tag{8.40}$$

where f is a correction factor to account for the eccentricity of the earth's orbit around the sun ($0.97 \leqslant f \leqslant 1.03$). S_c is the solar constant which can

Fig. 8.14 ● Solar
radiation and its
attenuation in
earth's atmosphere

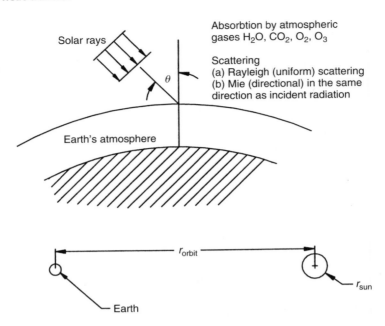

Solar rays

Absorbtion by atmospheric
gases H_2O, CO_2, O_2, O_3

Scattering
(a) Rayleigh (uniform) scattering
(b) Mie (directional) in the same
direction as incident radiation

θ

Earth's atmosphere

r_{orbit}

r_{sun}

Earth

be calculated from $E_{sun} = \sigma T_{sun}^4$, and since the intensity obeys the inverse square law

$$S_c = \left(\frac{r_{sun}}{r_{orbit}}\right)^2 E_{sun}$$

$$= \left(\frac{0.695 \times 10^9}{1.5 \times 10^{11}}\right)^2 \times 56.7 \times 10^{-9} \times 5800^4$$

$$= 1353 \text{ W/m}^2$$

As solar radiation passes through the atmosphere it is absorbed by the various atmospheric gases and also scattered. The actual radiation reaching the surface is referred to as G_s, the *solar irradiation*. This is therefore a combination of the direct radiation from the sun in addition to some of the scattered component. In the discussion here it is assumed that solar irradiation occurs from an omnidirectional, diffuse source. This assumption will be valid on overcast days but not in daytime under a clear sky.

As well as radiation from the sun, environmental radiation occurs from the surface of the earth and there is emission from certain atmospheric gases. The earth's emissivity is close to unity; water, for example, has $\varepsilon = 0.97$. Temperatures on the earth's surface are generally in the range $250 \leqslant T \leqslant 320$ K. Atmospheric emission occurs from carbon dioxide and water molecules, and although the physical mechanisms are complex, it may be conveniently expressed in the form

$$G_{atm} = \sigma T_{sky}^4 \tag{8.41}$$

where T_{sky} is the effective sky temperature. The value of T_{sky} depends on the atmospheric conditions and can range from 230 K for a cold clear sky to 285 K when it is warm and there is full cloud cover. At night, atmospheric emission is the only significant source of irradiation.

Example 8.6

In the desert at night the air temperature is 20°C and the effective sky temperature is −40°C. If the convective heat transfer coefficient from a shallow pool of water is 5 W/m²K, is this water likely to freeze?

Solution Let the temperature of the water be T_w, the ambient air temperature T_a and the effective sky temperature T_{sky}. At night $G_s = 0$, so from a simple heat balance,

$$h(T_w - T_a) = \sigma(T_{sky}^4 - T_w^4)$$

from which

$$hT_w + \sigma T_w^4 = 1632.1$$

which can be solved using Newton–Raphson:

$$T_w^{(n+1)} = T_w^{(n)} - \frac{f(T_w)}{f'(T_w)}$$

where $f = hT_w + \sigma T_w^4 - 1632.1$ and the derivative $f' = h + 4\sigma T_w^3$. Guess:

$$T_w^{(0)} = 0°C \ (273 \ K)$$

$$T_w^{(1)} = 0 - \frac{47.84}{9.61} = -4.9°C$$

$$T_w^{(2)} = -4.9 - \frac{0.39}{9.36} = -5.0°C$$

Comment It is therefore likely to freeze, due to the low value of T_{sky}.

An important consideration when dealing with solar radiation to and from surfaces at temperatures significantly different from the source of radiation is that $\varepsilon \neq \alpha$; (see Kirchhoff's law (8.19). Consider a surface absorbing radiation from the sun which, as we know, approximates to a black body at 5800 K. The surface also emits radiant energy but at a much lower temperature. This difference in temperatures is large enough for there to be a significant difference in the respective wavelengths of the absorbed and emitted radiation: $\lambda_{absorbed} \ll \lambda_{emitted}$ and so $\varepsilon \neq \alpha$. This may be summarised by the rule that α 'sees' T_{source}, but ε 'sees' T_{sink}. The absorptivity of a surface designed to receive radiation from the sun is usually referred to as the solar absorptivity α_s.

8.6.2 Solar heating panels

Figure 8.15 shows a schematic diagram of a simple flat-plate solar collector panel. Incident radiation from the sun and sky heats the absorber surface which backs onto a good insulator. Cool water enters the collector from the top, flows over the absorber surface and leaves at the bottom. Losses from the collector occur from the cover plate through convection to the surrounding air and by radiation. The efficiency of the collector, η, is defined in the normal sense of (useful heat out)/(solar irradiation onto the collector).

To analyse the performance of such a collector, consider an energy balance which for unit area becomes

$$q_{rad,in} + q_{rad,sky} = q_{useful} + q_{rad,out} + q_{conv} \tag{8.42}$$

where $q_{rad,in} = \alpha_s G_s$

$$q_{rad,sky} = \alpha_{sky} G_{sky} = \alpha_{sky} \sigma T_{sky}^4$$

$$q_{rad,out} = \varepsilon \sigma T_s^4$$

$$q_{conv} = h(T_s - T_a)$$

$$q_{useful} = \eta G_s$$

T_s is the surface temperature of the cover plate and T_a is the temperature of the surrounding air. The solar irradiation, G_s, is usually known (8.40), as is the effective sky temperature, T_{sky}. The heat transfer coefficient, h, can be predicted using correlations for either free or forced convection as appropriate. The solar absorptivity and emissivity are usually obtained from data on the absorber surface, and a good design will have a large value ($\geqslant 1$) of the ratio α_s/ε. Conversely, if it is required to reject heat then a good design will have small values ($\ll 1$) of this ratio. Since radiation from the sky occurs at roughly the same temperature as emission from the surface, it is possible to put $\alpha_{sky} = \varepsilon$.

Fig. 8.15 ● A simple solar collector panel

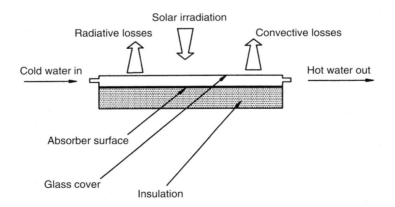

Solar irradiation

Radiative losses   Convective losses

Cold water in   Hot water out

Absorber surface

Glass cover

Insulation

Example 8.7

A solar panel comprises a square flat plate with sides $L = 0.7\,$m in length. Its surface has a solar absorptivity of $\alpha_s = 0.9$ and an emissivity of $\varepsilon = 0.2$. The panel operates in still air at 1 bar and 25°C ($\mu = 1.85 \times 10^{-5}\,$kg/m s and $k = 0.026\,$W/m K) and in still surrounding air the convective heat loss from the surface is given by $Nu_{av,L} = 0.5\,Gr_L^{1/4}$.

At a particular time of day, the solar irradiation is $0.5\,$kW/m² and the effective sky temperature is -3°C. If the efficiency of the panel is 50%, calculate the surface temperature of the cover plate.

Solution From an energy balance,

$$q_{rad,in} + q_{rad,sky} - q_{rad,out} - q_{conv} = q_{useful}$$

$$\alpha_s G_s + \alpha_{sky}\sigma T_{sky}^4 - \varepsilon\sigma T_s^4 - h(T_s - T_a) = \eta G_s$$

where T_s is the surface temperature of the cover plate.

The heat transfer coefficient is evaluated from the correlation $Nu_{av,L} = 0.5 Gr_L^{1/4}$:

$$h = \frac{0.5k}{L}\left\{\frac{g\beta(T_s - T_a)L^3\rho^2}{\mu^2}\right\}^{1/4}$$

$$p = \frac{\rho}{RT} = \frac{1 \times 10^5}{287 \times 298} = 1.17\,\text{kg/m}^3$$

$$h = 0.5 \times \left(\frac{0.026}{0.7}\right)\left\{\frac{9.81 \times (1/298) \times 0.7^3 \times 1.17^2}{(1.85 \times 10^{-5})^2}\right\}^{1/4}(T_s - T_a)^{1/4}$$

$$= 1.522\,(T_s - T_a)^{1/4}$$

Since $T_{sky} \approx T_s$ then $\alpha_{sky} = \varepsilon = 0.2$. Substituting appropriate values into the energy balance gives

$$\{0.9 \times 500\} + \{0.2 \times 56.7 \times 10^{-9} \times (270^4 - T_s^4)\} - 1.522(T_s - T_a)^{5/4}$$
$$= 0.5 \times 500$$

from which

$$260.3 - 1.134 \times 10^{-8}\,T_s^4 - 1.522\,(T_s - T_a)^{5/4} = 0$$

Solving by iteration, with $T_a = 25$°C, gives

$$T_s = 330\,\text{K} \quad (57°\text{C})$$

Comment The simple design of solar collector analysed above has its limitations: the cover plate has negligible thermal resistance and consequently the overall effectiveness of converting solar radiation into useful heat is low, owing to the convective loss from the outer surface. If the wind is blowing, this loss can be very large (and would depend on the Reynolds number not the

Grashof number). A more advanced design would use a double-glazed cover plate to reduce this convective loss: there is still a loss but it is now controlled by the free convective effects in the air gap inside the cover plate. To estimate this loss the designer needs to know about heat transfer in inclined air spaces (see Chapter 6). If the Grashof number can be reduced in this air space then so can the parasitic heat loss. An even more advanced design would reduce the Grashof number by reducing the pressure in the air space.

8.7 ● Closing comments

Radiative or infrared heat transfer occurs as a result of absolute temperature: anything with a temperature above absolute zero emits infrared radiation. This distinguishes thermal radiation as an electromagnetic phenomenon, and takes place in the region of wavelengths $0.1 < \lambda < 100\,\mu m$. Therefore, and unlike conduction and convection, thermal radiation can be propagated in the absence of any medium.

In addition to emission due to an object's absolute temperature, thermal radiation can be reflected, transmitted and absorbed. The physical properties known as emissivity, reflectivity, transmissivity and absorptivity determine the magnitude of these components. Values of these physical properties are easily obtained from tabulated data. The net radiation leaving a surface, termed the radiosity, therefore comprises separate components due to emission and reflection.

The idealised concept of a black body defines a perfect radiator and absorber. Real surfaces, which are termed grey surfaces, emit and absorb less radiation than a black body. The radiation view factor determines the fraction of radiation leaving one surface that is incident upon another. The general mathematical definition of the view factor is a complex double integration. In a few limiting cases this can be greatly simplified to provide simple analytical expressions. For other, more complex cases the reader is referred to the catalogue of radiation configuration factors by Howell (1982). View factor algebra is used to simplify the task of calculating view factors in a particular geometry.

A generalised methodology has been presented to deal with radiative exchange between grey surfaces in an enclosure. Most geometric configurations involving three or more separate surfaces can be made into an enclosure by the addition of one or more imaginary surfaces that behave as black bodies at the surrounding ambient conditions. For the interchange between two grey surfaces it is easier to use the formulae derived for the various limiting special cases.

A net radiative heat transfer is always present whenever there is a temperature difference. Consequently the first stage of any heat transfer analysis should be to establish its significance in relation to, for example, convection. In general, radiative heat transfer is likely to be significant in most free convection processes and in forced convection at high temperatures (for example, combustion, flames and film boiling).

8.8 ● References

Beiser, A. (1969). *Perspectives of Modern Physics*. Kogakusha, Japan: McGraw-Hill

Howell, J.R. (1982). *A Catalog of Radiation Configuration Factors*. New York: McGraw-Hill

Siegel, R. and Howell, J.R. (1992). *Thermal Radiation Heat Transfer* 3rd edn. Washington, DC: Hemisphere

Touloukian, Y.S. and DeWitt, D.P. (1972). *Thermophysical Properties of Matter*. Vol. 8: *Thermal Radiative Properties of Nonmetallic Solids* and Vol. 9: *Coatings*. New York: IFI/Plenum

8.9 ● End of chapter questions

8.1 Two large parallel plates are spaced a small distance apart and maintained at respective temperatures of 1000°C and 500°C.

(a) Assuming that the plates behave as black bodies, what is the net radiative heat flux from the hotter to colder plate?

[129 kW/m²]

(b) What is the value of this heat flux if the temperature of the cooler plate is increased to 700°C?

[98 kW/m²]

8.2 For radiative heat transfer between N black surfaces of an enclosure, show that the net radiation to or from surface i due to interactions with all the other surfaces is

$$\dot{Q}_{ij} = \sum_{j=1}^{n} \sigma F_{ij} A_i (T_i^4 - T_j^4)$$

8.3 Is a 'radiator' an appropriate name for a domestic central heating radiator? A radiator can be assumed to be a thin vertical plate positioned in a comparatively large room. The radiator is 0.5 m high and 2 m in length, and heat is emitted by free convection and radiation from both sides of the radiator. Assume that the radiator acts as a black body and the convective heat transfer coefficient from the radiator surface is given by

$$h = 1.6\Delta T^{1/3} \text{ W/m}^2\text{K}$$

(a) The surface temperature of the radiator is 60°C and the air temperature inside the room 20°C. What is the total heat transfer from the radiator; and the respective contributions from radiant and convective heat transfer?

[997 W, 44% convective and 56% radiative]

(b) What are the total heat transfer and respective contributions if the surface temperature is increased to 80°C?

[1677 W, 45% convective and 55% radiative]

8.4 In a bolier, heat is radiated from the burning fuel bed to the side walls and the boiler tubes at the top. The temperatures of the fuel and the tubes are T_1 and T_2 respectively and their areas are A_1 and A_2.

(a) Assuming that the side walls (denoted by the subscript 3) are perfectly insulated, show that the temperature of the side walls is given by

$$T_3 = \left(\frac{A_1 F_{13} T_1^4 + A_2 F_{23} T_2^4}{A_2 F_{23} + A_1 F_{13}}\right)^{1/4}$$

where F_{13} and F_{23} are the appropriate view factors.

(b) Show that the total radiative heat transfer to the tubes $\dot{Q}_2$ is given by

$$\dot{Q}_2 = \left(A_1 F_{12} + \frac{A_1 F_{13} A_2 F_{23}}{A_2 F_{23} + A_1 F_{13}}\right)\sigma (T_1^4 - T_2^4)$$

(c) Calculate the radiative heat transfer to the tubes if $T_1 = 1700°C$, $T_2 = 300°C$, $A_1 = A_2 = 12 \text{ m}^2$ and the view factors are each 0.5.

[7.68 MW]

8.5 The temperature of the fuel bed in the boiler above is to be measured using a simple high-temperature thermometer. This comprises a copper disc of diameter D painted with a temperature-sensitive paint – the colour of the paint changes with temperature. A small hole of diameter $d = 0.025$ m is made in the side of the furnace; radiation falls on the copper disc which records a temperature.

(a) By considering an energy balance on the disc and neglecting convection show that the fuel bed temperature T_1 is

$$T_1^4 = 2\left\{\frac{(D/d)^2}{F_{1-\text{disc}}} - 1\right\}(T_\text{disc}^4 - T_\text{amb}^4) + T_\text{disc}^4$$

where the subscripts 1, disc and amb refer to the fuel bed, copper disc and the ambient surroundings respectively.

(b) The view factor, the fraction of radiation emitted from surface j which is received by surface i is given by the integral

$$F_{ij} = \frac{1}{\pi A_i} \int_{A_j} \int_{A_i} \frac{\cos \theta_i \cos \theta_j \, dA_i \, dA_j}{R^2}$$

by making the appropriate assumptions show that for the view factor $F_{1-\text{disc}}$ the above expression simplifies to

$$F_{1-\text{disc}} = 2 \int_{A_j} \frac{r \cos^2 \theta \, dr}{R^2}$$

where A_j is the area of the copper disc, r is a radial coordinate on the disc surface, θ is the angle between a normal to the disc surface and R.

(c) Evaluate the above view factor for the case of a disc of 0.2 m dia. located 0.3 m from the hole.

[0.1]

(d) The thermal paint on the copper disc registers that the disc temperature is 69°C when the ambient temperature is 30°C. Estimate the temperature of the fuel bed. What is the reason for the difference between this and the 'actual' temperature?

[1337°C]

(e) Now allow for the heat loss by free convection from the copper disc. Assume a value of average heat transfer coefficient of 10 W/m²K. Show that the disc now registers the actual temperature of the fuel bed.

[1712°C]

8.6 By considering the radiative heat transfer between N grey surfaces of an enclosure, show that

$$\sum_{j=1}^{N} A_i F_{ij}(J_i - J_j) = \frac{A_i(E_{b,i} - J_i)}{(1 - \varepsilon_i)/\varepsilon_i}$$

where the symbols have their usual meaning.

8.7 Figure 8.16 shows part of a combustion system. The ends are open and a central cylinder (denoted by surface 1) is within a concentric outer cylinder (denoted by surface 2).

The view factors F_{21} and F_{22} are 0.225 and 0.617, respectively. Complete the enclosure by defining the endcaps as surfaces 3 and 4 and deduce the remaining view factors.

[$F_{23} = 0.079$, $F_{24} = 0.079$, $F_{12} = 0.9$, $F_{11} = 0$, $F_{13} = 0.05$, $F_{14} = 0.05$]

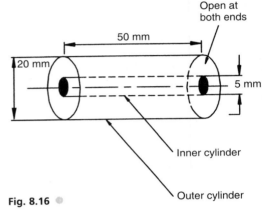

Fig. 8.16 ●

8.8 The inner cylinder of Figure 8.16 has a temperature of 1800 K and the ambient temperature T_amb is 300 K. The emissivity for the inner cylinder is $\varepsilon_1 = 0.22$; for the inner surface of the outer cylinder it is $\varepsilon_2 = 0.5$, and for the outer surface of the outer cylinder it is $\varepsilon_0 = 0.17$.

(a) Apply the formula given in Question 8.6 to surface 1 and show that the radiosity J_1 is

$$J_1 = 130\,996 + 0.702 J_2$$

(b) Apply it to surface 2 and show that the radiosity J_2 is

$$J_2 = \frac{\sigma T_2^4 + 0.225 J_1 + 72.56}{1.383}$$

(c) Neglect convection and consider a heat balance on the outer surface of the outer cylinder. Show that

$$J_2 = \sigma(\varepsilon_0 + 1) T_2^4 - \sigma \varepsilon_0 T_\text{amb}^4$$

(d) Now calculate the surface temperature of the outer cylinder.

[1050 K]

8.9 A hot-air airship is 50 m in length and the temperature of the air inside is 122°C. The airship is designed to fly at a speed of 10 m/s in air at 30°C where the solar irradiation is 1 kW/m² and there is an effective sky temperature of 270 K. The skin of the ship has negligible thermal resistance, a solar absorptivity of $\alpha_s = 0.9$ and an emissivity of $\varepsilon = 0.2$. In addition to solar radiation, heat is transferred through the skin by free convection on the inside and forced convection on the outside. The respective Nusselt numbers for the inside and outside are given by

$$Nu_{in} = 0.13Gr_{in}^{1/3} \quad \text{and} \quad Nu_{out} = 0.03Re_L^{0.8}$$

(a) Using tables for air at the appropriate temperatures show that the heat transfer coefficients on the inside and outside are:

$$h_{in} = 1.446\,(T_{in} - T_s)^{1/3}$$

$$h_{out} = 15.7\,\text{W/m}^2\text{K}$$

where T_s is the temperature of the skin.

(b) Assume that the rate of heat loss is equal over the whole surface of the airship. Apply a heat balance to unit area of the skin and verify that the skin temperature can be obtained from

$$(122 - T_s)^{4/3} = 7.84 \times 10^{-9}\,T_s^4$$
$$+ 10.84\,(T_s - 30) - 664$$

(c) What is the skin temperature?

[89°C]

(d) What is the corresponding rate of heat loss from the airship?

[153 W/m²]

8.10 A schematic diagram of an infrared heating lamp is shown in Figure 8.17. The filament has an effective surface area of 10 mm² and a black body surface temperature of 3000 K. The circular glass lens has a radius of 0.1 m. The reflector is assumed to be radiatively adiabatic and, inside the the lamp, heat is transferred by radiation to the lens; outside the lamp, heat is transferred by

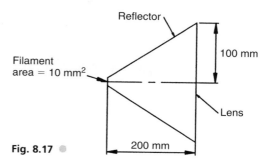

Fig. 8.17 ◉

radiation and convection from the lens to the surroundings which are at 20°C. Assuming that all surfaces behave as black bodies, calculate the temperature of the lens. Neglect the thermal resistance of the glass and, for the convective heat transfer coefficient on the outside of the lamp, take $h = 1.2\Delta T^{1/3}\,\text{W/m}^2\text{K}$.

[390 K]

8.11 A helicopter platform (16 m × 16 m) on an oil rig is subjected to an incident radiative heat flux from a nearby gas flame of 6 kW/m². The absorptivity, α, of the platform surface to thermal radiation at this wavelength is 0.7, and the average emissivity, ε, at the wavelength of emission is 0.6.

By considering an energy balance on the surface of the platform, show that for an ambient air temperature of 300 K, the surface temperature of the platform, T_s, is given by

$$4475.6 - 3.402 \times 10^{-8}T_s^4$$
$$- 1.551(T_s - 300)^{4/3} = 0$$

Now calculate the surface temperature.

[518 K]

For air, take $\mu = 1.846 \times 10^{-5}\,\text{kg/m s}$, $k = 0.026\,24\,\text{W/m K}$, $\rho = 1.177\,\text{kg/m}^3$ and $Pr = 0.707$. The following correlation may be used to predict the convective heat loss: $Nu_L = 0.13\,(Gr_L\,Pr)^{1/3}$, where the characteristic length, L, is the ratio of the area to the perimeter.

Chapter nine

Heat exchangers

9.1 ● Introduction

As the name implies, heat exchangers are thermal devices that exchange heat from one fluid stream to one or more others. The term 'heat exchanger' encompasses a broad range of devices that operate in one of three ways:

1. By recuperation, or recovery, of heat from the hot stream to the cold stream.
2. By regeneration, as the hot and cold streams alternately flow through a matrix.
3. By direct contact of one fluid stream with another.

This chapter focuses on recuperative heat exchangers, as these are the most common, and a general methodology can be developed without too many complications. Regenerative heat exchangers consist of either a moving (usually a rotating wheel) or a fixed matrix. The former is often found in gas turbines, where it is used to preheat combustion gases by the exhaust stream. Examples of the latter are the heat exchanger matrix in a Stirling cycle engine, and combustion beds for blast furnaces. Examples of direct contact heat exchanger are the cooling of sheet metal by water droplets or air as it leaves a rolling press, and cooling towers where water droplets are cooled by an updraught of air.

For recuperative heat exchangers, mixing and contamination between the streams is prevented by solid walls which constitute the heat exchanger surface area. The dominant mechanisms of heat transfer are convection from the hotter fluid stream to the walls, conduction through the separating wall, and convection from the wall to the other (cooler) fluid stream. Heat transfer by radiation will also be present, although the analysis and theory developed in this chapter assume that this is negligible (in practice this is generally true unless combustion processes are taking place within the heat exchanger). Similarly, the discussion is limited to the most common configuration of heat exchangers using two fluids, considering only the steady-state performance of single-phase fluids and focusing on only a few of the many possible flow arrangements. The reader is referred to the article by

Sekulić and Shah (1995) for a theoretical treatment of heat exchangers using three fluid streams.

Heat exchangers are extremely common devices and find application in a wide range of uses, for example:

Domestic	– refrigerator condenser, central heating boiler
Automotive	– automobile radiator and oil cooler
Medical	– kidney dialysis machine, cooling of blood during brain surgery
Industrial	– power generation equipment, process heating/cooling, air conditioning systems
Biological	– lungs

An understanding of heat exchanger theory, analysis and design is therefore of direct importance to mechanical engineers as well as serving as a good illustrative example for the application of convective heat transfer theory. In this respect, analysis of heat exchangers divides into either analysis for design or analysis of performance. In the former, the operating conditions are given and a particular heat exchanger type and flow arrangement is chosen – the task involves calculating the surface area. In the latter, the concern is to predict the outlet conditions (usually the temperatures) given a particular geometry, flow arrangement and the inlet conditions.

9.2 ● Classification

Heat exchangers can be classified by flow arrangement. Three types of arrangement in their simplest form, together with hypothetical temperature distributions are illustrated in Figure 9.1.

The concentric pipe arrangement shown in Figure 9.1a is a useful starting point as both a concept for more advanced analysis and also as the simplest practical form of recuperative heat exchanger. When the flows are in the same direction, the arrangement is referred to as *parallel flow*; if the flows are in opposite directions, as in Figure 9.1b, it is referred to as *counterflow*. In Figure 9.1c one fluid stream flows across the direction of the other, and this is termed *crossflow*. An important distinction to make in the crossflow heat exchanger is whether each fluid stream is mixed (where the entire stream is spread out across the heat exchanger) or unmixed (where a fluid stream is divided across the heat exchanger). These terms are illustrated in Figure 9.1c. Although not shown here, in terms of classification by flow arrangement, regenerative devices may be classified as periodic flow heat exchangers.

The internal and external surfaces of the heat exchanger tubes shown in Figures 9.1a and b are smooth. In practice, the heat transfer rates can be augmented by the use of internal and/or external fins (whose purpose is to increase the surface area), or inserts which promote turbulence and secondary flows.

Fig. 9.1 ● Simple heat exchanger geometries and temperature distributions: (a) concentric pipe with parallel flow; (b) concentric pipe with counterflow; (c) plate heat exchanger with crossflow

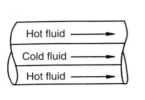

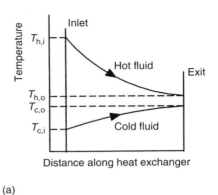

(a)

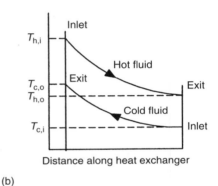

(b)

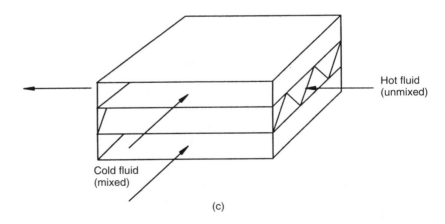

(c)

Another way to classify heat exchangers is by their construction. The double pipe arrangement discussed above (and shown in Figure 9.2a) is the simplest form, but in practice this will result in a large device to achieve a

Fig. 9.2 ● Heat exchanger types: (a) double pipe heat exchanger; (b) shell and tube heat exchanger; (c) plate fin heat exchanger; (d) tube fin heat exchanger

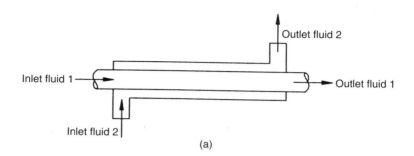

(a)

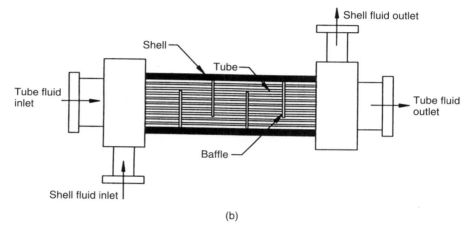

(b)

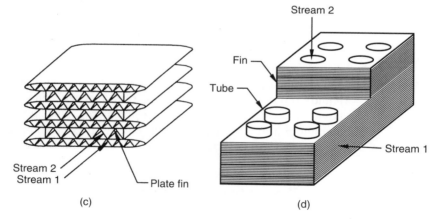

(c)　　　　　　　　　　(d)

required heat transfer duty. A derivative of this, which exploits the large volume of the outer pipe (or shell) to pack in a large number of tubes, is the shell and tube heat exchanger (Figure 9.2b). The inner stream in this case flows through not one, but several, even hundreds, of tubes. Internal baffles force the outer stream to cross the tubes, improving the convective heat

Fig. 9.3 ●
Classification of
heat exchangers by
surface area to
volume ratio

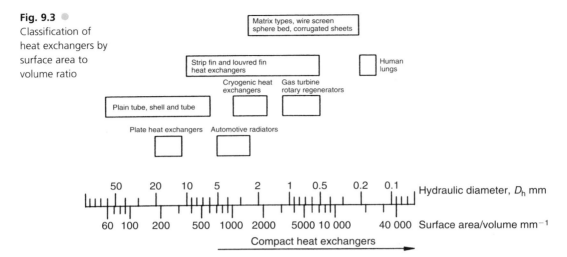

transfer. In a plate fin heat exchanger (Figure 9.2c) the flow channels are constructed from parallel plates separated by fins. Fins are used on both sides in gas-to-gas applications. For gas-to-liquid applications, fins are used on the gas side (where the heat transfer coefficient is lower). Tube fin heat exchangers (Figure 9.2d) are used in gas-to-liquid applications (such as automotive radiators) where the liquid flows through the tubes (which can withstand the higher pressure) and the gas over the fins.

Yet another means of classification is by the degree of compactness, expressed as the ratio of surface area to volume. For a number of different heat exchanger types, Figure 9.3 shows the ranges of their characteristic hydraulic diameter, D_h, and their surface area to volume ratio. Most of the examples in Figure 9.3 are self-explanatory. Compact heat exchangers are devices offering high surface area densities; these high densities can be achieved in a number of ways. They tend to be used in gaseous, rather than liquid, applications where the heat transfer coefficients are lower than for liquids and consequently the surface area needs to be large to transfer a given amount of heat. Figure 9.2c depicts an example of a compact heat exchanger. An excellent text covering this topic of is Kays and London (1984).

9.3 ● Overall heat transfer coefficient

The concept of an overall heat transfer coefficient was discussed in Chapter 2, where it was applied to composite walls and pipes.

The heat flow, $\dot{Q}$, through the wall of the pipe of length L, shown in Figure 9.4, is given by

$$\frac{\dot{Q}}{L} = 2\pi h_i r_i (T_i - T_1) = -2\pi r k \frac{\mathrm{d}T}{\mathrm{d}r} = 2\pi h_o r_o (T_2 - T_o) \tag{9.1}$$

Fig. 9.4 ● Overall heat transfer coefficient for a pipe

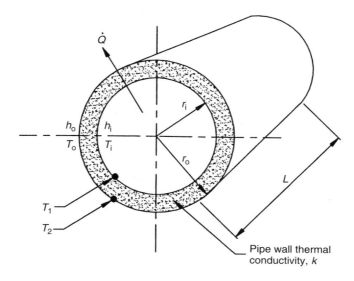

From equating the first and third terms in Equation (9.1),

$$dT = -\left(\frac{\dot{Q}}{2\pi k L}\right)\frac{dr}{r} \tag{9.2}$$

which gives, after integration,

$$T_1 - T_2 = \frac{\dot{Q}}{2\pi k L}\ln(r_o/r_i) \tag{9.3}$$

From the first and second terms in Equation (9.1),

$$T_i - T_1 = \frac{\dot{Q}}{2\pi L h_i r_i} \tag{9.4}$$

Likewise, from the first and last terms,

$$T_2 - T_o = \frac{\dot{Q}}{2\pi L h_o r_o} \tag{9.5}$$

By adding Equations (9.3), (9.4) and (9.5)

$$T_i - T_o = \frac{\dot{Q}}{2\pi L}\left\{\frac{\ln(r_o/r_i)}{k} + \frac{1}{h_i r_i} + \frac{1}{h_o r_o}\right\}$$

and rearranging,

$$\frac{\dot{Q}}{2\pi L r_o} = \frac{T_i - T_o}{\dfrac{r_o}{k}\ln\left(\dfrac{r_o}{r_i}\right) + \dfrac{r_o}{h_i r_i} + \dfrac{1}{h_o}} \tag{9.6a}$$

$$= U_o(T_i - T_o) \tag{9.6b}$$

Table 9.1 ●
Orders of
magnitude for
U, the overall
heat transfer
coefficient

Hot fluid	Cold fluid	$U\,(\text{W/m}^2\text{K})$
Water	Water	1000–2500
Ammonia	Water	1000–2500
Gases	Water	10–250
Light organics[a]	Water	370–730
Heavy organics[b]	Water	20–370
Steam	Water	1000–3500
Steam	Ammonia	1000–3500
Steam	Gases	25–250
Steam	Light organics	500–1000
Steam	Heavy organics	30–300
Light organics	Light organics	200–400
Heavy organics	Heavy organics	50–200
Light organics	Heavy organics	50–200
Heavy organics	Light organics	150–300

[a] $\mu < 5 \times 10^{-4}$ kg/m s
[b] $\mu > 1 \times 10^{-3}$ kg/m s
Source: Adapted from Bejan (1993)

where, by dividing by the area of the outer surface, the overall heat transfer coefficient U_o is said to be referred to the outer surface and is given by

$$U_o = \left\{ \left[\frac{r_o}{k} \ln\left(\frac{r_o}{r_i}\right) \right] + \left[\frac{r_o}{h_i r_i} \right] + \left[\frac{1}{h_o} \right] \right\}^{-1} \tag{9.7}$$

It can be useful to think of the analogy between Equation (9.6b) and Ohm's law ($I = V/R$). In the context of heat exchange by convection and conduction, the temperature difference ($T_i - T_o$) provides the potential to 'drive' a current, in this case the heat flux ($\dot{Q}/2\pi L r_o$), and overcome the thermal resistance, $1/U_o$.

Similarly, the overall heat transfer coefficient U_i referred to the inside surface is

$$U_i = \left\{ \left[\frac{r_i}{k} \ln\left(\frac{r_o}{r_i}\right) \right] + \left[\frac{1}{h_i} \right] + \left[\frac{r_i}{r_o h_o} \right] \right\}^{-1} \tag{9.8}$$

Some typical values of overall heat transfer coefficient are given in Table 9.1. They are presented as guidelines only; it is always preferable to calculate values using standard convection correlations for plain tubes and other passageways (see Chapter 5) or the data of Kays and London (1984), for more geometrically complex designs.

Very often $r_o/r_i \approx 1$ and the tubes are made from a good conductor whose thermal resistance can be neglected, so Equations (9.7) and (9.8) become

$$U_i \approx U_o = \left\{ \frac{1}{h_i} + \frac{1}{h_o} \right\}^{-1}$$

$$= \frac{h_o h_i}{h_o + h_i} = U \tag{9.9}$$

An important practical implication of Equation (9.9) is that if $h_i \gg h_o$, then $U_o \approx h_o$, or if $h_o \gg h_i$, then $U_o \approx h_i$, which may be summarised as the heat transfer will be controlled by the lower of the two heat transfer coefficients. To illustrate this, consider a water-to-air heat exchanger, where the heat transfer coefficient on the water side is $1000\,W/m^2 K$ and that on the air side $40\,W/m^2 K$. Neglecting the thermal resistance of the tube walls, the overall heat transfer coefficient will be

$$U = \frac{h_o h_i}{h_o + h_i}$$

$$= \frac{4 \times 10^4}{1040} = 38.46\,W/m^2 K$$

Doubling the water-side heat transfer coefficient will increase the value of the overall heat transfer coefficient to

$$U = \frac{h_o h_i}{h_o + h_i}$$

$$= \frac{8 \times 10^4}{2040} = 39.21\,W/m^2 K$$

an increase of 2%. Doubling the air-side heat transfer coefficient (lower value and hence controlling) will lead to an overall heat transfer coefficient of

$$U = \frac{h_o h_i}{h_o + h_i} = \frac{8 \times 10^4}{1080} = 74.07\,W/m^2 K$$

an increase of 93%. Clearly, if the overall heat transfer coefficient needs to be increased, then it is more worthwhile to focus attention on the side with the lower value of heat transfer coefficient.

In addition, the thermal resistance will be increased owing to the action of deposits formed from salts, solid particles, chemical reactions, corrosion and biological organisms. The general term for this action is fouling and, in practice, fouling is handled by an extra resistance term known as the fouling factor or fouling resistance ($R_{f,i}$ for the inside, $R_{f,o}$ for the outside). Some typical representative values of fouling resistance are listed in Table 9.2.

If such fouling resistances are included, then Equations (9.7) and (9.8) for the overall heat transfer coefficient become

$$U_o = \left\{ \left[\frac{r_o}{k} \ln\left(\frac{r_o}{r_i}\right) \right] + \left[\frac{r_o}{r_i} R_{f,i} \right] + \left[\frac{r_o}{h_i r_i} \right] + \left[\frac{1}{h_o} \right] + R_{f,o} \right\}^{-1} \tag{9.10a}$$

$$U_i = \left\{ \left[\frac{r_i}{k} \ln\left(\frac{r_o}{r_i}\right) \right] + \left[R_{f,i} \right] + \left[\frac{1}{h_i} \right] + \left[\frac{r_i}{r_o h_o} \right] + \left[\frac{r_i}{r_o} R_{f,o} \right] \right\}^{-1} \tag{9.10b}$$

Table 9.2 ●
Representative
values of fouling
resistance, R_f

Fluid	R_f (m^{2}K/W)
Water	
Distilled	0.0001
Sea water	0.0001–0.0002
Hard water	0.0005–0.0009
Liquids	
Liquid gasoline, oil and LPG	0.0002–0.0004
Vegetable oils	0.0005
Caustic solutions	0.0004
Refrigerants, ammonia	0.0002
Methanol, ethanol and ethylene glycol solutions	0.0004
Gases	
Natural gas	0.0002–0.0004
Acid gas	0.0004–0.0005
Solvent vapours	0.0002
Steam (non-oil-bearing)	0.0001
Steam (oil-bearing)	0.0003–0.0004
Compressed air	0.0002
Ammonia	0.0002

Source: Adapted from Bejan (1993)

The above analysis has been developed for circular tubes. For a heat exchanger where the channels are constructed from plates of thickness t and thermal conductivity, k, with the outside and inside areas being equal, it is relatively easy to show that the overall heat transfer coefficient is given by

$$U = \left\{ \left[\frac{t}{k}\right] + R_{f,i} + \left[\frac{1}{h_i}\right] + \left[\frac{1}{h_o}\right] + R_{f,o} \right\}^{-1} \qquad (9.10c)$$

Example 9.1

A concentric tube heat exchanger is used to cool the lubricating oil for a large marine Diesel engine. The inner tube is constructed from 2 mm wall thickness stainless steel, having $k = 16$ W/m K. The flow rate of cooling water through the inner tube ($r_i = 30$ mm) is 0.3 kg/s, the flow rate of oil through the outer tube ($r_o = 50$ mm) is 0.15 kg/s. Assuming fully developed flow and taking account of fouling resistances, calculate the overall heat transfer coefficient referred to the inside surface. Evaluate properties at 80°C for the oil and at 35°C for the water.

Solution   Step 1: Obtain relevant data from tables

For oil at 80°C, $C_p = 2131$ J/kg K, $\mu = 3.25 \times 10^{-2}$ kg/m s, $k = 0.138$ W/m K, and from Table 9.2, $R_f = 0.0004$ m^{2}K/W.

For water at 35°C, $C_p = 4178$ J/kg K, $\mu = 725 \times 10^{-6}$ kg/m s, $k = 0.625$ W/m K, and from Table 9.2, $R_f = 0.0001$ m^{2}K/W.

Step 2: Evaluate the Reynolds and Nusselt numbers

$$Re = \rho V_m D_h / \mu$$

where the mean velocity, V_m, is obtained from continuity $V_m = \dot{m}/\rho A$. Hence for a tube with area $A = \pi D^2/4$,

$$Re = 4\dot{m}/\pi D \mu$$

For water in the inner tube,

$$Re = \frac{4 \times 0.3}{\pi \times 0.06 \times 725 \times 10^{-6}} = 8781$$

$$Pr = \frac{\mu C_p}{k} = \frac{725 \times 10^{-6} \times 4178}{0.625} = 4.85$$

Since $Re > 2300$, the flow is therefore turbulent and the Nusselt number can be obtained from the Dittus–Boelter equation (5.42) with $n = 0.4$:

$$Nu_D = 0.023 Re_D^{0.8} \, Pr^{0.4} = 0.023 \times (8781)^{0.8} \times (4.85)^{0.4} = 62$$

From which

$$h_i = \frac{Nu_D k}{D} = \frac{62 \times 0.625}{0.06} = 643 \text{ W/m}^2\text{K}$$

For oil in the outer annulus of inner radius r_a and outer radius r_b, the hydraulic diameter, D_h, is given by

$$\begin{aligned} D_h &= \frac{4 \times \text{area}}{\text{perimeter}} \\ &= \frac{4\pi(r_b^2 - r_a^2)}{2\pi(r_b + r_a)} \\ &= 2(r_b - r_a) \\ &= 2(0.05 - 0.032) = 0.036 \text{ m} \end{aligned}$$

and

$$\begin{aligned} Re &= \frac{\rho V_m D_h}{\mu} \\ &= \frac{2\dot{m}(r_b - r_a)}{\pi(r_b^2 - r_a^2)\mu} \\ &= \frac{2\dot{m}}{\pi(r_b + r_a)\mu} \\ &= \frac{2 \times 0.15}{\pi \times (0.05 + 0.032) \times 3.25 \times 10^{-2}} = 36 \end{aligned}$$

The flow is therefore laminar, the Nusselt number is constant (see Section 5.3). Although we could use a correlation for a plain tube, for an annulus of this radius ratio ($r_a/r_b = 0.64$), a better correlation is given by Kays and Perkins (1973) as $Nu_o (= h_o D_h/k) = 5.6$, from which

$$h_o = \frac{Nu_o k}{D_h}$$

$$= \frac{5.6 \times 0.138}{0.036} = 21.5 \, \text{W/m}^2\text{K}$$

Step 3: Evaluate overall heat transfer coefficient
From Equation (9.10b),

$$U_i = \left\{ \left[\frac{r_i}{k} \ln \left(\frac{r_o}{r_i} \right) \right] + [R_{f,i}] + \left[\frac{1}{h_i} \right] + \left[\frac{r_i}{r_o h_o} \right] + \left[\frac{r_i}{r_o} R_{f,o} \right] \right\}^{-1}$$

$$= \left\{ \left[\frac{0.030}{16} \ln \left(\frac{32}{30} \right) \right] + [0.0001] + \left[\frac{1}{643} \right] + \left[\frac{0.030}{0.032 \times 21.5} \right] \right.$$

$$\left. + \left[\frac{0.030}{0.032} \times 0.0004 \right] \right\}^{-1}$$

$$= 21.9 \, \text{W/m}^2\text{K}$$

Comment   The overall heat transfer coefficient U_i is of similar magnitude to the lower heat transfer coefficient h_o. Since h_o is significantly lower than h_i, its value more or less governs the overall heat transfer coefficient. Ignoring fouling resistances and the thermal resistance of the tube walls, we have $U = h_o h_i/(h_o + h_i)$ which gives $U = 20.8 \, \text{W/m}^2\text{K}$, a difference of 5% which is safe enough to neglect for the purpose of a first approximation. Because the tube walls are of finite thickness, $r_i/r_o < 1$, and consequently $U_i > h_o$.

One way of improving the heat transfer performance is to augment the surface area on the side with the controlling heat transfer coefficient. This can be done with the addition of fins, which can be an integral part of the surface, or attached by adhesive, solder, welding or brazing. The method of attachment does, however, impose limitations on the maximum operating temperature. The analysis of fin heat transfer is covered in Section 2.8.

Example 9.2

The performance of the heat exchanger tube in Example 9.1 is to be improved by the addition of eight stainless steel radial fins in the annular gap which conveys the oil, as shown in Figure 9.5. The fins are made of

Fig. 9.5 ● Finned concentric tube heat exchanger (Example 9.2)

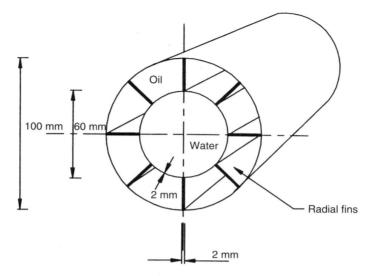

the same material as the tubes ($k = 16\,\text{W/m K}$) and have a thickness of $t = 2\,\text{mm}$. Neglecting fouling resistances and also the thermal resistance of the tube walls, what now is the overall heat transfer coefficient referred to the inside surface?

Solution

Step 1. Analysis to develop the appropriate expression for U_i

Let the surface temperature of the tube be T_s. Since the tube has negligible thermal resistance, T_s is uniform. The outside heat transfer coefficient is assumed to be unchanged by the addition of fins; this and the inside heat transfer coefficient therefore have the same values as evaluated for Example 9.1. Let the temperature of the oil be T_{oil} and that of the water be T_{water}. Let the surface area of the inside be A_i and that on the outside A_o. Since the fins are attached to the outside, $A_o = A_{o,f} + A_{o,u}$ where $A_{o,f}$ and $A_{o,u}$ are the finned and unfinned surface areas, respectively. Although (as shown in Example 9.1) it is reasonable to ignore the thermal resistance created by the tube walls, conduction cannot be ignored in the fins. In particular, a temperature gradient along the length of each fin is necessary to conduct the heat through it. This may be accounted for using the definition of fin efficiency (2.56):

$$\eta_{fin} = \frac{\text{Actual heat flow}}{\text{Ideal heat flow}}$$

and the introduction of the mean fin temperature, $T_{fin,m}$. Using Equation (2.56) it can be shown that

$$\eta_{fin} = \frac{T_{fin,m} - T_{oil}}{T_s - T_{oil}}$$

The heat transferred from the inner surface to the water is given by

$$\dot{Q} = A_i h_i (T_{\text{water}} - T_s) \tag{9.11}$$

which is equal to that transferred from the oil to the outer surface

$$\dot{Q} = A_{o,u} h_o (T_s - T_{\text{oil}}) + A_{o,f} h_o (T_{\text{fin,m}} - T_{\text{oil}}) \tag{9.12}$$

Following from the definition of fin efficiency, Equation (9.12) may be rewritten as

$$\dot{Q} = A_{o,u} h_o (T_s - T_{\text{oil}}) \left[1 + \left(\frac{A_{o,f} \eta_{\text{fin}}}{A_{o,u}} \right) \right] \tag{9.13}$$

Using Equations (9.11) and (9.13) to express the heat flow in terms of temperatures T_{oil} and T_{water} gives

$$\dot{Q} = A_i U_i (T_{\text{oil}} - T_{\text{water}})$$

where

$$U_i = \left\{ \frac{1}{h_i} + \frac{A_i}{A_{o,u} h_o} \left[1 + \left(\frac{A_{o,f} \eta_{\text{fin}}}{A_{o,u}} \right) \right]^{-1} \right\}^{-1}$$

Step 2: Obtain fin efficiency and calculate areas
For a rectangular fin with an adiabatic tip (2.49) and (2.56),

$$\eta_{\text{fin}} = \frac{\tanh mL}{mL}$$

where $m = (2h_o/kt)^{1/2}$ and the fin length $L = 0.050 - 0.032 = 0.018\,\text{m}$. Hence

$$mL = \left(\frac{2 \times 21.5}{16 \times 0.002} \right)^{1/2} \times 0.018 = 0.66$$

and $\eta_{\text{fin}} = \dfrac{\tanh(0.66)}{0.66} = 0.875$

Hence the three relevant areas are

$$A_i = 2\pi r_i = 0.1885\,\text{m}^2/\text{metre length}$$

$$A_{o,u} = 2\pi r_o - 8 \times 0.002 = 0.1851\,\text{m}^2/\text{metre length}$$

$$A_{o,f} = 8 \times 0.018 \times 2 = 0.288\,\text{m}^2/\text{metre length}$$

Step 3. Evaluate U_i

$$U_i = \left\{ \frac{1}{h_i} + \frac{A_i}{A_{o,u} h_o} \left[1 + \left(\frac{A_{o,f} \eta_{fin}}{A_{o,u}} \right) \right]^{-1} \right\}^{-1}$$

$$= \left\{ \frac{1}{643} + \frac{0.1885}{0.1851 \times 21.5} \left[1 + \left(\frac{0.288 \times 0.875}{0.1851} \right) \right]^{-1} \right\}^{-1}$$

$$= 46.3 \text{ W/m}^2\text{K}$$

Comment   Comparison with Example 9.1 shows that the overall heat transfer coefficient has been approximately doubled (previously, $U_i = 21.9 \text{ W/m}^2\text{K}$) by the addition of the fins. The analysis assumes that $h_o = 21.5 \text{ W/m}^2\text{K}$ (a value obtained from considering flow of the oil in a plane annulus) also acts along the length of the fin. In reality, the addition of fins protruding into the flow may bring about changes in the flow structure which also modify the heat transfer coefficient. There may, for example, be variations around the sector between adjacent fins. Such detail is clearly beyond the scope of this text; the flow and metal temperature distribution in the vicinity of the fins is likely to be complex, and for detailed design purposes a combination of experimental data and numerical techniques is probably the best way ahead.

9.4 ● Analysis of thermal performance: the logarithmic mean temperature difference

In accordance with the arguments developed so far it is both customary and convenient to express the overall heat transfer, $\dot{Q}$, from one fluid to another as

$$\dot{Q} = UA\Delta T_m \qquad \text{(watts)} \tag{9.14}$$

The meaning of the overall heat transfer coefficient, U, has already been discussed, A is the surface area for heat exchange, and ΔT_m some, and as yet undefined, representative mean temperature difference. Equation (9.14) and the following reasoning for determining ΔT_m makes use of a number of assumptions:

1. There are no external losses from the heat exchanger.
2. There is negligible axial conduction along the length of the tubes.
3. Changes in potential and kinetic energy are negligible.

4. The heat transfer coefficients are constant along the length of the tubes.

5. The specific heats of the fluids do not vary with temperature.

Consider, as shown in Figure 9.6, the flow of two fluids in a parallel flow heat exchanger. By applying an energy balance to differential elements of these fluids and with the above assumptions in mind, it follows that

$$d\dot{Q}_h = -d\dot{Q}_c = d\dot{Q} \tag{9.15}$$

where $\quad d\dot{Q}_h = -\dot{m}_h C_{p,h}\, dT_h \tag{9.16a}$

$$d\dot{Q}_c = \dot{m}_c C_{p,c}\, dT_c \tag{9.16b}$$

and dT_h and dT_c are the respective temperature differences of the hot and cold fluids as they flow across the boundaries of the element. Equations (9.16a) and (9.16b) can be combined, using Equation (9.15) and rearranged to give the overall change in temperature over the element, $d\Delta T$ as

$$(dT_h - dT_c) = d\Delta T = -d\dot{Q}\left\{\frac{1}{(\dot{m}_h C_{p,h})} + \frac{1}{(\dot{m}_c C_{p,c})}\right\} \tag{9.17}$$

Equation (9.14) in differential form is

$$d\dot{Q} = U\Delta T\, dA \tag{9.18}$$

where ΔT is the local temperature difference between the hot and cold fluids $\Delta T = (T_h - T_c)$. Combining Equations (9.17) and (9.18) and inserting the

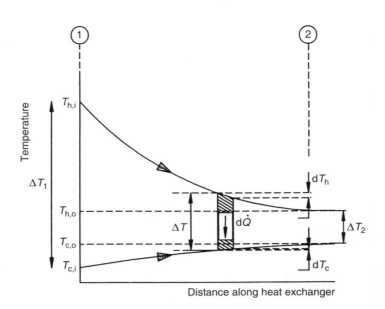

Fig. 9.6 ●
Logarithmic mean temperature difference applied to a parallel flow heat exchanger

Distance along heat exchanger

limits of integration to be consistent with the nomenclature on Figure 9.6 gives

$$\int_1^2 \frac{d\Delta T}{\Delta T} = -U \left\{ \frac{1}{\dot{m}_h C_{p,h}} + \frac{1}{\dot{m}_c C_{p,c}} \right\} \int_1^2 dA \qquad (9.19)$$

Integrating gives

$$\ln\left(\frac{\Delta T_2}{\Delta T_1}\right) = -UA \left\{ \frac{1}{\dot{m}_h C_{p,h}} + \frac{1}{\dot{m}_c C_{p,c}} \right\} \qquad (9.20)$$

This expression provides a hint as to the definition of ΔT_m in Equation (9.14). The complete form is arrived at by recognising that since

$$\dot{Q} = \dot{m}_h C_{p,h}(T_{h,i} - T_{h,o}) = \dot{m}_c C_{p,c}(T_{c,o} - T_{c,i}) \qquad (9.21)$$

and

$$\Delta T_1 = T_{h,i} - T_{c,i} \quad \text{and} \quad \Delta T_2 = T_{h,o} - T_{c,o} \qquad (9.22)$$

then, by substituting for the heat capacity rates (mass flow × specific heat capacity) in Equation (9.20), we have

$$\ln\left(\frac{\Delta T_2}{\Delta T_1}\right) = -\frac{UA}{\dot{Q}}(T_{h,i} - T_{h,o} + T_{c,o} - T_{c,i})$$

$$= -\frac{UA}{\dot{Q}}(\Delta T_1 - \Delta T_2) \qquad (9.23)$$

which, on rearrangement, becomes

$$\dot{Q} = UA\left(\frac{\Delta T_2 - \Delta T_1}{\ln(\Delta T_2/\Delta T_1)}\right)$$

$$\dot{Q} = UA\Delta T_{lm} \qquad (9.24)$$

where

$$\Delta T_{lm} = \frac{\Delta T_2 - \Delta T_1}{\ln(\Delta T_2/\Delta T_1)} \qquad (9.25)$$

is the logarithmic mean temperature difference (LMTD). It is left to the reader to verify that an identical expression arises for the case of a counterflow heat exchanger where $\Delta T_1 = T_{h,i} - T_{c,o}$ and $\Delta T_2 = T_{h,o} - T_{c,i}$.

For crossflow and other heat exchanger geometries, Equation (9.24) can still be used but the logarithmic mean temperature difference (derived for counterflow conditions) $\Delta T_{lm,cf}$ is multiplied by a correction factor F, which is a function of two other dimensionless variables, denoted by P and R. Graphs of these correction factors, which are derived from analytical

Fig. 9.7 ● Correction factor, F, for single-pass crossflow heat exchanger with both streams unmixed

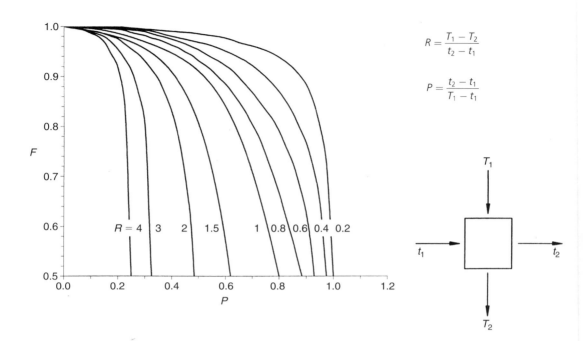

$$R = \frac{T_1 - T_2}{t_2 - t_1}$$

$$P = \frac{t_2 - t_1}{T_1 - t_1}$$

expressions, are given in Kays and London (1984). One such graph for the case of a single-pass crossflow heat exchanger with both fluids unmixed, including the definitions of P and R, is given in Figure 9.7. The parameter R represents the ratio of the heat capacity rates (product of mass flow and specific heat capacity, $\dot{m}C_p$) of the two streams. It is a simple matter to prove that $P = 0$ or $R = 0$ corresponds to the case where one of the streams experiences heat transfer without a temperature change, as in an evaporator or condenser.

Example 9.3

The oil cooler geometry of Example 9.1 is to be used to cool oil from 90°C to 50°C using water available at 10°C. Calculate:

(a) The length of the tube required for parallel flow.

(b) The length of the tube required for counterflow.

(c) The area required for a single-pass crossflow heat exchanger with both streams unmixed, operating at the same temperatures and flow rates and with the same value of overall heat transfer coefficient as in (a) and (b).

Solution   **Step 1: Obtain relevant data and calculate heat flow $\dot{Q}$**

For the oil (hot stream), $\dot{m}_h = 0.15\,\text{kg/s}$, $C_{p,h} = 2131\,\text{J/kg K}$, $T_{h,i} = 90°C$, $T_{h,o} = 50°C$.

For the water (cold stream), $\dot{m}_c = 0.3\,\text{kg/s}$, $C_{p,c} = 4178\,\text{J/kg K}$, $T_{c,i} = 10°C$, $T_{c,o} = ?$

$$\dot{Q} = \dot{m}_h C_{p,h}(T_{h,i} - T_{h,o}) = 12\,786\,\text{W}$$

The unknown exit temperature $T_{c,o}$ can be evaluated from an overall energy balance (assuming no losses, etc.) since

$$\dot{m}_h C_{p,h}(T_{h,i} - T_{h,o}) = \dot{m}_c C_{p,c}(T_{c,o} - T_{c,i})$$

$$T_{c,o} = \left[\frac{\dot{m}_h C_{p,h}(T_{h,i} - T_{h,o})}{\dot{m}_c C_{p,c}}\right] + T_{c,i}$$

$$= \left[\frac{0.15 \times 2131 \times (90 - 50)}{0.3 \times 4178}\right] + 10$$

$$= 20.2°C$$

Step 2: Evaluate ΔT_{lm}

For parallel flow,

$$\Delta T_1 = 90 - 10 = 80°C$$

$$\Delta T_2 = 50 - 20.2 = 29.8°C$$

$$\Delta T_{lm} = \frac{\Delta T_2 - \Delta T_1}{\ln(\Delta T_2/\Delta T_1)} = \frac{29.8 - 80}{\ln(29.8/80)} = 50.83°C$$

For counterflow,

$$\Delta T_1 = 90 - 20.2 = 69.8°C$$

$$\Delta T_2 = 50 - 10 = 40°C$$

$$\Delta T_{lm} = \frac{\Delta T_2 - \Delta T_1}{\ln(\Delta T_2/\Delta T_1)} = \frac{40 - 69.8}{\ln(40/69.8)} = 53.52°C$$

For crossflow with both streams unmixed, in terms of the nomenclature indicated in Figure 9.7,

$$t_i = T_{h,i} = 90°C;\ t_2 = T_{h,o} = 50°C;\ T_1 = T_{c,i} = 10°C;\ T_2 = T_{c,o} = 20.2°C$$

$$R = \frac{T_1 - T_2}{t_2 - t_1} = 0.255$$

$$P = \frac{t_2 - t_1}{T_1 - t_1} = 0.5$$

From Figure 9.7, for $P = 0.5$ and $R = 0.255$, $F \approx 0.98$, and

$$\Delta T_m = \Delta T_{lm,cf} F = 53.52 \times 0.98 = 52.45°C$$

Step 3. Evaluate areas and tube lengths

$$\dot{Q} = UA\Delta T_{lm}$$

as $\dot{Q}$, U and ΔT_{lm} are all known then

$$A = (2\pi r_i L) = \dot{Q}/U\Delta T_{lm}$$

(a) For parallel flow, $A = \dfrac{12\,786}{21.9 \times 50.83} = 11.5\,\text{m}^2$ and for $r_i = 30\,\text{mm}$, $L = 61\,\text{m}$.

(b) For counterflow, $A = \dfrac{12\,786}{21.9 \times 53.52} = 10.9\,\text{m}^2$ and $L = 57.9\,\text{m}$.

(c) For crossflow, $A = \dfrac{12\,786}{21.9 \times 52.45} = 11.13\,\text{m}^2$.

Comment For the same inlet and outlet temperatures and overall heat transfer coefficients, the counterflow heat exchanger geometry has a higher logarithmic mean temperature difference and consequently requires less surface area than any other heat exchanger. The counterflow heat exchanger therefore has the highest thermodynamic efficiency of heat exchanger types. For a double pipe arrangement operating in either counterflow or parallel flow, the areas lead to impractical tube lengths. This can be overcome using a shell and tube design.

As illustrated in Example 9.3, providing that the fluid inlet and outlet temperatures are known, it is a relatively simple matter to apply the logarithmic mean temperature difference method to *design* a heat exchanger (i.e. to specify either the area or the overall heat transfer coefficient). The following example highlights some of the difficulties encountered when this method is applied to evaluating heat exchanger *performance* (i.e. when the area, the overall heat transfer coefficient and the capacity rates are known plus only two of the fluid temperatures).

Example 9.4

The double pipe, parallel flow heat exchanger geometry of Example 9.1 is to be used to cool oil (0.15 kg/s at 90°C), using sea water (0.3 kg/s at 10°C). The area of the heat exchanger is 11.5 m² and the overall heat transfer coefficient is 21.9 W/m²K. What are the exit states of the water and oil from the heat exchanger? (For oil $C_p = 2131$ J/kg K; for water $C_p = 4178$ J/kg K).

Solution   Before calculating the heat transfer rate we need to evaluate ΔT_{lm} but this involves knowing all the temperatures, so the following iterative cycle is applied

1. Assume a value for $T_{h,o}$.
2. Calculate $T_{c,o}$ and ΔT_{lm}.
3. Knowing these, calculate the heat transfer, $\dot{Q}$.
4. From $\dot{Q}$ and the capacity rates, recalculate $T_{h,o}$.
5. Compare the latest value of $T_{h,o}$ with the initial estimate and if the two do not agree (to within, say, $\pm 0.5°C$), the iterative cycle is repeated until they do.

Step 1
$T_{h,i} = 90°C$, $T_{c,i} = 10°C$ and assume $T_{h,o} = 70°C$.

Step 2
The corresponding cold stream exit temperature $T_{c,o}$ can be evaluated from an overall energy balance (assuming no losses, etc.):

$$\dot{m}_h C_{p,h}(T_{h,i} - T_{h,o}) = \dot{m}_c C_{p,c}(T_{c,o} - T_{c,i})$$

$$T_{c,o} = \left[\frac{\dot{m}_h C_{p,h}(T_{h,i} - T_{h,o})}{\dot{m}_c C_{p,c}} \right] + T_{c,i}$$

$$= \left[\frac{0.15 \times 2131 \times (90 - 70)}{0.3 \times 4178} \right] + 10$$

$$= 15.1°C$$

Evaluate T_{lm}:

$$\Delta T_1 = 90 - 10 = 80°C$$

$$\Delta T_2 = 70 - 15.1 = 54.9°C$$

$$\Delta T_{lm} = \frac{\Delta T_2 - \Delta T_1}{\ln(\Delta T_2/\Delta T_1)}$$

$$= \frac{54.9 - 80}{\ln(54.9/80)} = 66.66°C$$

Step 3
Evaluate $\dot{Q}$:

$$\dot{Q} = UA\Delta T_{lm}$$

$$= 21.9 \times 11.5 \times 66.66$$

$$= 16\,788 \text{ W}$$

Step 4
Recalculate $T_{h,o}$:

$$\dot{Q} = \dot{m}_h C_{p,h}(T_{h,i} - T_{h,o})$$

$$T_{h,o} = T_{h,i} - \left(\frac{\dot{Q}}{\dot{m}_h C_{p,h}}\right)$$

$$= 90 - \frac{16\,788}{0.15 \times 2131}$$

$$= 37.5°C$$

Step 5
Compare with initial guess:

$$\text{Difference} = 70 - 37.5 = 32.5°C$$

Make a second guess.

A reasonable choice for a second guess at $T_{h,o}$ is to take the average of the initially assumed value and the recalculated value. A more elegant method is to plot $(T_{h,o})_{\text{assumed}}$ against $(T_{h,o})_{\text{recalculated}}$; this allows the convergence history to be seen graphically and a more enlightened choice to be made for $T_{h,o}$.

$$(T_{h,o})_{\text{assumed}} = \frac{70 + 37.5}{2} = 53.7°C$$

From which

$$T_{c,o} = 19.24°C$$

$$\Delta T_{\text{lm}} = 54.09°C$$

$$\dot{Q} = 13\,623\text{ W}$$

$$T_{h,o} = 47.38°C$$

$$\text{Difference} = 53.7 - 47.4 = 6.3°C$$

Make a third guess:

$$T_{h,o} = \frac{47.38 + 53.75}{2} = 50.56°C$$

From which

$$T_{c,o} = 20.05°C$$

$$\Delta T_{\text{lm}} = 51.33°C$$

$$\dot{Q} = 12\,365\text{ W}$$

$$T_{h,o} = 51.32°C$$

$$\text{Difference} = 50.6 - 51.3 = -0.7°C$$

Comment   It is left to the reader to continue this iterative cycle until the hot stream outlet temperatures (assumed and recalculated) are suitably converged. Nonetheless, the above exercise serves to demonstrate that using the logarithmic mean temperature difference method to evaluate heat exchanger performance can be a tedious process. The procedure is even more involved when applied to a crossflow or shell and tube heat exchanger, since the correction factor F depends on all four fluid temperatures. Furthermore, in some applications the specific heat capacity of one or more of the fluid streams can change significantly across the temperature range and this must be brought into the iterative cycle.

9.5 ● Analysis of thermal performance: effectiveness and NTU

The effectiveness/NTU (number of transfer units) method of performance analysis leads directly to the unknown temperatures, avoiding the iterative process of the logarithmic mean temperature difference method. Effectiveness and NTU are non-dimensional groups. The NTU is defined as follows:

$$NTU = \frac{UA}{C_{min}} \tag{9.26}$$

where C_{min} is the minimum heat capacity rate, that is,

$$C_{min} = \min(\dot{m}_h C_{p,h}, \dot{m}_c C_{p,c})$$

The effectiveness, ε, is defined as the ratio of the actual heat transfer rate, $\dot{Q}$, to the maximum possible heat transfer rate, $\dot{Q}_{max}$:

$$\varepsilon = \frac{\dot{Q}}{\dot{Q}_{max}} \tag{9.27}$$

where

$$\dot{Q} = \dot{m}_h C_{p,h}(T_{h,i} - T_{h,o})$$
$$= \dot{m}_c C_{p,c}(T_{c,o} - T_{c,i}) \tag{9.28}$$

or, using the capacity rate C to replace the product $\dot{m}C_p$:

$$\dot{Q} = C_h(T_{h,i} - T_{h,o})$$
$$= C_c(T_{c,o} - T_{c,i}) \tag{9.29}$$

To obtain an expression for the maximum possible heat transfer rate, consider a counterflow heat exchanger. To increase the heat transfer the area will be increased while maintaining both inlet temperatures at fixed values. As the area (or the product UA) is made infinitely large, the

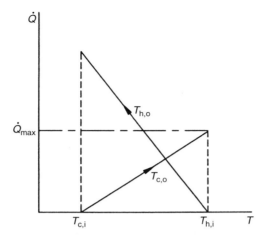

stream-to-stream temperature differences are reduced everywhere; in particular, the cold stream outlet temperature $T_{c,o}$ approaches the hot stream inlet temperature, $T_{h,i}$ and the hot stream outlet temperature $T_{h,o}$ approaches the cold stream inlet temperature $T_{c,i}$. This is illustrated graphically in Figure 9.8. The limit is reached when the outlet temperature of the stream having the smaller capacity rate becomes equal to the inlet temperature of the other stream. For the sake of argument let us assume this occurs when $T_{c,o} = T_{h,i}$ (which implies $C_c = C_{min}$). According to the limitations imposed by the Second Law of Thermodynamics, the heat transfer in this state is the maximum possible heat transfer; it is given by

$$\dot{Q}_{max} = C_{min}(T_{c,o} - T_{c,i})$$

$$= C_{min}(T_{h,i} - T_{c,i}) \tag{9.30}$$

Consequently we are able to write

$$\varepsilon = \frac{\dot{Q}}{\dot{Q}_{max}}$$

$$= \frac{C_h(T_{h,i} - T_{h,o})}{C_{min}(T_{h,i} - T_{c,i})} \tag{9.31a}$$

$$= \frac{C_c(T_{c,o} - T_{c,i})}{C_{min}(T_{h,i} - T_{c,i})} \tag{9.31b}$$

Although derived from considering a counterflow device, Equation (9.31b) for the effectiveness is valid for all other heat exchanger configurations, since the temperature difference in the denominator is always the largest temperature difference between the two streams.

Now let us restate the heat transfer rate in terms of these two new parameters for a parallel flow heat exchanger, and for argument assume $C_{min} = C_c$. First, from Equation (9.24),

$$\dot{Q} = UA\left(\frac{\Delta T_2 - \Delta T_1}{\ln(\Delta T_2/\Delta T_1)}\right)$$

it is apparent that

$$\ln\left(\frac{\Delta T_2}{\Delta T_1}\right) = \frac{UA}{\dot{Q}}(\Delta T_2 - \Delta T_1) \tag{9.32}$$

Since $\Delta T_1 = T_{h,i} - T_{c,i}$ and $\Delta T_2 = T_{h,o} - T_{c,o}$ the temperature differences on the right-hand side of Equation (9.32) may be expanded out to give

$$\ln\left(\frac{\Delta T_2}{\Delta T_1}\right) = -\frac{UA}{\dot{Q}}(T_{c,o} - T_{c,i})\left\{1 - \left(\frac{T_{h,i} - T_{h,o}}{T_{c,o} - T_{c,i}}\right)\right\} \tag{9.33}$$

From Equation (9.29) $\dot{Q} = C_c(T_{c,o} - T_{c,i})$ and from the definition of NTU (9.26) Equation (9.33) becomes

$$\ln\left(\frac{\Delta T_2}{\Delta T_1}\right) = -\text{NTU}\left\{1 - \left(\frac{T_{h,i} - T_{h,o}}{T_{c,o} - T_{c,i}}\right)\right\} \tag{9.34}$$

or

$$\ln\left(\frac{\Delta T_2}{\Delta T_1}\right) = -\text{NTU}\left\{1 + \left(\frac{C_{min}}{C_{max}}\right)\right\} \tag{9.35}$$

By manipulation of the temperature differences on the left-hand side of Equation (9.35), and using the definition of the effectiveness ε (9.31), it can be shown that

$$\frac{\Delta T_2}{\Delta T_1} = 1 + \varepsilon\left\{-1 + \left(\frac{T_{h,o} - T_{h,i}}{T_{c,o} - T_{c,i}}\right)\right\}$$

$$= 1 + \varepsilon\left\{-1 - \frac{C_{min}}{C_{max}}\right\} \tag{9.36}$$

From which, by substitution into Equation (9.35),

$$\varepsilon_{\text{parallel flow}} = \frac{1 - \exp[-\text{NTU}\{1 + (C_{min}/C_{max})\}]}{1 + (C_{min}/C_{max})} \tag{9.37}$$

and, from inverting this expressio

$$NTU_{parallel\,flow} = \frac{-\ln[1 - \varepsilon\{1 + (C_{min}/C_{max})\}]}{1 + (C_{min}/C_{max})} \tag{9.38a}$$

Although the above assumes that $C_{min} = C_c$, an identical expression will be obtained with $C_{min} = C_h$. The cases of $C_{max} = C_{min}$ and $C_{min}/C_{max} = 0$ (when the fluid C_{max} changes phase at constant pressure), are given by

$$\varepsilon_{parallel\,flow} = \frac{1}{2}\{1 - \exp(-2NTU)\} \qquad C_{max} = C_{min} \tag{9.38b}$$

$$\varepsilon_{parallel\,flow} = \{1 - \exp(-NTU)\} \qquad \frac{C_{min}}{C_{max}} = 0 \tag{9.38c}$$

A similar analysis applied to a counterflow heat exchanger yields the following relations:

$$\varepsilon_{counterflow} = \frac{1 - \exp\left[-NTU\left(1 - \dfrac{C_{min}}{C_{max}}\right)\right]}{\left(1 - \dfrac{C_{min}}{C_{max}}\right)\exp\left[-NTU\left(1 - \dfrac{C_{min}}{C_{max}}\right)\right]} \tag{9.39a}$$

$$\varepsilon_{counterflow} = \frac{NTU}{1 + NTU} \qquad C_{max} = C_{min} \tag{9.39b}$$

$$\varepsilon_{counterflow} = 1 - \exp(-NTU) \qquad \frac{C_{min}}{C_{max}} = 0 \tag{9.39c}$$

$$NTU_{counterflow} = \frac{-\ln\left[\left(1 - \dfrac{\varepsilon\,C_{min}}{C_{max}}\right)\Big/(1 - \varepsilon)\right]}{1 - \dfrac{C_{min}}{C_{max}}} \tag{9.40}$$

The above effectiveness/NTU relations are shown graphically in Figures 9.9 and 9.10, for parallel flow and counterflow respectively. The effectiveness/NTU relations for most other types of heat exchangers can be obtained from Kays and London (1984).

The merits of this technique can best be illustrated with an example.

Example 9.5

Use the effectiveness/NTU method to carry out the performance analysis of Example 9.4.

Fig. 9.9 ●
Effectiveness/NTU
relation for a
parallel flow heat
exchanger

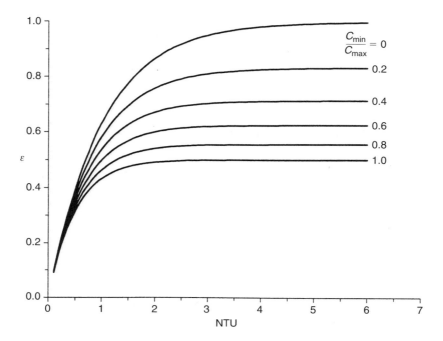

Fig. 9.10 ●
Effectiveness/NTU
relation for a
counterflow heat
exchanger

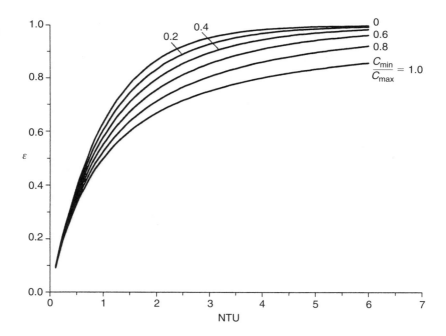

Solution   *Step 1: Identify C_{min}*

$$C_h = 0.15 \times 2131 = 319.65 \, \text{W/K}$$

$$C_c = 0.30 \times 4178 = 1253.4 \, \text{W/K}$$

Hence $C_{min} = C_h$

Step 2: Calculate NTU

$$\text{NTU} = \frac{UA}{C_{min}}$$

$$= \frac{21.9 \times 11.5}{319.65}$$

$$= 0.788$$

Step 3: Calculate effectiveness

$$\varepsilon_{\text{parallel flow}} = \frac{1 - \exp\left[-\text{NTU}\{1 + (C_{min}/C_{max})\}\right]}{1 + (C_{min}/C_{max})}$$

$$= \frac{0.628}{1.255}$$

$$= 0.5$$

Step 4: Evaluate $\dot{Q}$

$$\dot{Q} = \varepsilon \dot{Q}_{max}$$

$$= \varepsilon C_{min}(T_{h,i} - T_{c,i})$$

$$= 0.5 \times 319.65 \times (90 - 10)$$

$$= 12\,796 \, \text{W}$$

Step 5: Apply an energy balance and calculate outlet temperatures

$$\dot{Q} = C_h(T_{h,i} - T_{h,o})$$

so

$$T_{h,o} = T_{h,i} - \frac{\dot{Q}}{C_h}$$

$$= 90 - \left(\frac{12\,796}{319.65}\right)$$

$$= 50°\text{C}$$

Similarly,

$$T_{c,o} = T_{c,i} + \frac{\dot{Q}}{C_c}$$

$$= 10 + \left(\frac{12\,796}{1253.4}\right)$$

$$= 20.2°C$$

Comment   Using the effectiveness/NTU method, a direct solution is obtained without recourse to lengthy iterative calculations which can lead to errors. Although the derivation of the ε and NTU relations is complicated, the results are simple to apply.

9.6 ● Pressure losses

The discussion so far has considered only the thermal performance of heat exchangers. However, the Reynolds analogy (4.89) and (4.90) which links heat transfer and shear stress tells us that increased thermal performance is always associated with an increased pressure loss.

By way of illustration, in forced convection $Nu \propto Re^n$, which implies that the heat flux, q, varies with the average velocity, V^n (where $n < 1$). Since the friction factor, f, varies as $f \propto Re^m$, the pressure drop ΔP will vary as V^{m+2}. It is obvious that $m + 2 > n$, so an increase in the velocity will bring about a larger increase in the pressure drop than in the heat flux. If this pressure drop is too large, the *number* of channels can be increased which will reduce both the heat flux and the pressure drop, but the pressure drop will be reduced more severely. The reduction in heat flux can then be regained by increasing the length of the channels, which also increases ΔP, but in the same proportion as q.

In general, the losses in a heat exchanger are made up of entrance and exit losses and the pressure loss in the passages. These are depicted in Figure 9.11. The loss in the *passageways*, ΔP_p, due to friction is characterised by the friction factor, f, and

$$\Delta P_p = f\left(\frac{L}{D_h}\right)\frac{1}{2}\rho V^2 \tag{9.41}$$

$$f = \frac{64}{Re_{D,h}} \qquad \text{laminar flow, } Re_{D,h} < 2000$$

$$f = 0.3164\,Re_{D,h}^{-1/4} \qquad 2 \times 10^3 < Re_{D,h} < 2 \times 10^4$$

$$f = 0.184\,Re_{D,h}^{-1/5} \qquad 2 \times 10^4 < Re_{D,h} < 10^6$$

Fig. 9.11 ●
Pressure
distribution in a
shell and tube heat
exchanger

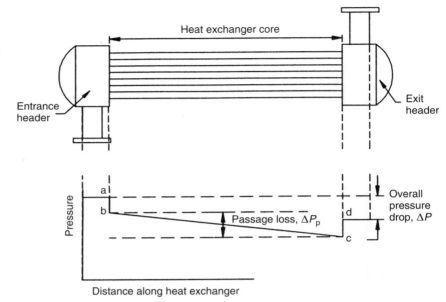

where D_h is the hydraulic diameter of the passageway ($4 \times$ area/perimeter), ρ the fluid density and V (used here to avoid confusion with U), the bulk average velocity of the fluid. Correlations for f for smooth tubes are given above; roughness will affect these results (see the Moody chart presented in Figure 5.4), and for a relative roughness of greater than 0.01, f becomes virtually independent of the Reynolds number.

The entrance and exit losses are characterised by contraction and expansion coefficients, K_c and K_e, respectively. The actual pressure loss, due to irreversibility, during contraction or expansion being the product of K_c or K_e and the dynamic pressure $\frac{1}{2}\rho V^2$.

Considering the *contraction* from a to b (entrance header to tubes) in Figure 9.11 and applying the modified form of Bernoulli's equation to take account of the loss in total pressure,

$$\left(P_a + \frac{1}{2}\rho_a V_a^2\right) - \left(P_b + \frac{1}{2}\rho_b V_b^2\right) = K_c + \frac{1}{2}\rho_b V_b^2 \tag{9.42}$$

From continuity with $\rho_a = \rho_b$, it is apparent that $A_a V_a = A_b V_b$, and Equation (9.42) may be more conveniently rewritten as

$$P_a - P_b = \left(1 - \sigma_{a-b}^2\right)\frac{1}{2}\rho_b V_b^2 + K_c \frac{1}{2}\rho_b V_b^2 \tag{9.43}$$

where σ_{a-b} is an area ratio which for a contraction in the direction of flow is defined as

$$\sigma_{a-b} = \frac{A_b}{A_a} \tag{9.44}$$

Fig. 9.12 ●

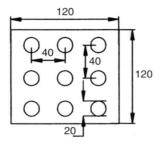

or more generally, the ratio of flow cross-sectional area to the frontal area. For example, Figure 9.12 shows a 3×3 array of equispaced 20 mm dia. tubes pitched 40 mm apart (centreline to centreline) in a square section of dimensions 120 mm $\times$ 120 mm

$$A_b = \frac{9 \times \pi \times 20^2}{4} = 2827\,\text{mm}^2$$

$$A_a = 120 \times 120 = 14\,400\,\text{mm}^2$$

$$\sigma_{a-b} = \frac{2827}{14\,400} = 0.196$$

The pressure rise for a fluid undergoing an *enlargement*, as from the heat exchanger tubes to the exit header (c to d in Figure 9.11), is given in a similar way by the expression

$$P_d - P_c = \left(1 - \sigma_{c-d}^2\right)\frac{1}{2}\rho_c V_c^2 - K_e \frac{1}{2}\rho_c V_c^2 \qquad (9.45)$$

It is worth noting that Equations (9.43) and (9.45) make use of a single velocity to evaluate the respective pressure differences (in this case the velocity in the passegeways, $V_b = V_c$). This velocity is easily obtained, from continuity and knowing the flow rate.

Example 9.6

A tubular heat exchanger comprises 120 smooth tubes, each 8 mm dia., 1.8 m long and set in a cylindrical shell of 0.18 m dia. Find and compare with the overall pressure loss, the entrance, exit and passageway pressure losses for the following two situations:

(a) A total flow of 1 kg/s of air at 200°C and 4 bar

(b) A total flow of 20 kg/s of water at 30°C

For these conditions, take $K_c = 0.35$ and $K_e = 0.55$.

Solution *Step 1: Obtain relevant data, evaluate Reynolds numbers*

$$N = 120, \ L = 1.8 \,\mathrm{m}, \ D_\mathrm{h} = 0.008 \,\mathrm{m}$$

$$\text{Flow area} = \frac{120 \times \pi D_\mathrm{h}^2}{4} = 6032 \,\mathrm{mm}^2$$

$$\text{Frontal area} = \frac{180^2 \times \pi}{4} = 25\,447 \,\mathrm{mm}^2$$

$$\sigma = \frac{6032}{25\,447} = 0.24$$

For air,

$$\rho = \frac{P}{RT} = \frac{4 \times 10^5}{287 \times (273 + 200)} = 2.95 \,\mathrm{kg/m}^3$$

$$\mu = 2.5775 \times 10^{-5} \,\mathrm{kg/m\,s} \quad \text{(tables)}$$

$$\text{Mass flow through each tube} = \dot{m} = \frac{1}{120} \,\mathrm{kg/s}$$

$$\text{Bulk average velocity} = V = \frac{\dot{m}}{\rho \pi D_\mathrm{h}^2/4}$$

$$= \frac{1/120}{2.95 \times \pi \times 0.008^2/4} = 56.2 \,\mathrm{m/s}$$

$$Re = \frac{\rho V D_\mathrm{h}}{\mu} = \frac{2.95 \times 56.2 \times 0.008}{2.5775 \times 10^{-5}} = 51\,456$$

For water,

$$\rho = 1000 \,\mathrm{kg/m}^3 \text{ and } \mu = 0.797 \times 10^{-3} \,\mathrm{kg/m\,s} \quad \text{(tables)}$$

$$\dot{m} = \frac{20}{120} \,\mathrm{kg/s}, \text{ from which } V = 3.3 \,\mathrm{m/s}$$

$$Re = 33\,280$$

Step 2: Evaluate pressure loss in passageways
Both flows exceed the transition Reynolds number of 2000, so a turbulent correlation (9.41), $f = 0.184 Re_D^{-0.2}$, is used to evaluate the friction factor f.
For air,

$$f = 0.184(51\,456)^{-0.2} = 0.0210$$

$$\Delta P_\mathrm{p} = f\left(\frac{L}{D_\mathrm{h}}\right)\frac{1}{2}\rho V^2$$

$$= 0.021 \times \left(\frac{1.8}{0.008}\right) \times 0.5 \times 2.95 \times 56.2^2$$

$$= 2.2 \times 10^4 \,\mathrm{N/m}^2$$

For water,

$$f = 0.184(33\,280)^{-0.2} = 0.0229$$

$$\Delta P_{\mathrm{p}} = f\left(\frac{L}{D_{\mathrm{h}}}\right)\frac{1}{2}\rho V^2$$

$$= 0.0229 \times \left(\frac{1.8}{0.008}\right) \times 0.5 \times 1000 \times 3.3^2$$

$$= 2.8 \times 10^4 \,\mathrm{N/m^2}$$

Step 3: Evaluate inlet and exit losses
Inlet loss:

$$P_{\mathrm{a}} - P_{\mathrm{b}} = (1 - \sigma_{\mathrm{a-b}}^2)\tfrac{1}{2}\rho_{\mathrm{b}}V_{\mathrm{b}}^2 + K_{\mathrm{c}}\tfrac{1}{2}\rho_{\mathrm{b}}V_{\mathrm{b}}^2$$

For air,

$$P_{\mathrm{a}} - P_{\mathrm{b}} = [(1 - 0.24^2) \times 0.5 \times 2.95 \times 56.2^2] + [0.35 \times 0.5 \times 2.95 \times 56.2^2]$$

$$= 6021 \,\mathrm{N/m^2}$$

For water,

$$P_{\mathrm{a}} - P_{\mathrm{b}} = [(1 - 0.24^2) \times 0.5 \times 1000 \times 3.33^2] + [0.35 \times 0.5 \times 1000 \times 3.33^2]$$

$$= 7166 \,\mathrm{N/m^2}$$

Exit loss:

$$P_{\mathrm{d}} - P_{\mathrm{c}} = (1 - \sigma_{\mathrm{c-d}}^2)\tfrac{1}{2}\rho_{\mathrm{c}}V_{\mathrm{c}}^2 - K_{\mathrm{e}}\tfrac{1}{2}\rho_{\mathrm{c}}V_{\mathrm{c}}^2$$

For air,

$$P_{\mathrm{d}} - P_{\mathrm{c}} = [(1 - 0.24^2) \times 0.5 \times 2.95 \times 56.2^2] - [0.55 \times 0.5 \times 2.95 \times 56.2^2]$$

$$= 1828 \,\mathrm{N/m^2}$$

For water,

$$P_{\mathrm{d}} - P_{\mathrm{c}} = [(1 - 0.24^2) \times 0.5 \times 1000 \times 3.33^2] - [0.55 \times 0.5 \times 1000 \times 3.33^2]$$

$$= 2176 \,\mathrm{N/m^2}$$

The total loss is $\Delta P = \Delta P_{\mathrm{p}} + \Delta P_{\mathrm{a-b}} - \Delta P_{\mathrm{d-c}}$:

For air,

$$\Delta P = 22\,000 + 6021 - 1828 = 26\,193 \,\mathrm{N/m^2}$$

For water,

$$\Delta P = 28\,000 + 7166 - 2176 = 32\,990 \,\mathrm{N/m^2}$$

Comments It is seen that the overall loss is dominated by the frictional pressure loss in the passageways. It accounts for 82%, of the overall pressure loss in the case of air and 85% in the case of water. The example also illustrates that the density of a fluid has a profound effect on the overall pressure loss: for roughly the same overall ΔP, the mass flow of water is 20 times the mass flow of air. It is for this reason that the pressure loss on the gas side of a heat exchanger is likely to be the most crucial factor in determining the overall pressure drop (just as the gas-side heat transfer coefficient can be the governing factor of the overall heat transfer coefficient).

9.7 ● Closing comments

Heat exchangers are used to transfer heat from one fluid stream to one or more others. The word itself describes a broad spectrum of devices, which may be classified as either recuperative, regenerative or direct contact heat exchangers. Heat exchangers may also be classified according to flow configuration (parallel flow, counterflow, crossflow or periodic flow); size (ratio of heat exchange surface area to volume); or type (double pipe, shell and tube, plate fin, tube fin, etc.).

This chapter has focused on recuperative heat exchangers. Two questions are most likely to arise:

1. The design question: What heat exchange surface area is required?

2. The performance question: What are the outlet temperatures?

The logarithmic mean temperature difference method is the favoured method to permit analysis for design. The effectiveness/NTU method is preferred for analysis for performance. In order to maintain simplicity, and to illustrate principles rather than discuss specific detail, much of the analysis in this chapter has concentrated on simple heat exchanger types and flow configurations. Correction factors to the logarithmic mean temperature difference (derived for counterflow conditions) and effectiveness/NTU characteristics are available (usually as charts but also as software) for other heat exchanger types and flow configurations.

In a recuperative heat exchanger, heat is transferred from the hot fluid stream to the cold fluid stream by convection in the streams and conduction in the solid walls. The effect of these two processes is combined in the overall heat transfer coefficient, which should also include an allowance for fouling of the passages due to the accumulation of deposits (for example, dissolved salts, debris). For simple geometries, such as pipes and passageways of regular cross-section, the overall heat transfer coefficient can be estimated using forced convection correlations such as those in Chapter 5, and conduction relations such as those in Chapter 2. For more complex geometries, it will probably be necessary to use experimental data.

Fins, extended surfaces and turbulence promoters may be incorporated in a heat exchanger to enhance the overall thermal performance. However, since the overall heat transfer coefficient is dominated by the highest thermal resistance (the stream with the lowest heat transfer coefficient), little advantage is gained by adding fins to the heat transfer surfaces adjacent to the fluid stream having the largest heat transfer coefficient.

The thermal performance of a heat exchanger is inextricably linked to pressure losses. Any judgement about the suitability of a heat exchanger must be made with information concerning the pressure losses as well as the thermal performance.

9.8 ● References

Bejan, A. (1993). *Heat Transfer*. New York: Wiley

Kays, W.M. and London, A.L. (1984). *Compact Heat Exchangers* 3rd edn. New York: McGraw-Hill

Kays, W.M. and Perkins, H.C. (1973). *Handbook of Heat Transfer*. New York: McGraw-Hill

Sekulic, D.P. and Shah, R.K. (1995). *Advances in Heat Transfer*. Vol. **26**: *Thermal Design Theory of Three-fluid Heat Exchangers*. New York: Academic Press.

9.9 ● End of chapter questions

9.1 A heat exchanger consists of numerous rectangular channels, each 18 mm wide and 2.25 mm high. In an adjacent pair of channels, there are two streams: water $k = 0.625$ W/m K and air $k = 0.0371$ W/m K, separated by an 18 mm wide and 0.5 mm thick stainless steel plate of $k = 16$ W/m K. The fouling resistances for air and water are 2×10^{-4} m^{2}K/W and 5×10^{-4} m^{2}K/W, respectively, and the Nusselt number is given by $Nu_{D,h} = 5.95$ where the subscript 'D,h' refers to the hydraulic diameter.

(a) Calculate the overall heat transfer coefficient ignoring both the thermal resistance of the separating wall and the two fouling resistances.

[52.1 W/m²K]

(b) Calculate the overall heat transfer coefficient with these resistances.

[50.2 W/m²K]

(c) Which is the controlling heat transfer coefficient?

9.2 Figure 9.13 shows the layout of a surface to be used in a gas-to-liquid heat exchanger, where it is proposed to enhance the overall heat transfer by the addition of cylindrical pin-fins each of 5 mm dia. and 20 mm length. The base material and the

pin-fins both have a thermal conductivity of 30 W/m K. The heat transfer coefficient on the gas side is 30 W/m²K, and that on the liquid side 180 W/m²K.

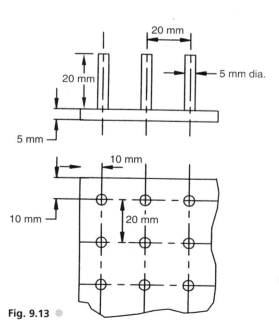

Fig. 9.13 ●

(a) Briefly explain why the fins should be attached to the air side.

By considering a 1 m × 1 m section of the surface:

(b) Calculate the overall heat transfer coefficient without the fins.

$$[25.6 \, \text{W/m}^2\text{K}]$$

(c) Calculate the overall heat transfer coefficient with the fins. (Ignore any contribution to the area or heat transfer from the fin tip.)

$$[38.8 \, \text{W/m}^2\text{K}]$$

9.3 A heat exchanger tube of $D = 20$ mm dia. conveys 0.0983 kg/s of water ($Pr = 4.3$, $k = 0.632$ W/m K, $\rho = 1000$ kg/m^3, $\mu = 0.651 \times 10^{-3}$ kg/m s) on the inside which is used to cool a stream of air on the outside where the external heat transfer coefficient has a value of $h_o = 100$ W/m^{2}K. Ignoring the thermal resistance of the tube walls, evaluate the overall heat transfer coefficient, U, assuming that the internal heat transfer coefficient is given by the Dittus–Boelter relation for fully developed pipe flow: $Nu_D = 0.023Re_D^{0.8}Pr^{0.4}$.

$$[95.2 \, \text{W/m}^2\text{K}]$$

9.4 In a tubular heat exchanger design, the overall heat transfer coefficient is to be increased (by a factor ϕ) by the addition of annular fins (of thickness t and outer diameter D), having a fin efficiency of η. If the spacing between the fins is s (i.e. when $s = 0$ adjacent fins touch) and the outer diameter of the unfinned (plane) tube is d (neglect the contributions of the peripheral area of the fin), show that the ratio s/d is given by

$$\frac{s}{d} = \frac{\left(\frac{h_o\eta}{2}\right)\left(\frac{D^2}{d^2} - 1\right)\left(\frac{1}{\phi U_{\text{plane tube}}} - \frac{1}{h_i}\right) - \frac{t}{d}}{1 - \left(\frac{h_o}{\phi U_{\text{plane tube}}} + \frac{h_o}{h_i}\right)}$$

where the inside heat transfer coefficient is h_i, the outside heat transfer coefficient h_o and the overall heat transfer coefficient for the plane tube, $U_{\text{plane tube}}$.

9.5 Use the data for Question 9.3 to estimate the fin spacing required to double the plane tube heat transfer coefficient using 50 mm dia. and 3 mm thick fins when $\eta = 0.95$.

$$[39.5 \, \text{mm}]$$

9.6 A heat exchanger is to be used as a water-cooled oil cooler. In order to determine its thermal performance in this role, a test is carried out using water and air in parallel flow. The flow rates of the water and air are set to 3 kg/s and 2 kg/s, respectively, the measured air inlet temperature is 90°C, the water inlet temperature 10°C and the water exit temperature 18°C.

Calculate the oil exit temperature when this heat exchanger is used as an oil cooler and the flow rates of oil and water are 3 kg/s and 3.3 kg/s, respectively. Assume that the original overall heat transfer coefficient is halved, the oil inlet temperature is 110°C and the water inlet temperature 10°C.

Take $C_{p,\text{air}} = 1$ kJ/kg K, $C_{p,\text{oil}} = 2.2$ kJ/kg K and $C_{p,\text{water}} = 4$ kJ/kg K.

$$[96°C]$$

9.7 A heat exchanger is required to cool 1 kg/s of compressed air from 250°C to 40°C using 5 kg/s of water at 20°C. Assuming a value for the overall heat transfer coefficient of 100 W/m^{2}K, and taking $C_{p,\text{air}} = 1$ kJ/kg K, $C_{p,\text{water}} = 4$ kJ/kg K, calculate the area required for (a) parallel flow, and (b) counterflow.

$$[(a) \ 30.35 \, \text{m}^2; \ (b) \ 25.22 \, \text{m}^2]$$

9.8 The coolant flow for the heat exchanger in Question 9.7a is reduced to 2 kg/s and the overall heat transfer coefficient remains unchanged. Calculate, using an iterative LMTD method, the new air exit temperature. As an aid, start with an initial guess of $T_{\text{air,out}} = 60°C$ and plot a graph of the guessed and recalculated air exit temperatures.

$$[52.5°C, \text{ approximately }]$$

9.9 Verify the answer to Question 9.8, using the effectiveness/NTU method.

$$[52.3°C]$$

9.10 A single-pass crossflow heat exchanger is to be used to cool air from 250°C to 40°C, using air at 20°C. Taking a value for the overall heat transfer coefficient of 100 W/m^{2}K, calculate the heat exchanger area required for cooling air flow rates of 2, 4, 8 and 10 kg/s. Which would be the most acceptable choice?

$$[52.3, \underline{34.3}, 28.8 \text{ and } 27.5 \, \text{m}^2]$$

9.11 A heat exchanger design is required to cool 10 kg/s of gaseous methane (CH_4) at an absolute pressure of 10 bar, from 100°C to 20°C using 25 kg/s of sea water which is available at 5°C. The methane is to flow through a number of 10 mm dia. tubes, and the water over the outside.

(a) Explain why this will not be able to operate in parallel flow.

(b) Calculate the tube pressure loss per metre length for a stack of 2000, 1000 and 500 tubes.

[9021, 2592, 744 Pa/m]

(c) Assuming that the tube-side heat transfer coefficient is the controlling heat transfer coefficient, calculate U and the heat exchanger area required for each of the configurations in part (b).

[1165, 670, 384 W/m²K and 40, 70, 122 m²]

(d) Estimate the length of tubing required and the overall tube-side pressure loss.

[2.55, 2.23, 1.94 m and 23 000, 5780, 1400 Pa]

(e) Comment on the effectiveness of reducing the pressure loss by increasing the number of tubes.

(f) The design with 1000 tubes is to be chosen, but owing to variations in operating conditions, the methane inlet temperature is increased to 160°C. Use an appropriate method to calculate the new gas exit temperature.

[29.5°C]

For $Re_D > 2000$ take $Nu_D = 0.023Re_D^{0.8}Pr^{0.4}$ and $f = 0.184Re_D^{-0.2}$.

Take R_0 (the universal gas constant) as 8.214 kJ/kmol K.

For methane, take $C_p = 2.226$ kJ/kg K, $\mu = 1.3 \times 10^{-5}$ kg/m s, $k = 0.03$ W/mK
For water, take $C_p = 4$ kJ/kg K.

Chapter ten

Heat transfer instrumentation

10.1 ● Introduction

In the previous chapters we have seen how surface temperatures and heat fluxes can be predicted from an understanding of the physical processes and by applying suitable approximations. But it is also desirable to *measure* temperature and heat flux directly. Theory and experiment both have their place and limitations. Theory can guide experimental work by providing insight into what are, and what are not, influential parameters; experiments can provide evidence to validate theoretical models. In certain circumstances, the combination of flow regime and geometry may be beyond the capability of analytical and numerical modelling, and experimental measurements are the only method in which one has confidence.

This chapter aims to review some of the more common temperature and heat flux measurement techniques. The three main sections cover temperature measurement, heat flux measurement and finally errors in temperature measurement. Naturally in the limited space available in this text, it has not been possible to cover all aspects of this extensive topic. In particular, I have chosen to omit any reference to the more complex optical techniques such as CARS (coherent anti-Stokes Raman spectroscopy) and LIF (laser-induced fluorescence), which the average user is unlikely to encounter. There are several books giving adequate coverage on the subject of heat transfer instrumentation, many comprehensive review articles in journals and countless journal and conference papers. A list of references is included at the end of this chapter and the interested reader can consult these publications for further information. The two review articles (one on heat flux and the other on temperature instrumentation) by Childs *et al.* (1999a, b) are also useful in this respect and contain a large and up-to-date list of references.

As a fundamental quantity, temperature is relatively straightforward to measure. However, although it may be easy to measure a temperature, if care is not taken with the installation of a temperature probe, it is also easy to measure the wrong temperature. A gas temperature probe, for example, may measure an erroneous temperature owing to radiation to nearby pipe walls; a surface temperature thermocouple installed in a slot may not record the temperature that would exist in the absence of the slot, thermocouple and adhesive. Uncertainties can also arise from calibration errors and use of inappropriate temperature sensors.

It is possible to buy ready-made heat transfer gauges, to design and makes one's own, or by careful design of temperature instrumentation, to make the test surface of the experiment itself a heat transfer gauge. Measurements of heat flux are usually derived from temperature measurements; for example, from a temperature difference across a material with a known thermal conductivity and thickness, or given the thermal capacity of a material, from the rate of change of temperature with time. Accurate temperature measurement is therefore a prerequisite to accurate measurement of heat flux.

10.2 ● Temperature measurement

A thorough discussion of the scientific principles which underpin the definition of temperature and temperature scales can be found in Rogers and Mayhew (1992). To *define* a temperature scale it is necessary to take two fixed points. For example, the Celsius (or centigrade) scale is based on taking the difference in temperature between the melting point of ice and the boiling point of water and dividing this into one hundred equal increments. For reference, Table 10.1 presents a list of some of the fixed points used in the international practical temperature scales, given at conditions of standard atmospheric pressure (101 325 Pa).

To *measure* temperature, use is made of some physical effect. For example, the expansion of a liquid or gas, change of electrical resistance with temperature, changes in crystal structure with temperature, or the thermoelectric effect. If the physical effect has a linear variation with temperature, then there will be a direct linear correspondence between the output of the instrument and the measured temperature. A variety of relatively common devices which transduce temperature to a physical effect are described below.

Table 10.1 ●
Selection of fixed points used in international practical temperature scales

Fixed point	°C
Boiling point of oxygen	−182.962
Sublimation point of carbon dioxide	−78.476C
Ice point of pure water	0
Triple point of pure water	0.01
Boiling point of pure water	100
Freezing point of bismuth	271.442
Boiling point of sulphur	444.674
Freezing point of silver	961.93
Freezing point of gold	1064.43

10.2.1 Liquid-in-glass thermometer

The liquid-in-glass thermometer is probably the most familiar and common form of temperature measuring instrument. The liquid is usually either alcohol or mercury and is used mostly either as a quick check, or for an accurate calibration of, for example, a thermocouple. Its advantages are simplicity in use and, for reference standard instruments, it can be very accurate. The disadvantages are inconvenience if many temperatures need to be measured, and the output is visual and cannot be readily interfaced to a computer.

10.2.2 Thermocouples

Thermocouples rely on the Seebeck effect, which occurs when two dissimilar metals are joined together and an electromotive force (e.m.f.) is generated. The magnitude of this e.m.f. depends on the temperature difference between the ends, or junctions, of the thermocouple. It is usual to refer to the junctions of a thermocouple probe as the hot junction, temperature T_h, and the cold junction, temperature T_c. The Peltier effect describes the reverse process, where a temperature difference is created by supplying an e.m.f. to two dissimilar materials. See Benedict (1984) or Doebelin (1976) for further details of the thermoelectric effect, thermocouple 'laws', calibration, and so on.

Thermocouples, therefore, require no external power supply. Many different materials are available for use as thermocouples. Some of the more common are listed in Table 10.2, along with their designation code, output (or sensitivity) and useful range (excluding any additional limitations imposed by insulating materials).

An idealised thermocouple circuit for $T_c = 0°C$ is shown in Figure 10.1. Over a relatively narrow range of, say, 0 to 100°C it is usual to represent the temperature–voltage relationship as a quadratic equation of the form

$$T_h = aV^2 + bV \tag{10.1}$$

where a and b are coefficients found from calibration. For a chromel/alumel thermocouple (see Table 10.2), $b \approx 25°C/mV$, and $a \approx -0.283°C/mV^2$. Equation (10.1) is unlikely to be acceptable for accurate measurements

Table 10.2 ●
Some common
thermocouple
materials

Thermocouple material	Designation code	Typical output (0–100°C)	Useful range (continuous use[a])
Chromel/constantan	Type E	68 $\mu V/°C$	$0 \leqslant T \leqslant 800°C$
Iron/constantan	Type J	46 $\mu V/°C$	$20 \leqslant T \leqslant 700°C$
Copper/constantan	Type T	46 $\mu V/°C$	$-185 \leqslant T \leqslant 300°C$
Chromel/alumel	Type K	42 $\mu V/°C$	$0 \leqslant T \leqslant 1100°C$

[a] Excluding any additional limitations imposed by insulating materials

Fig. 10.1 ●
An idealised
thermocouple
circuit and
hypothetical
calibration curve

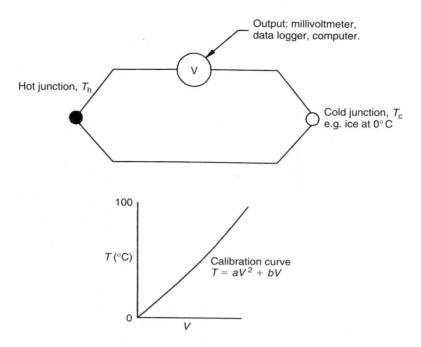

across a wider range, say 0 to 600°C, so a higher-order polynomial should be used. A word of warning is required here. A polynomial of degree n has $n - 1$ turning points. So using a high-order polynomial to calibrate an instrument in a relatively narrow range such as 0–100°C, may result in (physically unrealistic) turning points in the calibration curve when the range is extended.

Figure 10.2 shows a typical calibration set-up for thermocouples, which may also be adapted for the calibration of other temperature sensors. Since water is used to heat the thermocouple and reference instrument, the range of calibration is restricted to that between the freezing point and boiling point of water. For higher temperatures, a calibration oven could be used, or alternatively some of the fixed points (Table 10.1)

The hot junction of the thermocouple is taped (commercially available copper-backed adhesive tape is available for this) to a suitable reference instrument. A liquid-in-glass thermometer is indicated but a platinum resistance thermometer could also be used. The thermocouple and reference instrument are placed in a heated water bath containing pure (distilled) water. The cold junctions are arranged as shown (oil is used to create an even temperature at the cold junctions), and placed in a vacuum flask containing crushed ice also made from distilled water. The water bath, which should have a built-in mixer to circulate the water, is heated using a thermostatically controlled electric heater. Readings of voltage output from the thermocouple and the temperature indicated by the

Fig. 10.2 ●
Thermocouple
calibration
apparatus
(for 0–100°C)

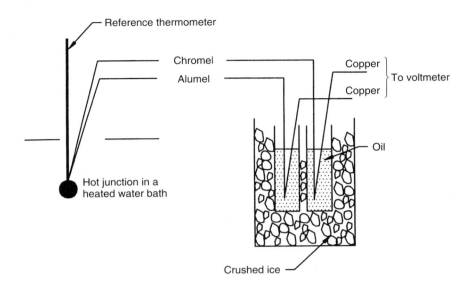

reference thermometer are taken at suitable (≈ 3 to 5°C) intervals. A least-squares fit of temperature against voltage provides the values of the coefficients in Equation (10.1). If high accuracy is required, it is worth repeating this process, both on an individual thermocouple in order to judge systematic errors, and on different thermocouples from the same batch of wire in order to gauge variations in calibration owing to variations in wire composition.

To obtain adequate time response, thermocouple probe wires are usually very thin (0.25 mm dia. is not uncommon), and are made of high quality wire which is relatively expensive. It is not always practicable to extend this thin, expensive wire all the way to a convenient measuring instrument, such as a data logger or computer. In these circumstances, compensating cable is used. This has the same thermoelectric characteristic as the thermocouple wire, but over a limited range of temperature. It may also not be practicable to maintain the cold junction(s) at 0°C. Figure 10.3 illustrates the use of compensating cable and a variable temperature cold junction, or reference junction.

The voltage, V, is generated by the temperature difference $T_h - T_c$, and for a quadratic fit, or higher order, of temperature against voltage, the magnitude of V will also depend on the temperature T_c. Since thermocouples are calibrated with respect to the ice point ($T_c = 0°C$) the following procedure must be applied to obtain the correct temperature when $T_c \neq 0°C$:

1. Measure T_c by some independent method (e.g. an electronic sensor).
2. Convert this to an equivalent signal for the thermocouple used.
3. Add this to the actual signal voltage.
4. Convert to °C.

Fig. 10.3 ● (a) Practical thermocouple measuring circuit;
(b) thermocouple output characteristic; (c) isothermal box

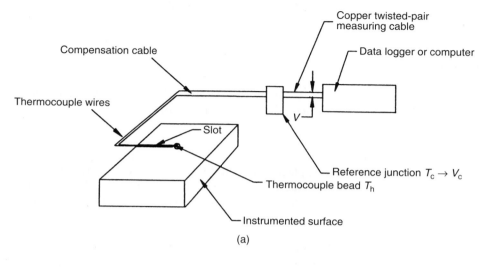

Copper twisted-pair
measuring cable

Compensation cable

Data logger or computer

Thermocouple wires

Slot

V

Reference junction $T_c \rightarrow V_c$

Thermocouple bead T_h

Instrumented surface

(a)

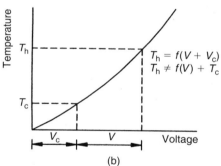

Temperature

T_h

T_c

$T_h = f(V + V_c)$
$T_h \neq f(V) + T_c$

V_c V Voltage

(b)

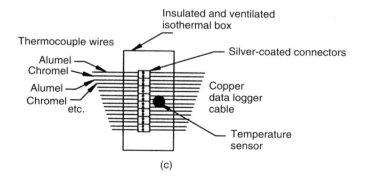

Insulated and ventilated
isothermal box

Thermocouple wires

Silver-coated connectors

Alumel
Chromel

Alumel
Chromel
etc.

Copper
data logger
cable

Temperature
sensor

(c)

In commercially available hand-held instruments, such as the thermocouple probe with a digital readout, this four-step procedure is carried out inside the instrument. The principle of cold junction compensation is illustrated in Example 10.1.

Example 10.1

A calibration on a thermocouple gives the following temperature/voltage characteristic, where T is in °C and V in mV:

$$T = 17V - 0.2V^2$$

The thermocouple is used to measure the surface temperature of a heated cylinder which is placed in a wind tunnel. The temperature of the reference junction is measured by an electronic sensor as 30°C. The voltage generated by the thermocouple is 5.6 mV. What is the temperature of the cylinder?

Solution　For the thermocouple,

$$T = aV^2 + bV$$

hence (discarding the negative root),

$$V = \frac{-b + (b^2 - 4ac)^{1/2}}{2a}$$

where $a = -0.2$, $b = 17$, $c = -T$.
　　The voltage equivalent for a cold junction at 30°C is

$$V = \frac{-17 + (17^2 - 4 \times -0.2 \times -30)^{1/2}}{2 \times -0.2}$$

$$= 1.803 \, \text{mV}$$

Total signal = Measured + Cold junction equivalent

$$= 5.6 + 1.803$$

$$= 7.403 \, \text{mV}$$

Surface temperature from the thermocouple equation $= (17 \times 7.403) - (0.2 \times 7.403^2) = 114.9°C$

Comment　The actual measurement of 114.9°C should be compared with the incorrect method of adding the cold junction temperature to the measured tempera-ture:

$$T = 30 + \{(17 \times 5.6) - (0.2 \times 5.6^2)\} = 118.9°C$$

So the error is 4°C.

When, as is often the case, many thermocouples are used, it is important that there are no variations of temperature across the cold junctions. Uniformity can be achieved by locating the cold junctions in an isothermal box, which through the use of insulation and ventilation maintains all the cold junctions at the same temperature. A schematic diagram of an isothermal box is depicted in Figure 10.3c.

10.2.3 Resistance thermometers

The operation of resistance thermometers depends on measuring the change of electrical resistance with temperature, which ideally may be expressed as

$$R \approx R_0(1 + \alpha T) \tag{10.2}$$

where R is the resistance at temperature T, R_0 is a reference resistance and α is the temperature coefficient of resistance. In general, the relationship is non-linear, so

$$R = R_0(1 + \alpha T + \beta T^2 + \gamma T^3 + \ldots)$$

Platinum, however, has a linear characteristic, $R \approx R_0(1 + \alpha T) \pm 4\%$ for the range $-200°C \leqslant T \leqslant 40°C$. This, combined with the relatively high resistivity $(10^{-7}\,\Omega m)$, makes it particularly suitable for resistance thermometry. For platinum, $\alpha \approx 38 \times 10^{-4}\,K^{-1}$ and typically $10\,\Omega \leqslant R_0 \leqslant 25\,k\Omega$. A high resistance is desirable as the effect of other resistances in the circuit is reduced and the relatively large e.m.f. produced is much greater than voltages induced by thermoelectric effects at junctions between dissimilar materials elsewhere in the circuit. Unlike thermo-couples, these devices require an external power supply (or energisation); they are usually arranged as part of a bridge circuit.

10.2.4 Thermistors

Thermistors are solid-state devices which exhibit a marked, but non-linear, change of resistance with temperature. Their temperature/resistance char-acteristic is usually exponential and can be expressed as

$$R = R_0 \exp \beta \left(\frac{1}{T} - \frac{1}{T_0} \right) \tag{10.3}$$

where β is the temperature coefficient of resistance, R is the resistance at temperature T and R_0, the resistance at T_0. The change in resistance is much greater than for pure resistance thermometers for example, in the range 20°C to 150°C the ratio R/R_0 varies from 1 to 0.001 (using the data for platinum resistance thermometers given in Section 10.2.3, the corresponding variation is from 1 to 1.5). This gives them a very high sensitivity.

10.2.5 Solid-state devices

Solid-state devices operate using the forward voltage/temperature coefficient of a silicon diode, which has a value of approximately $-2\,\mathrm{mV}/^{\circ}\mathrm{C}$. They require separate energisation, but the circuitry is relatively simple; see, for example, Usher (1985) for further details. The output is linear; however, the constant of proportionality should be established by calibration.

Solid-state devices are useful in practical thermocouple measuring circuits, in particular for measuring the cold junction temperature (see Figure 10.3). The signal from the solid-state sensor together with those from the thermocouples can be fed directly into a PC or data logger and the cold junction compensation carried out in software.

10.2.6 Thermal paints and liquid crystals

Thermal paints and liquid crystals undergo a change in crystal structure at a particular temperature, which manifests itself as a change in colour. For thermal paints, this is usually an irreversible process. The colour change with liquid crystals is reversible but liquid crystals generally cannot be used above 120°C.

Although paints were developed in the 1950s and thermochromic liquid crystals in the 1960s, their use was limited because they provided a purely visual indication of temperature. However, it is now possible to buy a relatively inexpensive video camera and a 'frame grabber' board for a PC. This allows electronic images of the painted test surface to be captured and stored on a computer. Since the colour displayed by the liquid crystal surface is directly related to temperature, software analysis of stored images converts the red, green and blue attributes of each pixel to the physical attribute of hue or colour. The temperature can then be inferred from a previous calibration of hue value against known temperatures. Further details on the selection, calibration and use of liquid crystals is given by Jones (1992) and Camci (1996). The significant advantage of using these methods is that they are non-intrusive (there is no physical presence of a probe creating local disturbances of heat flow and hence temperatures). They also make it possible to obtain a surface temperature 'map' rather than measurements at discrete locations.

10.2.7 Radiation pyrometer

A radiation pyrometer measures the thermal radiation emitted by an object ($q = \sigma T^4$ for a black body). A photon detector is used to detect the amount of incident thermal radiation, q. Signal processing converts this to a temperature, and most pyrometers include internal switches to allow for measurements from surfaces of different emissivity. Further detail on radiation thermometry is available in Morris (1988).

Radiation pyrometers can be used to measure temperature at a point or, as in the case of thermographic systems used in hospitals to detect tumours, to obtain a complete surface temperature map. As with liquid crystals, they provide a non-intrusive form of surface temperature measurement. The disadvantage is the relatively high cost. Care needs to be taken, however, when viewing through optically transparent 'windows' as most conventional glasses are virtually opaque to thermal radiation at wavelengths greater than $3\,\mu$m, corresponding to the peak radiation emitted from a black body at approximately $800\,$K, and it is often necessary to use sapphire, germanium or zinc selenide glass.

10.3 ● Measurement of heat flux

Heat fluxes and heat transfer coefficients can be obtained experimentally by measuring temperatures at discrete points of a solid. These are used as boundary conditions to solve the appropriate conduction equation and the temperature field inside the solid. Numerical differentiation of the interior temperature field then provides the heat flux at the surface.

Heat fluxes can also be measured using the rate of change of the temperature of an object with time, either as a lumped mass (that is, the low Biot number approximation) or as a semi-infinite solid (when the Biot number is large). The reader is referred to the article by Schultz and Jones (1973) for a comprehensive review of transient techniques in experimental heat transfer. In addition, it is possible to purchase ready-made heat flux gauges. Diller (1987) describes a number of devices made using micro-manufacturing techniques.

It is important to note that in all cases the heat flux measured is a *total* heat flux. That is, the sum of the *convective and radiative* contributions. If, as is often the case, an experimental investigation is concerned with measuring and correlating convective heat flux, then the contribution from radiation must be subtracted from the measurements. In practice this is best achieved either by minimising the radiative component (reducing the emissivity by polishing the gauge or test surface), or by using a surface with an emissivity that is close to unity, assuming black body radiation and making the appropriate radiation calculations.

10.3.1 Simple one-dimensional steady-state conduction solution

The simplest form of conduction solution, the one-dimensional steady-state method is illustrated in Figure 10.4.

To obtain a measurement of the convective heat flux, q, the surface temperatures T_1 and T_2 are measured in a thermal steady state, and

$$q_{\text{tot}} = \frac{k(T_1 - T_2)}{L}$$

Fig. 10.4 ●
One-dimensional
steady-state
conduction
solution

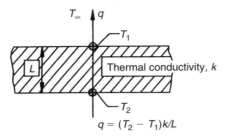

$$q = (T_2 - T_1)k/L$$

$$q_{tot} = q_{rad} + q$$

After estimating the radiative heat flux, q_{rad}, the heat transfer coefficient is obtained from

$$h = \frac{q}{(T_1 - T_\infty)} \tag{10.4}$$

There are three limitations of this technique:

1. Achieving a thermal steady state may take several hours.
2. One-dimensional conduction implies that the heat flux is constant along the surface, and this is usually not the case.
3. It is difficult to measure T_1 and T_2 accurately.

This last point can be demonstrated by considering Example 10.2, which is of considerable practical significance (for simplicity, the effects of radiation are ignored).

Example 10.2

A steady-state heat transfer experiment is carried out on a test surface with $L = 10$ mm and $k = 10$ W/m K. Neglecting radiative heat transfer, calculate the convective heat transfer coefficient when $T_\infty = 20°C$, $T_1 = 60°C$ and $T_2 = 58.8°C$.

Solution Since the radiative heat transfer is neglected,

$$q_{tot} = q = \frac{k(T_1 - T_2)}{L}$$

$$= \frac{10 \times (60 - 58.8)}{0.01}$$

$$= 1200 \text{ W/m}^2$$

$$h = \frac{q}{T_1 - T_\infty}$$

$$= \frac{1200}{60 - 20}$$

$$= 30\,\text{W/m}^2\text{K}$$

Now suppose there is an error of 0.5°C in the measurement of T_2, i.e. $T_2 = 58.8 + 0.5 = 59.3°C$. What is now the 'measured' value of the heat transfer coefficient?

$$h = \frac{k(T_1 - T_2)}{L(T_1 - T_\infty)}$$

$$= \frac{10.0 \times (60 - 59.3)}{0.01 \times (60 - 20)}$$

$$= 17.5\,\text{W/m}^2\text{K}$$

Comment   A small error, $0.5/(58.8 - 20) = 1.3\%$, in a surface temperature measurement gives rise to a very large error (-42%) in the measured heat transfer coefficient. Such sensitivity to small errors is inherent in any thermal gradient method such as a conduction solution technique.

10.3.2   Two-dimensional transient conduction solution

The many hours to attain a thermal steady state and the variations of heat flux along the surface, limitations inherent with the one-dimensional conduction solution, can be both overcome by using a two-dimensional transient conduction solution. For the slab $(0 \leqslant x \leqslant L_x;\ 0 \leqslant y \leqslant L_y)$ shown in Figure 10.5, the relevant time-dependent conduction equation is as follows:

$$\frac{\partial^2 T}{\partial x^2} + \frac{\partial^2 T}{\partial y^2} = \frac{1}{\alpha}\frac{\partial T}{\partial t} \tag{10.5}$$

Fig. 10.5 ●
Two-dimensional
conduction
solution

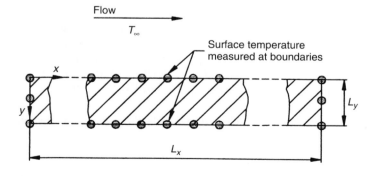

Fig. 10.6 ●
Calculation of local
heat transfer
coefficient from
two-dimensional
conduction
solution results
(Example 10.3)

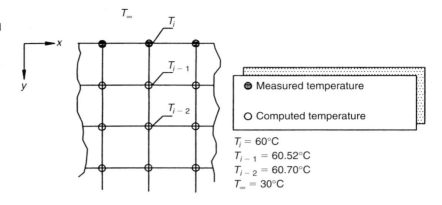

⊕ Measured temperature

○ Computed temperature

$T_i = 60°C$
$T_{i-1} = 60.52°C$
$T_{i-2} = 60.70°C$
$T_\infty = 30°C$

The surface heat transfer rates are obtained in three steps:

1. Measuring the surface temperature distribution at known times; this gives the boundary conditions for Equation (10.5).
2. Solving the conduction equation using finite differences or finite elements, to obtain the internal temperature field at each time step.
3. The total (radiative and convective) local heat flux is obtained from

$$q_{tot} = -k \left(\frac{\partial T}{\partial y} \right)_{surface}$$

where the gradient, $\partial T/\partial y$, can be calculated numerically using the second-order backward difference formula:

$$\frac{\partial T}{\partial y} \approx \frac{1.5T_i - 2T_{i-1} + 0.5T_{i-2}}{\Delta y} \tag{10.6}$$

The application of this to a single point on the surface at one time step is illustrated in Figure 10.6 and Example 10.3.

Example 10.3

Figure 10.6 shows results of a conduction solution where $k = 10$ W/m K and $\Delta y = 1$ mm, the measured surface temperature is $T_i = 60°C$; and the computed internal temperatures (from a finite difference solution) are $T_{i-1} = 60.52°C$ and $T_{i-2} = 60.7°C$. Calculate the value of the local convective heat transfer coefficient for $T_\infty = 30°C$.

Solution   Using Equation (10.6) to evaluate the gradient at the surface

$$q_{tot} = -k \left(\frac{\partial T}{\partial y} \right)$$

$$= \frac{-10[(1.5 \times 60) - (2.0 \times 60.52) + (0.5 \times 60.7)]}{0.001}$$

$$= 6900 \text{ W/m}^2$$

Assuming black body radiative behaviour to surroundings at 30°C,

$$q_{rad} = 56.7 \times 10^{-9} \times ((273 + 60)^4 - (273 + 30)^4)$$
$$= 219 \text{ W/m}^2$$

The convective heat flux, q, is then

$$q = q_{tot} - q_{rad}$$
$$= 6900 - 219$$
$$= 6681 \text{ W/m}^2$$

$$h = \frac{q}{T_i - T_\infty}$$

$$= \frac{6681}{60 - 30}$$

$$= 228 \text{ W/m}^2\text{K}$$

Comment   The contribution of radiative heat transfer is relatively small in this example because of the large value of convective heat transfer coefficient and the small temperature difference involved (30–60°C). In free convection, where the magnitude of h is smaller, the radiative (assuming black body) and convective contributions to the total heat flux are likely to be of comparable magnitude.

10.3.3   Low Biot number method

It is often possible to design an experiment to take advantage of various simplifications to the general heat conduction equation. If $Bi \ll 1$ we can use the low Biot number, lumped mass or lumped capacitance method; see Chapter 2 and in particular Equations (2.84a) and (2.84b). Then

$$\frac{T - T_\infty}{T_{initial} - T_\infty} = \exp(-Bi \cdot Fo) \tag{10.7}$$

and: $Bi = hL/k$ is the Biot number

$F_o = \alpha t / L^2$ is the Fourier number

L = volume/surface area

$\alpha = k/\rho C$ is the thermal diffusivity

Rewriting Equation (10.7) as: $\theta = \exp(-\lambda h t)$, where $\lambda = 1/L\rho C$, and taking the natural logarithm of both sides gives

$$\ln \theta = -\lambda h t \tag{10.8}$$

Fig. 10.7 ● How the low Biot number method can be used to obtain heat transfer coefficients

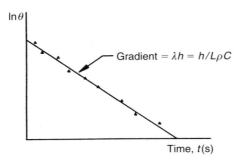

which is a form of the equation for a straight line. So, from measuring the rate of change of surface temperature with time, it is possible to obtain the heat transfer coefficient from the above equation. As shown in Figure 10.7, a convenient method for doing this is to plot a graph of $\ln\theta$ against time, t, and to calculate the gradient. Since L, ρ and C are known, the gradient is a direct measure of the heat transfer coefficient.

Example 10.4

A polished, rotating copper disc of thickness 6 mm and diameter 0.15 m is heated to a surface temperature of 100°C. Whilst rotating at a constant speed, both surfaces of the disc are allowed to cool in air which is at 20°C. From the measurements made of the surface temperature at different times, calculate the convective heat transfer coefficient. For copper, take $k = 401$ W/m K, $\rho = 8933$ kg/m³ and $C = 385$ J/kg K, and since the surface is polished, assume that the contribution of radiative heat transfer is negligible.

Solution   Draw up a table, as below. The times and temperatures are given, the value of θ at each time is calculated and entered in the table, then plot a graph of $\ln\theta$ against time, t (Figure 10.8).

Time (s)	Temperature, T (°C)	$\theta = \dfrac{T - T_\infty}{T_{initial} - T_\infty}$	$\ln\theta$
0	100.0	1	0
2	99.0	0.992	−0.00803
4	98.8	0.985	−0.0151
8	97.6	0.970	−0.0305
20	94.1	0.927	−0.0758
40	88.8	0.860	−0.151
60	83.8	0.797	−0.227
100	74.8	0.685	−0.378
200	57.6	0.470	−0.755

Fig. 10.8 ●

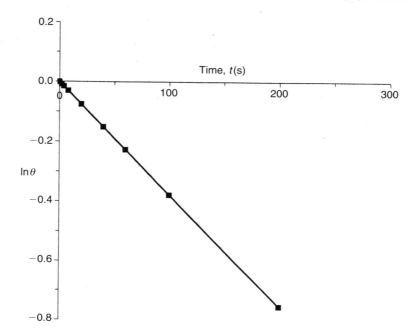

From the graph, gradient $= -0.378/100 = -3.78 \times 10^{-3}\,\mathrm{s}^{-1} = -\lambda h$ so

$$h/L\rho C = -3.78 \times 10^{-3}\,\mathrm{s}^{-1}$$

$$L = 3\,\mathrm{mm} \quad \text{(volume/area)}$$

So

$$h = -\lambda C\rho L$$
$$= 3.78 \times 10^{-3} \times 385 \times 8933 \times 0.003$$
$$= 39\,\mathrm{W/m^2 K}$$

Comment A final check on the magnitude of the Biot number ($hL/k = 0.0003$) indicates that the analysis is valid. In fact, because copper has been used for the experiment, the analysis could be used for (corresponding to $Bi = 0.1$) $h < 1.3 \times 10^4\,\mathrm{W/m^2 K}$, a value very unlikely to occur with air and a spinning disc of this diameter.

The implicit assumption with the small Biot number, lumped mass analysis presented in Example 10.4 is that there are no spatial variations of the heat transfer coefficient (the surface behaves as a lumped mass, after all), so the value measured represents an averaged value of heat transfer coefficient. It is possible, however, to divide a surface into a number of lumped mass regions, separated by sections of insulation, and to apply the lumped mass analysis to each region to obtain local values of the heat transfer coefficient. Such devices are often referred to as calorimeters.

10.3.4 Semi-infinite solid method

For large Biot numbers, $Bi \gg 1$, and for small Fourier numbers, $Fo < 1$, the temperature field inside a one-dimensional region can be approximated as a semi-infinite solid. This is illustrated in Figure 10.9 (and has been referred to previously in Section 2.10.3).

A solution to this equation for the case of a constant surface heat flux, q, requires one boundary condition at $x = 0$, the measured variation of surface temperature with time. The other boundary at $x = \infty$, is assumed to be a constant temperature. The solution is given in Equation (2.88) and can be written as follows:

$$T(x,t) - T_{\text{initial}} = \left\{ \left[\frac{q}{k} (4\alpha t/\pi)^{1/2} \right] \exp\left(\frac{-x^2}{4\alpha t} \right) \right\} - \frac{qx}{k} \operatorname{erfc} \left(\frac{x}{(4\alpha t)^{1/2}} \right)$$

(10.9)

where erfc is the complementary error function obtained from tables (a listing of a computer program to calculate error functions Erf.bas is given in Appendix C) and

$$\operatorname{erfc}(x) = 1 - \operatorname{erf}(x) = 1 - \frac{2}{\sqrt{\pi}} \int_{u=0}^{u=x} \exp(-u^2) \, du$$

(10.10)

From Equation (10.9), since $\exp(0) = 1$ and $\operatorname{erfc}(0) = 0$, then at $x = 0$ the surface temperature variation with time is given by

$$T_{(x=0,t)} - T_{\text{initial}} = \frac{q}{k} \left(\frac{4\alpha t}{\pi} \right)^{1/2}$$

(10.11)

Rearranging Equation (10.11) provides a means for measuring the heat flux from the surface temperature/time history:

Fig. 10.9 ● The semi-infinite solid (constant heat flux at surface)

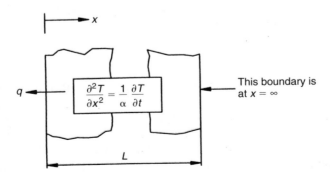

This boundary is at $x = \infty$

$$q = k(T - T_{\text{initial}})\left(\frac{\pi}{4\alpha t}\right)^{1/2} \tag{10.12}$$

One question to be considered when using this technique to measure heat transfer rates is how deep should the material be, or how long can the experiment run so as not to invalidate the semi-infinite assumption?

To answer this we define $x^* = x/(4\alpha t)^{1/2}$ as a dimensionless penetration depth, and from Equations (10.9) and (10.11)

$$(T_x - T_{\text{initial}})/(T_{x=0} - T_{\text{initial}}) = \exp{-(x^*)^2} - \{\sqrt{\pi}x^* \operatorname{erfc}(x^*)\} \tag{10.13}$$

A value of $(T_x - T_{\text{initial}})/(T_{x=0} - T_{\text{initial}}) = 0.01$ occurs when

$$x^* = \frac{x}{(4\alpha t)^{1/2}} \approx 1.6 \tag{10.14}$$

Polycarbonate (a commonly used material for this technique), which has a thermal diffusivity of $\alpha = 10^{-7}\,\text{m}^2/\text{s}$, gives $x \approx 10\,\text{mm}$ with $t = 100\,\text{s}$. Metals have a larger thermal diffusivity and consequently need to be thicker or require a shorter experimental time. The semi-infinite technique has been used extensively at the University of Oxford (Schultz and Jones, 1973; Jones 1992) to measure heat transfer rates on turbine blades under rapid transient conditions.

10.3.5 Heat flux gauges

It is possible to purchase or to make heat flux gauges. The simplest is the plug gauge (or fluxmeter), shown in Figure 10.10.

Fig. 10.10 ●
Schematic diagram
of a plug fluxmeter

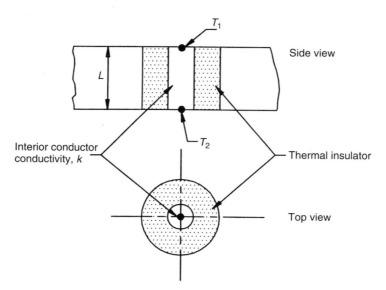

T_1

Side view

L

T_2

Interior conductor
conductivity, k

Thermal insulator

Top view

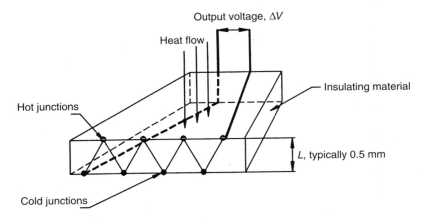

The conducting element is thermally insulated from the surroundings, therefore conduction is locally one-dimensional. The temperatures T_1 and T_2 are measured and, knowing k and L, the heat flux can be calculated from the usual relation, $q = k(T_1 - T_2)/L$. When using these gauges it is advisable to carry out calibrations (and/or numerical modelling) in order to assess conduction losses and errors in the surface temperature measurement.

The thermopile fluxmeter shown in Figure 10.11 comprises a number (four pairs are shown here) of thermocouple junctions in series either side of a thin layer of insulating material. Commercial versions use Kapton polyimide, $k \approx 0.3 \, \text{W/m K}$, and are about the same size as strain gauges. The voltage output is related to the temperature difference across the gauge which, for one-dimensional conduction, is directly proportional to the heat flux through it.

For a single thermocouple, assuming a linear relation between temperature and voltage,

$$T = bV_1 \tag{10.15}$$

where V_1 is the voltage output from a single thermocouple and b is the thermocouple calibration constant as in Equation (10.1).

For n thermocouple junction pairs in series,

$$T = \frac{bV_n}{n} \tag{10.16}$$

Hence

$$T_\text{h} = \frac{bV_{n,\text{h}}}{n}$$

$$T_\text{c} = \frac{bV_{n,\text{c}}}{n}$$

$$T_\text{h} - T_\text{c} = \frac{b\Delta V}{n} \tag{10.17}$$

where ΔV is the voltage output from the thermopile.

Since

$$q = -k\left(\frac{dT}{dx}\right)_{x=0} = \frac{k(T_h - T_c)}{L}$$

the output characteristic is given by

$$q = \left(\frac{kb}{Ln}\right)\Delta V \tag{10.18}$$

Substituting some typical values, $b \approx 25°C/mV$, $n = 4$, $L = 0.5\,mm$ and $k = 0.3\,W/m\,K$, allows us to estimate the sensitivity as

$$q/\Delta V = 3750\,W/m^2\,mV$$

Diller (1987) describes a number of gauges which use this principle. In order to increase the sensitivity, while remaining physically small, the gauges are made using micromanufacturing techniques and incorporate hundreds of junctions.

10.4 ● **Errors in temperature measurement**

A full analysis of errors in temperature measurement is complicated and beyond the scope of this chapter. It is rather the intention here to impress upon the reader that such errors can occur and to illustrate this with one or two examples.

10.4.1 **Dynamic error**

A thermocouple (or any other temperature probe) bead has a finite size and therefore an associated mass and thermal response time. Consider the spherical bead of diameter D shown in Figure 10.12, and assume this

Fig. 10.12 ●
Dynamic error for
a thermocouple
bead

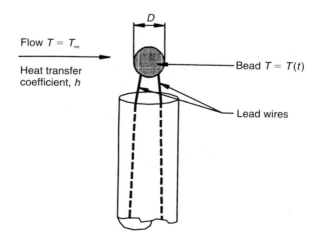

behaves as a lumped mass. For a sphere (area $= \pi D^2$, volume $= \pi D^3/6$), the response to a step change in temperature is given by

$$\frac{T - T_\infty}{T_{initial} - T_\infty} = \exp(-Bi \cdot Fo)$$

$$= \exp - \left(\frac{6h}{D\rho C}\right) t \tag{10.19}$$

It will read 95% of the 'true' value when $(T - T_\infty)/ (T_{initial} - T_\infty) = 0.95$; this occurs at time $t = \tau$.

Rearranging Equation (10.19) for τ gives

$$\tau = -\left(\frac{D\rho C}{6h}\right) \ln (0.95) \tag{10.20}$$

Substituting some typical values in Equation (10.20), $\rho = 8000\,\text{kg/m}^3$, $D = 2\,\text{mm}$, $C = 400\,\text{J/kg K}$ and $h = 50\,\text{W/m}^2\text{K}$, gives $\tau = 1.1\,\text{s}$. And for a probe of $D = 0.4\,\text{mm}$, $\tau = 0.22\,\text{s}$. Consequently, depending upon the probe size, physical properties and the magnitude of the heat transfer coefficient, the response time can be significant; it should therefore be calculated and, if necessary, allowed for when making measurements.

10.4.2 Errors due to radiation effects

Figure 10.13 shows a spherical thermocouple probe mounted in a duct, with the objective of measuring the temperature of the fluid in the duct T_f. For $T_f > T_s$, and in a thermal steady state, the probe will radiate thermal energy to its surroundings and consequently the temperature *recorded* by the probe, T_p, is different from T_f. This heat loss by radiation is balanced by the convective heat gain from the fluid, so we can find T_f given T_p and the temperature of the surroundings T_s from the energy balance

$$h(T_f - T_p) = \sigma\varepsilon (T_p^4 - T_s^4) \tag{10.21}$$

Fig. 10.13 ● Error due to radiation effects

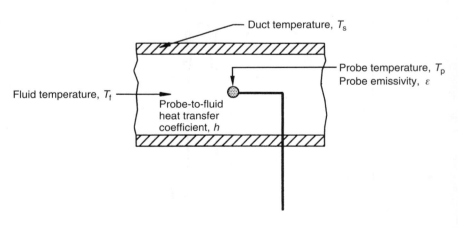

Duct temperature, T_s

Probe temperature, T_p
Probe emissivity, ε

Fluid temperature, T_f

Probe-to-fluid heat transfer coefficient, h

The heat transfer coefficient, h, can be estimated from a suitable correlation; in the example shown here, forced convection from a sphere would be an obvious choice.

Radiation effects can be minimised by fitting one or more radiation shields around the probe, but care must be taken to ensure that the fluid temperature, not the shield temperature, is measured.

Example 10.5

Exhaust gas from a Diesel engine flows at a rate of $0.066\,\text{m}^3/\text{s}$ along an exhaust duct with a diameter of 38 mm. The temperature of the gas is measured using a spherical probe of diameter 3 mm and emissivity 0.5. If the pipe walls are at 500 K and the thermocouple records a temperature of 777 K, what is the actual temperature of the exhaust gas?

For a sphere of diameter D, the Nusselt number is given by Equation (5.92):

$$Nu_{D,\text{av}} = 2 + (0.4Re_D^{1/2} + 0.06Re_D^{2/3})Pr^{0.4}\left(\frac{\mu_\infty}{\mu_w}\right)$$

Take $k = 0.0577\,\text{W/m K}$, $C_p = 1.1\,\text{kJ/kg K}$, $R = 287\,\text{J/kg K}$, $p = 1$ bar, $\rho = 0.435\,\text{kg/m}^3$ and approximate

$$\mu_\infty = \mu_w = 3.6 \times 10^{-5}\,\text{kg/m s}$$

Solution

$$Pr = \frac{\mu C_p}{k} = \frac{3.6 \times 10^{-5} \times 1100}{0.0577} = 0.686$$

$$U = \frac{\text{Flow rate}}{\text{Area}} = \frac{0.066}{(\pi \times 0.038^2/4)} = 58\,\text{m/s}$$

$$Re_D = \frac{\rho U D}{\mu} = \frac{0.435 \times 58 \times 0.003}{3.6 \times 10^{-5}} = 2103$$

From the correlation for heat transfer from a sphere:

$$Nu_{D,\text{av}} = 2 + (0.4Re_D^{1/2} + 0.06Re_D^{2/3})Pr^{0.4}\left(\frac{\mu_\infty}{\mu_w}\right)$$

$$= 2 + (0.4 \times (2103)^{1/2} + 0.06 \times (2103)^{2/3}) \times 0.686^{0.4}$$

$$= 26$$

$$h = \frac{Nu_{D,\text{av}} k}{D} = \frac{26 \times 0.0577}{0.003} = 500\,\text{W/m}^2\text{K}$$

Applying Equation (10.21), $h(T_f - T_p) = \sigma\varepsilon(T_p^4 - T_s^4)$ and solving for T_f gives

$$T_f = \left(\frac{\sigma\varepsilon}{h}\right)(T_p^4 - T_s^4) + T_p$$

$$= \left(\frac{56.7 \times 10^{-9} \times 0.5}{500}\right)(777^4 - 500^4) + 777$$

$$= 794\,\text{K} \quad (\text{error} = 17\,\text{K})$$

Comment   The example demonstrates that without a measurement of T_s it would be difficult to estimate the error. The magnitude of error can be reduced by using a radiation shield around the probe head. Also the emissivity of the probe should be as low as possible.

10.4.3   Errors due to conduction along lead wires

Figure 10.14a depicts an installation of a surface temperature measuring probe where the lead wires are taken out from the surface and are exposed to the flow. Since there is a temperature difference between the

Fig. 10.14 ●
(a) Error caused
by conduction
along lead wires;
(b, c) how to
reduce it

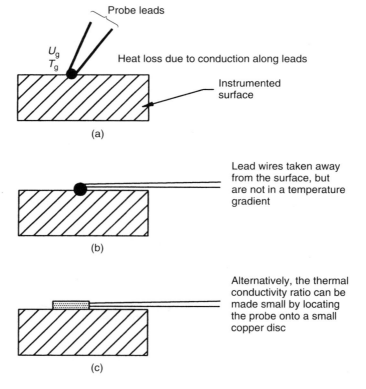

Probe leads

U_g
T_g

Heat loss due to conduction along leads

Instrumented
surface

(a)

Lead wires taken away
from the surface, but
are not in a temperature
gradient

(b)

Alternatively, the thermal
conductivity ratio can be
made small by locating
the probe onto a small
copper disc

(c)

surface and surrounding fluid, the lead wires are in a temperature gradient. Consequently, heat will be conducted along lead wires and dissipated into the surroundings by convection. To obtain an estimate of the error we may equate the heat loss to the probe leads with the heat taken from surface.

The probe leads, of diameter D, can be considered as an infinite fin of thermal conductivity k_t, where the measured temperature is T_t, and the 'true' temperature T_s, perimeter P and area A, and where the heat flow into the surrounding fluid at temperature T_a is given by Equation (2.53):

$$\dot{Q} = (hPk_tA)^{1/2}(T_t - T_a) \tag{10.22a}$$

which, for $P = \pi D$ and $A = \pi D^2/4$ becomes

$$\dot{Q} = \frac{\pi}{2}(T_t - T_a)(hD^3k_t)^{1/2} \tag{10.22b}$$

The heat flow in Equation (10.22) needs to be supplied from inside the surface where the probe is attached. Considering this as a semi-infinite solid of thermal conductivity k_s, which at some distance from the surface is maintained at a temperature T_s, the heat flow can be approximated using a semi-infinite solution (no proof is given here), from which

$$\dot{Q} = 2Dk_s(T_s - T_t) \tag{10.23}$$

where the effective diameter of contact between the probe and surface is equal to the probe lead diameter. Combining Equations (10.22) and (10.23) gives the result for the temperature error:

$$\frac{T_s - T_t}{T_s - T_a} = \frac{\dfrac{\pi}{4k_s}(k_tDh)^{1/2}}{1 + \dfrac{\pi}{4k_s}(k_tDh)^{1/2}} \tag{10.24}$$

Lead wire errors may be reduced by installation design, as shown in Figure 10.14b and c.

10.4.4 Embedding errors

Embedding errors occur when a thermocouple or other temperature measuring device is located in a slot cut into a 'substrate' material using an adhesive with a quite different thermal conductivity. The presence of adhesive and thermocouple material in a slot creates a local anomaly in the thermal conductivity (for example, $k_{adhesive} \approx 0.3\,\text{W/m K}$, $k_{steel} \approx 30\,\text{W/m K}$ and $k_{copper} \approx 400\,\text{W/m K}$). This leads to a disturbance in the isotherms in the neighbourhood of the slot. The result is that the temperature 'measured' by the sensor is not the temperature that would have existed at that point in the absence of the slot, sensor, insulation and adhesive (Figure 10.15).

Fig. 10.15 ●
Thermocouple
embedding error

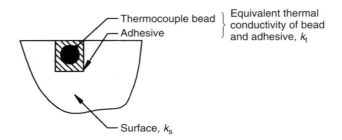

Finite element/finite difference methods are usually required to accurately estimate the error accurately; see for example Long (1985). However, errors can be minimised using a surface coating or 'instrumentation mat' which covers the surface and has a similar thermal conductivity to the adhesive used.

10.5 ● Closing comments

An understanding of heat transfer instrumentation is an essential prerequisite to accurate heat transfer measurements. This chapter has presented a brief review of some of the more common hardware and techniques in temperature and heat transfer measurement. The review is not meant to be exclusive, and for supplementary information the reader is referred to the list of references.

Temperatures can be measured using a variety of different instruments and techniques. These may be considered as either intrusive or non-intrusive. Examples of intrusive instrumentation include thermocouples and resistance thermometers. Radiation thermometers and liquid crystals are non-intrusive techniques. Thermocouples are relatively inexpensive, require no external energisation and are easily interfaced to a computer or data logger. Their main disadvantages are that the output is relatively small, and the cold junction temperature needs to be measured. Platinum resistance thermometers (PRTs) are capable of very high accuracy, yet the probes tend to be relatively bulky, and they are expensive. They are most useful as a reference standard for calibration. Solid-state devices are cheap and relatively accurate and are ideal for use in monitoring cold junction temperatures. Radiation thermometers (also called pyrometers) can be used to obtain point temperature measurement or, in a more sophisticated form, to build up a complete temperature map of the surface; they are relatively expensive. Liquid crystals can also be used to build up a temperature map and are non-intrusive; however, their main limitation is that the range is restricted to below approximately 100°C.

Heat flux measurements are obtained from either the variation of temperature across a distance, or with time. A number of techniques and commercial devices are available; the choice depends on the specific requirements of the user. For example, is the Biot number sufficiently small that the lumped mass technique can be used? Are measurements in the variation of heat transfer coefficient across the surface important? Is the experimental equipment capable of generating a rapid transient, and are the data acquisition facilities capable of recording it? And so on.

Finally, the chapter has discussed some of the main sources of error in temperature instrumentation: dynamic errors, radiation errors, lead wire errors and embedding errors. The first three errors are relatively straightforward to quantify; an investigation of embedding errors is probably best carried out using numerical modelling.

10.6 ● References

Benedict. R.P. (1984). *Fundamentals of Temperature, Pressure and Flow Measurement*. Wiley: New York

Camci, C. (1996). Liquid crystal thermography. In *Temperature Measurements*, chapters 1–3. Von Kármán Institute Lecture Series 1996-07

Chapman, A.J. (1984). *Heat Transfer* 4th edn. New York: Macmillan

Childs, P.R.N., Greenwood, J.R. and Long, C.A. (1999a). Temperature measurement techniques and selection. *I. Mech. E. J. Mech. Engng Sci.* (in press)

Childs, P.R.N., Greenwood, J.R. and Long, C.A. (1999b). Heat flux measurement techniques. *I. Mech. E. J. Mech. Engng Sci.* (in press)

Diller, T.E. (1987). *Advances in Heat Transfer*. Vol. 23: *Advances in Heat Flux Measurements* pp. 279–368. New York: Academic Press

Doebelin, E.O. (1976). *Measurement Systems Application and Design*. New York: McGraw-Hill

Jones, T.V. (1992). Liquid crystal techniques. In *International Centre for Heat and Mass Transfer Int. Symp. Heat Transfer in Turbomachinery*, Athens, Greece

Kaye, G.W.C. and Laby, T.H. (1978). *Tables of Physical and Chemical Constants* 14th edn. London: Longman

Long, C.A. (1985). The effect of thermocouple disturbance errors on the measurement of local heat transfer coefficients. *Proc. Test and Transducer Conference*, **3**, 73–104

Morris, A.S. (1988). *Principles of Measurement and Instrumentation*. Hemel Hempstead: Prentice Hall

Rogers, G.F.C. and Mayhew, Y.R. (1992) *Engineering Thermodynamics: Work and Heat Transfer*. 4th edn. Harlow: Longman

Schultz, D.L. and Jones, T.V. (1973). Heat Transfer Measurements in Short Duration Hypersonic Facilities. *AGARD-AG-165*

Usher, M.J. (1985). *Sensors and Transducers*. London: Macmillan

10.7 ● End of chapter questions

10.1 Sketch a practical thermocouple calibration circuit which uses an ice cold-junction. What precautions should be taken when carrying out a calibration?

10.2 The table opposite gives measured temperatures (°C) and output voltage (mV) for a type E (chromel–constantan) thermocouple. Using an appropriate method (software, spreadsheet or hand-drawn graph) verify that a quadratic least-squares fit gives the calibration equation

$$T = aV^2 + bV + C \quad (°C)$$

where $a = -0.154\,21°C/mV^2$, $b = 16.945°C/mV$, $c = 0.194\,54°C$.

If you are unable to do this on a computer then plot the experimental points and verify that the curve is a good fit. What is the physical significance of coefficient c?

T (°C)	V (mV)
14.3	0.836
20.8	1.23
31.85	1.902
40.4	2.428
48	2.895
55.05	3.341
58.5	3.559
65.3	3.993
69	4.23
74.6	4.582
79.5	4.882
85.6	5.287
93.5	5.81
97.4	6.073
100	6.256

10.3 Sketch a practical circuit for measuring surface or fluid temperatures using a reference junction and compensation cable.

10.4 A thermocouple used to measure the temperature of air from a compressor exit has the following calibration curve:

$$T = -0.3V^2 + 20.5V$$

The reference junction temperature is measured using a thermistor and $T_{ref} = 28°C$. If the output from the thermocouple is 7.3 mV, what is the temperature of the air?

[155.5°C]

10.5 A resistance thermometer probe is used to measure the ambient air temperature in the vicinity of a large, heated steel pipe. The pipe and the probe are both painted matt black. The surface temperature of the pipe is measured to be 100°C the probe records an air temperature of 57.5°C. Consider an energy balance between convective heat transfer from the probe and radiation to it and use this to obtain the true air temperature. The convective heat transfer from the probe may be estimated from

$$Nu_D = \frac{0.387 Gr_D^{1/3} Pr^{1/6}}{[1 + (0.559/Pr)^{9/16}]^{8/27}}$$

For air take $\beta = 1/300\,K^{-1}$, $k = 0.026\,24\,W/m\,K$, $v = 1.5 \times 10^{-5}\,m^2/s$ and $\alpha = 2.14 \times 10^{-5}\,m^2/s$. The Stefan–Boltzmann constant is $56.7 \times 10^{-9}\,W/m^2\,K^4$

[$T_{air} = 26°C$]

10.6 The measured surface temperature distribution on a titanium compressor disc is used to solve the full conduction equation for the complete compressor disc. Part of the finite element mesh is illustrated in Figure 10.16. The numbers are the computed temperatures (in kelvin) at the indicated node point.

Fig. 10.16 ●

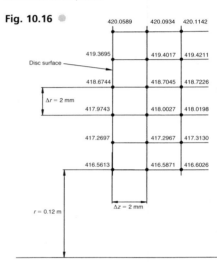

(a) Using a second-order backward difference method, obtain the local surface heat fluxes.

[269, 251, 235, 221, 210, 201 W/m²]

(b) Assume black body radiation to a nearby surface at 410 K and estimate the local radiative heat fluxes.

[163, 152, 140, 128, 117, 105 W/m²]

(c) For an ambient air temperature of 375 K obtain the local convective Nusselt numbers (use the radius as the characteristic length scale).

[9.5, 8.9, 8.5, 8.4, 8.4, 8.7]

Use the following property values: for titanium $k = 13\,W/m\,K$ and for air $k = 0.032\,W/m\,K$; take the Stefan–Boltzmann constant to be $56.7 \times 10^{-9}\,W/m^2\,K^4$.

10.7 A thermopile heat flux gauge comprises n hot–cold junction pairs separated by a thin layer of thickness L and of thermal conductivity k.

(a) Assuming that the thermocouple material used in the construction of the gauge has a linear temperature–voltage characteristic of the form

$$T = C_1 V \quad (°C) \text{ where } V \text{ is in mV.}$$

show that the output of the gauge ΔV for a heat flux q is given by

$$\Delta V = \frac{qLn}{kC_1}$$

(b) Assuming a quadratic temperature–voltage characteristic for the thermocouple,

$$T = C_0 V^2 + C_1 V$$

show that the gauge sensitivity is given by

$$\frac{q}{\Delta V} = A_0\{[A_1 + (4A_2 T)]^{1/2} + A_2 \Delta V\}$$

$$(W/m^2)$$

where $A_0 = k/L\,(W/m^2 K)$, $A_1 = (C_1/n)^2$ $(°C/mV)^2$ and $A_2 = C_0/n^2$ $(°C/mV^2)$.

10.8 In free convection the heat transfer coefficient depends on the surface-to-air temperature difference as $h \sim (T_s - T_{air})^n$, where $n = 1/4$ for laminar flow and $1/3$ for turbulent flow.

(a) Using the lumped mass or low Biot number method,

$$mC\frac{d(T_s - T_{air})}{dt} = -hA(T_s - T_{air})$$

show that the variation of temperature ratio with time, t, is given by

$$\left(\frac{T_s - T_{air}}{T_{s,initial} - T_{air}}\right)^{-n} = 1 + \left\{\frac{nAh_{initial}}{mC}\right\}t$$

where $T_{s,initial}$ is the surface temperature, and $h_{initial}$ is the heat transfer coefficient at $t = 0$.

(b) An experiment to measure the heat transfer coefficient in free convection is carried out on a polished circular aluminium block of 6 mm thickness and 75 mm dia., $\rho = 2702\,kg/m^3$, $C = 903\,J/kg\,K$. The table

below gives the recorded values of surface temperature taken at different times throughout the test; the air temperature remained constant at 23°C.

Time, t (s)	Surface temperature (°C)
0	74
60	72
180	68
300	65
600	58
1200	48
1500	44
1800	40
2400	36

Assuming laminar flow where $n = 1/4$, use the result from part (a) to calculate the heat transfer coefficient at $t = 0$.

[10 W/m²K]

(c) Compare the answer from part (b) with that obtained from the correlation

$$Nu_L = 0.54(Gr_L Pr)^{1/4}$$

Take the characteristic length as the ratio of the area to the perimeter and evaluate all properties at the mean film temperature.

[9.8 W/m²K]

10.9 A plate is 0.3 m long, 0.1 m wide and has a thickness of 12 mm. It is made from stainless steel with a thermal conductivity of $k = 16\,W/m\,K$ and is placed in an airstream of temperature 20°C. In an experiment, the plate is heated by an electrical heater (also 0.3 m × 0.1 m) positioned on the underside of the plate and the temperature of the plate next to the heater is maintained at 100°C. A voltmeter and ammeter are connected to the heater, and these read 200 V and 0.25 A. What is the *convective* heat transfer coefficient?

[12.7 W/m² K]

10.10 A plug-type fluxmeter is located a distance 0.2 m from the leading edge of a heat transfer model. The model, which is heated from underneath, is placed in a wind tunnel using air at 20°C, and a series of experiments are carried out for different values of the free stream velocity U_∞ The fluxmeter is instrumented with two thermocouples; one on the top surface (T_1) and one on the bottom (T_2) the distance between these two measurement points is 0.01 m and the fluxmeter has a thermal conductivity of $k = 9\,\text{W/m\,K}$.

The table below shows the variation of measured temperatures T_1 and T_2 with velocity U_∞.

U_∞ (m/s)	T_1 (°C)	T_2 (°C)
5	100.0	101.5
10	100.0	101.8
15	100.0	102.0
20	100.0	102.2
25	100.0	105.4
30	100.0	106.2
40	100.0	107.5
50	100.0	108.8

(a) Using the above data evaluate:

(i) The total heat flux.

(ii) The radiative heat flux (assume black body radiation from the fluxmeter to 20°C).

(iii) The convective heat flux.

(b) Plot a graph of the variation of local Nusselt number (at $x = 0.2$ m) with Reynolds number.

(c) Correlate the experimental results (separately for $Re > 3 \times 10^5$ and $Re < 3 \times 10^5$) in the form

$$Nu_x = C \cdot Re_x^n$$

where C and n are constants whose values are to be obtained from the graph.

(d) Comment on the form of the graph obtained.

Take the Stefan–Boltzmann constant as $\sigma = 56.7 \times 10^{-9}\,\text{W/m}^2\text{K}^4$, and for air at 20°C take $\mu = 1.8 \times 10^{-5}\,\text{kg/m\,s}$, $k = 0.02\,\text{W/m\,K}$ and $\rho = 1.2\,\text{kg/m}^3$.

Thermophysical properties of matter

A.1 ● Physical constants

Universal gas constant:
$$\mathcal{R} = \forall.\in/\triangledown \times \infty/^{-\in} \text{ m}^3 \text{ atm/kmol K}$$
$$= 8.314 \times 10^{-2} \text{ m}^3 \text{ bar/kmol K}$$
$$= 8.315 \text{ kJ/kmol K}$$
$$= 1545 \text{ ft lb}_f/\text{lbmole } ^\circ\text{R}$$
$$= 1.986 \text{ Btu/lbmole } ^\circ\text{R}$$

Avogadro's number:
$$\mathcal{N} = 6.024 \times 10^{23} \text{ molecules/mol}$$

Planck's constant:
$$h = 6.625 \times 10^{-34} \text{ J s/molecule}$$

Boltzmann's constant:
$$k = 1.380 \times 10^{-23} \text{ J/K molecule}$$

Speed of light in vacuum:
$$c = 2.998 \times 10^8 \text{ m/s}$$

Stefan–Boltzmann constant:
$$\sigma = 5.670 \times 10^{-8} \text{ W/m}^2 \text{ K}^4$$
$$= 0.1714 \times 10^{-8} \text{ Btu/h ft}^2 \, ^\circ\text{R}^4$$

Gravitational acceleration (sea level):
$$g = 9.807 \text{ m/s}^2$$

Normal atmospheric pressure
$$\rho = 101\,325 \text{ N/m}^2$$

A.2 ● Conversion factors

Acceleration	1 m/s^2	$= 4.2520 \times 10^7 \text{ ft/h}^2$
Area	1 m^2	$= 1550.0 \text{ in.}^2$
		$= 10.764 \text{ ft}^2$
Energy	1 J	$= 9.4787 \times 10^{-4} \text{ Bu}$
Force	1 N	$= 0.224\,81 \text{ lb}_f$
Heat transfer rate	1 W	$= 3.4123 \text{ Btu/h}$
Heat flux	1 W/m^2	$= 0.3171 \text{ Btu/h ft}^2$
Heat generation rate	1 W/m^3	$= 0.096\,65 \text{ Btu/h ft}^3$

Heat transfer coefficient	$1 \, \text{W/m}^2 \, \text{K}$	$= 0.176 \, 12 \, \text{Btu/h ft}^2 \, {}^\circ\text{F}$
Kinematic viscosity	$1 \, \text{m}^2/\text{s}$	$= 3.875 \times 10^4 \, \text{ft}^2/\text{h}$
and diffusivities		
Latent heat	$1 \, \text{J/kg}$	$= 4.2995 \times 10^{-4} \, \text{Btu/lb}_\text{m}$
Length	$1 \, \text{m}$	$= 39.370 \, \text{in.}$
		$= 3.2808 \, \text{ft.}$
	$1 \, \text{km}$	$= 0.621 \, 37 \, \text{mile}$
Mass	$1 \, \text{kg}$	$= 2.2046 \, \text{lb}_\text{m}$
Mass density	$1 \, \text{kg/m}^3$	$= 0.062 \, 428 \, \text{lb}_\text{m}/\text{ft}^3$
Mass flow rate	$1 \, \text{kg/s}$	$= 7936.6 \, \text{lb}_\text{m}/\text{h}$
Mass transfer coefficient	$1 \, \text{m/s}$	$= 1.1811 \times 10^4 \, \text{ft/h}$
Pressure and stress[1]	$1 \, \text{N/m}^2$	$= 0.020 \, 886 \, \text{lb}_\text{f}/\text{ft}^2$
		$= 1.4504 \times 10^{-4} \, \text{lb}_\text{f}/\text{in.}^2$
		$= 4.015 \times 10^{-3} \, \text{in. water}$
		$= 2.953 \times 10^{-4} \, \text{in. Hg}$
	$1.0133 \times 10^5 \, \text{N/m}^2$	$= 1 \, \text{standard atmosphere}$
	$1 \times 10^5 \, \text{N/m}^2$	$= 1 \, \text{bar}$
Specific heat	$1 \, \text{J/kg K}$	$= 2.3886 \times 10^{-4} \, \text{Btu/lb}_\text{m} \, {}^\circ\text{F}$
Temperature	K	$= (5/9){}^\circ\text{R}$
		$= (5/9)({}^\circ\text{F} + 459.67)$
		$= {}^\circ\text{C} + 273.15$
Temperature difference	$1 \, \text{K}$	$= 1{}^\circ\text{C}$
		$= (9/5){}^\circ\text{R} = (9/5){}^\circ\text{F}$
Thermal conductivity	$1 \, \text{W/m K}$	$= 0.577 \, 82 \, \text{Btu/h ft} \, {}^\circ\text{F}$
Thermal resistance	$1 \, \text{K/W}$	$= 0.527 \, 50 \, {}^\circ\text{F/h Btu}$
Viscosity (dynamic)[2]	$1 \, \text{N s/m}^2$	$= 2419.1 \, \text{lb}_\text{m}/\text{ft h}$
		$= 5.8016 \times 10^{-6} \, \text{lb}_\text{f} \, \text{h/ft}^2$
Volume	$1 \, \text{m}^3$	$= 6.1023 \times 10^4 \, \text{in.}^3$
		$= 35.314 \, \text{ft}^3$
		$= 264.17 \, \text{gal}$
Volume flow rate	$1 \, \text{m}^3/\text{s}$	$= 1.2713 \times 10^5 \, \text{ft}^3/\text{h}$
		$= 2.1189 \times 10^3 \, \text{ft}^3/\text{min}$
		$= 1.5850 \times 10^4 \, \text{gal/min}$

[1] The SI name for the quantity pressure is Pascal (Pa) having units N/m^2 or kg/m s^2.

[2] Also expressed in equivalent units of kg/s m.

A.3 ● Tables of thermophysical properties

Table A.1 ● Thermophysical properties of selected metallic solids

Composition	Melting point (K)	ρ (kg/m³)	Cp (J/kg K)	k (W/m K)	α × 10⁶ (m²/s)	100	200	400	600	800	1000	1200	1500	2000	2500
Aluminum															
Pure	933	2702	903	237	97.1	302	237	240	231	218					
						482	798	949	1033	1146					
Alloy 2024-T6 (4.5% Cu, 1.5% Mg, 0.6% Mn)	775	2770	875	177	73.0	65	163	186	186						
						473	787	925	1042						
Alloy 195, Cast (4.5% Cu)		2790	883	168	68.2			174	185						
								—	—						
Beryllium	1550	1850	1825	200	59.2	990	301	161	126	106	90.8	78.7			
						203	1114	2191	2604	2823	3018	3227	3519		
Bismuth	545	9780	122	7.86	6.59	16.5	9.69	7.04							
						112	120	127							
Boron	2573	2500	1107	27.0	9.76	190	55.5	16.8	10.6	9.60	9.85				
						128	600	1463	1892	2160	2338				
Cadmium	594	8650	231	96.8	48.4	203	99.3	94.7							
						198	222	242							
Chromium	2118	7160	449	93.7	29.1	159	111	90.9	80.7	71.3	65.4	61.9	57.2	49.4	
						192	384	484	542	581	616	682	779	937	
Cobalt	1769	8862	421	99.2	26.6	167	122	85.4	67.4	58.2	52.1	49.3	42.5		
						236	379	450	503	550	628	733	674		
Copper															
Pure	1358	8933	385	401	117	482	413	393	379	366	352	339			
						252	356	397	417	433	451	480			
Commercial bronze (90% Cu, 10% Al)	1293	8800	420	52	14		42	52	59						
						785	785	460	545						

Properties at 300 K / Properties at various temperatures (K) / k (W/m K)/Cp (J/kg K)

(continued)

Table A.1 ● Continued

Composition	Melting point (K)	Properties at 300K				Properties at various temperatures (K)									
		ρ (kg/m³)	C_p (J/kg K)	k (W/m K)	$\alpha \times 10^6$ (m²/s)	col				k (W/m K)/C_p (J/kg K)					
						100	200	400	600	800	1000	1200	1500	2000	2500
Phosphor gear bronze (89% Cu, 11% Sn)	1104	8780	355	54	17		41	65	74						
Cartridge brass (70% Cu, 30% Zn)	1188	8530	380	110	33.9	75	95	137	149						
							360	395	425						
Constantan (55% Cu, 45% Ni)	1493	8920	384	23	6.71	17	19								
						237	362								
Germanium	1211	5360	322	59.9	34.7	232	96.8	43.2	27.3	19.8	17.4	17.4			
						190	290	337	348	357	375	395			
Gold	1336	19300	129	317	127	327	323	311	298	284	270	255			
						109	124	131	135	140	145	155			
Iridium	2720	22500	130	147	50.3	172	153	144	138	132	126	120	111		
						90	122	133	138	144	153	161	172		
Iron Pure	1810	7870	447	80.2	23.1	134	94.0	69.5	54.7	43.3	32.8	28.3	32.1		
						216	384	490	574	680	975	609	654		
Armco (99.75% pure)		7870	447	72.7	20.7	95.6	80.6	65.7	53.1	42.2	32.3	28.7	31.4		
						215	384	490	574	680	975	609	654		
Carbon steels Plain carbon (Mn ≤ 1%, Si ≤ 0.1%)		7854	434	60.5	17.7			56.7	48.0	39.2	30.0				
								487	559	685	1169				
AISI 1010 (Mn ≤ 1%, Si ≤ 0.1%)		7832	434	63.9	18.8			58.7	48.8	39.2	31.3				
								487	559	685	1168				
Carbon–silicon (Mn ≤ 1%, 0.1% < Si ≤ 0.6%)		7854	446	51.9	14.9			49.8	44.0	37.4	29.3				
								501	582	699	971				

(continued)

Table A.1 ● *Continued*

Composition	Melting point (K)	ρ (kg/m³)	C_p (J/kg K)	k (W/m K)	$\alpha \times 10^6$ (m²/s)	100	200	400	600	800	1000	1200	1500	2000	2500
		Properties at 300 K				**Properties at various temperatures (K)** k (W/m K)/C_p (J/kg K)									
Carbon–manganese–silicon (1% < Mn ≤ 1.65%, 0.1% < Si ≤ 0.6%)		8131	434	41.0	11.6			42.2 / 487	39.7 / 559	35.0 / 685	27.6 / 1090				
Chromium (low) steels															
$\frac{1}{2}$Cr–$\frac{1}{4}$Mo–Si (0.18% C, 0.65% Cr, 0.23% Mo, 0.6% Si)		7822	444	37.7	10.9			38.2 / 492	36.7 / 575	33.3 / 688	26.9 / 969				
1 Cr–$\frac{1}{2}$ Mo (0.16% C, 1% Cr, 0.54% Mo, 0.39% Si)		7858	442	42.3	12.2			42.0 / 492	39.1 / 575	34.5 / 688	27.4 / 969				
1 Cr–V (0.2% C, 1.02% Cr, 0.15% V)		7836	443	48.9	14.1			46.8 / 492	42.1 / 575	36.3 / 688	28.2 / 969				
Stainless steels															
AISI 302		8055	480	15.1	3.91			17.3 / 512	20.0 / 559	22.8 / 585	25.4 / 606				
AISI 304	1670	7900	477	14.9	3.95	9.2 / 272	12.6 / 402	16.6 / 515	19.8 / 557	22.6 / 582	25.4 / 611	28.0 / 640	31.7 / 682		
AISI 316		8238	468	13.4	3.48			15.2 / 504	18.3 / 550	21.3 / 576	24.2 / 602				
AISI 347		7978	480	14.2	3.71			15.8 / 513	18.9 / 559	21.9 / 585	24.7 / 606				
Lead	601	11340	129	35.3	24.1	39.7 / 118	36.7 / 125	34.0 / 132	31.4 / 142						

(continued)

Table A.1 ● Continued

Composition	Melting point (K)	Properties at 300 K				Properties at various temperatures (K)									
		ρ (kg/m³)	C_p (J/kg K)	k (W/m K)	$\alpha \times 10^6$ (m²/s)	k (W/m K)/C_p (J/kg K)									
						100	200	400	600	800	1000	1200	1500	2000	2500
Magnesium	923	1740	1024	156	87.6	169 / 649	159 / 934	153 / 1074	149 / 1170	146 / 1267					
Molybdenum	2894	10240	251	138	53.7	179 / 141	143 / 224	134 / 261	126 / 275	118 / 285	112 / 295	105 / 308	98 / 330	90 / 380	86 / 459
Nickel Pure	1728	8900	444	90.7	23.0	164 / 232	107 / 383	80.2 / 485	65.6 / 592	67.6 / 530	71.8 / 562	76.2 / 594	82.6 / 616		
Nichrome (80% Ni, 20% Cr)	1672	8400	420	12	3.4			14 / 480	16 / 525	21 / 545					
Inconel X-750 (73% Ni, 15% Cr, 6.7% Fe)	1665	8510	439	11.7	3.1	8.7 / –	10.3 / 372	13.5 / 473	17.0 / 510	20.5 / 546	24.0 / 626	27.6 / –	33.0 / –		
Niobium	2741	8570	265	53.7	23.6	55.2 / 188	52.6 / 249	55.2 / 274	58.2 / 283	61.3 / 292	64.4 / 301	67.5 / 310	72.1 / 324	79.1 / 347	
Palladium	1827	12020	244	71.8	24.5	76.5 / 168	71.6 / 227	73.6 / 251	79.7 / 261	86.9 / 271	94.2 / 281	102 / 291	110 / 307		
Platinum Pure	2045	21450	133	71.6	25.1	77.5 / 100	72.6 / 125	71.8 / 136	73.2 / 141	75.6 / 146	78.7 / 152	82.6 / 157	89.5 / 165	99.4 / 179	
Alloy 60Pt–40Rh (60% Pt, 40% Rh)	1800	16630	162	47	17.4		52 / –	52 / –	59 / –	65 / –	69 / –	73 / –	76 / –		
Rhenium	3453	21100	136	47.9	16.7	58.9 / 97	51.0 / 127	46.1 / 139	44.2 / 145	44.1 / 151	44.6 / 156	45.7 / 162	47.8 / 171	51.9 / 186	
Rhodium	2236	12450	243	150	49.6	186 / 147	154 / 220	146 / 253	136 / 274	127 / 293	121 / 311	116 / 327	110 / 349	112 / 376	
Silicon	1685	2330	712	148	89.2	884 / 259	264 / 556	98.9 / 790	61.9 / 867	42.2 / 913	31.2 / 946	25.7 / 967	22.7 / 992		

(continued)

Table A.1 ● Continued

Composition	Melting point (K)	ρ (kg/m³)	C_p (J/kg K)	k (W/m K)	α × 10⁶ (m²/s)	Properties at various temperatures (K) k (W/m K)/C_p (J/kg K)									
						100	200	400	600	800	1000	1200	1500	2000	2500
Silver	1235	10500	235	429	174	444	430	425	412	396	379	361			
					187		225	239	250	262	277	292			
Tantalum	3269	16600	140	57.5	24.7	59.2	57.5	57.8	58.6	59.4	60.2	61.0	62.2	64.1	65.6
					110		133	144	146	149	152	155	160	172	189
Thorium	2023	11700	118	54.0	39.1	59.8	54.6	54.5	55.8	56.9	56.9	58.7			
						99	112	124	134	145	156	167			
Tin	505	7310	227	66.6	40.1	85.2	73.3	62.2							
						188	215	243							
Titanium	1953	4500	522	21.9	9.32	30.5	24.5	20.4	19.4	19.7	20.7	22.0	24.5		
						300	465	551	591	633	675	620	686 ·		
Tungsten	3660	19300	132	174	68.3	208	186	159	137	125	118	113	107	100	95
						87	122	137	142	145	148	152	157	167	176
Uranium	1406	19070	116	27.6	12.5	21.7	25.1	29.6	34.0	38.8	43.9	49.0			
						94	108	125	146	176	180	161			
Vanadium	2192	6100	489	30.7	10.3	35.8	31.3	31.3	33.3	35.7	38.2	40.8	44.6	50.9	
						258	430	515	540	563	597	645	714	867	
Zinc	693	7140	389	116	41.8	117	118	111	103						
						297	367	402	436						
Zirconium	2125	6570	278	22.7	12.4	33.2	25.2	21.6	20.7	21.6	23.7	26.0	28.8	33.0	
						205	264	300	322	342	362	344	344	344	

Source: Adapted from References 1–7.

Table A.2 ● Thermophysical properties of selected non-metallic solids

Composition	Melting point (K)	Properties at 300 K ρ (kg/m³)	C_p (J/kg K)	k (W/m K)	α × 10⁶ (m²/s)	Properties at various temperatures (K) k (W/m K)/C_p (J/kg K) 100	200	400	600	800	1000	1200	1500	2000	2500
Aluminum oxide, sapphire	2323	3970	765	46	15.1	450	82	32.4 / 940	18.9 / 1110	13.0 / 1180	10.5 / 1225	6.55	5.66	6.00	
Aluminum oxide, polycrystalline	2323	3970	765	36.0	11.9	133	55	26.4 / 940	15.8 / 1110	10.4 / 1180	7.85 / 1225				
Beryllium oxide	2725	3000	1030	272	88.0			196 / 1350	111 / 1690	70 / 1865	47 / 1975	33 / 2055	21.5 / 2145	15 / 2750	
Boron	2573	2500	1105	27.6	9.99	190	52.5	18.7 / 1490	11.3 / 1880	8.1 / 2135	6.3 / 2350	5.2 / 2555			
Boron fibre epoxy (30% vol) composite	590	2080													
k, ‖ to fibres				2.29		2.10	2.23	2.28							
k ⊥ to fibres				0.59		0.37	0.49	0.60							
C_p			1122			364	757	1431							
Carbon Amorphous	1500	1950	—	1.60	—	0.67	1.18	1.89	2.19	2.37	2.53	2.84	3.48		
Diamond, type IIa insulator	—	3500	509	2300	—	10000 / 21	4000 / 194	1540 / 853							
Graphite, pyrolytic	2273	2210													
k ‖ to layers				1950		4970	3230	1390	892	667	534	448	357	262	
k ⊥ to layers				5.70		16.8	9.23	4.09	2.68	2.01	1.60	1.34	1.08	0.81	
C_p			709			136	411	992	1406	1650	1793	1890	1974	2043	
Graphite fibre epoxy (25% vol) composite	450	1400													
k, heat flow ‖ to fibres				11.1		5.7	8.7	13.0							
k, heat flow ⊥ to fibres				0.87		0.46	0.68	1.1							
C_p			935			337	642	1216							
Pyroceram Corning 9606	1623	2600	808	3.98	1.89	5.25	4.78	3.64 / 908	3.28 / 1038	3.08 / 1122	2.96 / 1197	2.87 / 1264	2.79 / 1498		

(continued)

Table A.2 ● Continued

Composition	Melting point (K)	ρ (kg/m³)	Cp (J/kg K)	k (W/m K)	α × 10⁶ (m²/s)	\(k\) (W/m K)/\(C_p\) (J/kg K) — 100	200	400	600	800	1000	1200	1500	2000	2500
Silicon carbide — k	3100	3160	675	490	230						87	58	30		
Silicon carbide — C_p								880	1050	1135	1195	1243	1310		
Silicon dioxide, crystalline (quartz)	1883	2650													
k, ∥ to c axis				10.4		39	16.4	7.6	5.0	4.2					
k, ⊥ to c axis				6.21		20.8	9.5	4.70	3.4	3.1					
C_p			745			–	–	885	1075	1250					
Silicon dioxide, polycrystalline (fused silica) — k	1883	2220	745	1.38	0.834	0.69	1.14	1.51	1.75	2.17	2.87	4.00			
C_p						–	–	905	1040	1105	1155	1195			
Silicon nitride — k	2173	2400	691	16.0	9.65	–	–	13.9	11.3	9.88	8.76	8.00	7.16	6.20	
C_p						–	578	778	937	1063	1155	1226	1306	1377	
Sulphur — k	392	2070	708	0.206	0.141	0.165	0.185								
C_p						403	606								
Thorium dioxide — k	3573	9110	235	13	6.1			10.2	6.6	4.7	3.68	3.12	2.73	2.5	
C_p								255	274	285	295	303	315	330	
Titanium dioxide, polycrystalline — k	2133	4175	710	8.4	2.8			7.01	5.02	3.94	3.46	3.28			
C_p								805	880	910	930	945			

Source: Adapted from References 1, 2, 3 and 6

Table A.3 ● Thermophysical properties of common materials

Structural building materials

Description/composition	Density, ρ (kg/m³)	Typical properties at 300 K		
		Thermal conductivity, k (W/m K)	Specific heat C_p (J/kg K)	
Building boards				
Asbestos–cement board	1920	0.58	–	
Gypsum or plaster board	800	0.17	–	
Plywood	545	0.12	1215	
Sheathing, regular density	290	0.055	1300	
Acoustic tile	290	0.058	1340	
Hardboard, siding	640	0.094	1170	
Hardboard, high density	1010	0.15	1380	
Particle board, low density	590	0.078	1300	
Particle board, high density	1000	0.170	1300	
Woods				
Hardwoods (oak, maple)	720	0.16	1255	
Softwoods (fir, pine)	510	0.12	1380	
Masonry materials				
Cement mortar	1860	0.72	780	
Brick, common	1920	0.72	835	
Brick, face	2083	1.3	–	
Clay tile, hollow				
1 cell deep, 10 cm thick	–	0.52	–	
3 cells deep, 30 cm thick	–	0.69	–	
Concrete block, 3 oval cores				
Sand/gravel, 20 cm thick	–	1.0	–	
Cinder aggregate, 20 cm thick	–	0.67	–	
Concrete block rectangular core				
2 cores, 20 cm thick, 16 kg	–	1.1	–	
Same with filled cores	–	0.60	–	
Plastering materials				
Cement plaster, sand aggregate	1860	0.72	–	
Gypsum plaster, sand aggregate	1680	0.22	1085	
Gypsum plaster, vermiculite aggregate	720	0.25	–	

Table A.3 ● Continued

Insulating materials and systems

Description/composition	Density, ρ (kg/m³)	Typical properties at 300 K	
		Thermal conductivity, k (W/m K)	Specific heat C_p (J/kg K)
Blanket and batt			
Glass fibre, paper faced	16	0.046	–
	28	0.038	–
	40	0.035	–
Glass fibre, coated; duct liner	32	0.038	835
Board and slab			
Cellular glass	145	0.058	1000
Glass fibre, organic bonded	105	0.036	795
Polystyrene, expanded			
Extruded (R-12)	55	0.027	1210
Moulded beads	16	0.040	1210
Mineral fibreboard; roofing material	265	0.049	–
Wood, shredded/cemented	350	0.087	1590
Cork	120	0.039	1800
Loose fill			
Cork, granulated	160	0.045	–
Diatomaceous silica, coarse	350	0.069	–
Powder	400	0.091	–
Diatomaceous silica, fine powder	200	0.052	–
	275	0.061	–
Glass fibre, poured or blown	16	0.043	835
Vermiculite, flakes	80	0.068	835
	160	0.063	1000
Formed/foamed-in-place			
Mineral wool granules with asbestos/ inorganic binders, sprayed	190	0.046	–
Polyvinyl acetate cork mastic; sprayed or trowelled	–	0.100	–
Urethane, two-part mixture; rigid foam	70	0.026	1045
Reflective			
Aluminum foil separating fluffy glass mats; 10–12 layers, evacuated; for cyrogenic applications (150 K)	40	0.000 16	–
Aluminum foil and glass paper laminate; 75–150 layers; evacuated; for cyrogenic application (150 K)	120	0.000 017	–
Typical silica powder, evacuated	160	0.0017	–

Table A.3 ● Continued

Industrial insulation

Description/composition	Maximum service temperature (K)	Typical density (kg/m³)	Typical thermal conductivity, k (W/m K), at various temperatures (K)													
			200	215	230	240	255	270	285	300	310	365	420	530	645	750
Blankets																
Blanket, mineral fibre, metal reinforced	920	96–192									0.038	0.046	0.056	0.078		
Blanket, mineral fibre, glass; fine fibre, organic bonded	815	40–96									0.035	0.045	0.058	0.088		
	450	10				0.036	0.038	0.040	0.043	0.048	0.052	0.076				
		12						0.035	0.036	0.039	0.042	0.046	0.049	0.069		
		16						0.033	0.035	0.036	0.039	0.042	0.046	0.062		
		24						0.030	0.032	0.033	0.036	0.039	0.040	0.053		
		32						0.029	0.030	0.032	0.033	0.036	0.038	0.048		
		48						0.027	0.029	0.030	0.032	0.033	0.035	0.045		
Blanket, alumina–silica fiber	1530	48												0.071	0.105	0.150
		64												0.059	0.087	0.125
		96												0.052	0.076	0.100
		128												0.049	0.068	0.091
Felt, semirigid; organic bonded	480	50–125						0.035	0.036	0.038	0.039		0.051	0.063		
	730	50	0.023	0.025	0.026	0.027	0.029	0.030	0.032	0.033	0.035		0.051	0.079		
Felt, laminated; no binder	920	120											0.051	0.065	0.087	
Blocks, boards and pipe insulations																
Asbestos paper, laminated and corrugated																
4-ply	420	190								0.078	0.082	0.098				
6-ply	420	255								0.071	0.074	0.085				
8-ply	420	300								0.068	0.071	0.082				

(continued)

Table A.3 ● Continued

Industrial insulation (continued)

Description/composition	Maximum service temperature (K)	Typical density (kg/m³)	Typical thermal conductivity, k (W/m K), at various temperatures (K)													
			200	215	230	240	255	270	285	300	310	365	420	530	645	750
Magnesia, 85%	590	185									0.051	0.055	0.061			
Calcium silicate	920	190									0.055	0.059	0.063	0.075	0.089	0.104
Cellular glass	700	145			0.046	0.048	0.051	0.052	0.055	0.058	0.062	0.069	0.079			
Diatomaceous silica	1145	345												0.092	0.098	0.104
Diatomaceous silica	1310	385												0.101	0.100	0.115
Polystyrene, rigid																
Extruded (R-12)	350	56	0.023	0.023	0.022	0.023	0.023	0.025	0.026	0.027	0.029					
Extruded (R-12)	350	35	0.023	0.023	0.023	0.025	0.025	0.026	0.027	0.029						
Moulded beads	350	16	0.026	0.029	0.030	0.033	0.035	0.036	0.038	0.040						
Rubber, rigid foamed	340	70						0.029	0.030	0.032	0.033					
Insulating cement																
Mineral fibre (rock, slag or glass)																
With clay binder	1255	430									0.071	0.079	0.088	0.105	0.123	
With hydraulic setting binder	922	560									0.108	0.115	0.123	0.137		
Loose fill																
Cellulose, wood or paper pulp	–	45							0.038	0.039	0.042					
Perlite, expanded	–	105	0.036	0.039	0.042	0.043	0.046	0.049	0.051	0.053	0.056					
Vermiculite, expanded	–	122			0.056	0.058	0.061	0.063	0.065	0.068	0.071					
Vermiculite, expanded		80			0.049	0.051	0.055	0.058	0.061	0.063	0.066					

Table A.3 ● Continued

Other materials

Description/composition	Temperature (K)	Density, ρ (kg/m³)	Thermal conductivity, k (W/m K)	Specific heat, Cₚ (J/kg K)
Asphalt	300	2115	0.062	920
Bakelite	300	1300	1.4	1465
Brick, refractory				
Carborundum	872	–	18.5	–
	1672	–	11.0	–
Chrome brick	473	3010	2.3	835
	823		2.5	
	1173		2.0	
Diatomaceous	478	–	0.25	–
silica, fired	1145	–	0.30	
Fire clay, burnt 1600 K	773	2050	1.0	960
	1073	–	1.1	
	1373	–	1.1	
Fire clay, burnt 1725 K	773	2325	1.3	960
	1073		1.4	
	1373		1.4	
Fire clay brick	478	2645	1.0	960
	922		1.5	
	1478		1.8	
Magnesite	478	–	3.8	1130
	922	–	2.8	
	1478		1.9	
Clay	300	1460	1.3	880
Coal, anthracite	300	1350	0.26	1260
Concrete (stone mix)	300	2300	1.4	880
Cotton	300	80	0.06	1300
Foodstuffs				
Banana (75.7% water content)	300	980	0.481	3350
Apple, red (75% water content)	300	840	0.513	3600
Cake, batter	300	720	0.223	–
Cake, fully baked	300	280	0.121	–
Chicken meat, white	198	–	1.60	–
(74.4% water content)	233	–	1.49	
	253		1.35	
	263		1.20	
	273		0.476	
	283		0.480	
	293		0.489	
Glass				
Plate (soda lime)	300	2500	1.4	750
Pyrex	300	2225	1.4	835
Ice	273	920	1.88	2040
	253	–	2.03	1945

(continued)

Table A.3 ● Continued

Other materials (continued)

Description/composition	Temperature (K)	Density, ρ (kg/m³)	Thermal conductivity, k (W/m K)	Specific heat, Cₚ (J/kg K)
Leather (sole)	300	998	0.159	–
Paper	300	930	0.180	1340
Paraffin	300	900	0.240	2890
Rock				
Granite, Barre	300	2630	2.79	775
Limestone, Salem	300	2320	2.15	810
Marble, Halston	300	2680	2.80	830
Quartzite, Sioux	300	2640	5.38	1105
Sandstone, Berea	300	2150	2.90	745
Rubber, vulcanised				
Soft	300	1100	0.13	2010
Hard	300	1190	0.16	–
Sand	300	1515	0.27	800
Soil	300	2050	0.52	1840
Snow	273	110	0.049	–
		500	0.190	–
Teflon	300	2200	0.35	–
	400		0.45	–
Tissue, human				
Skin	300	–	0.37	–
Fat layer (adipose)	300	–	0.2	–
Muscle	300	–	0.41	–
Wood, cross grain				
Balsa	300	140	0.055	–
Cypress	300	465	0.097	–
Fir	300	415	0.11	2720
Oak	300	545	0.17	2385
Yellow pine	300	640	0.15	2805
White pine	300	435	0.11	–
Wood, radial				
Oak	300	545	0.19	2385
Fir	300	420	0.14	2720

Source: Adapted from References 1 and 8–13

Table A.4 ● Thermophysical properties of gases at atmospheric pressure

T (K)	ρ (kg/m³)	C_p (kJ/kg K)	$\mu \times 10^7$ (N s/m²)	$\nu \times 10^6$ (m²/s)	$k \times 10^3$ (W/m K)	$\alpha \times 10^6$ (m²/s)	Pr
Air							
100	3.5562	1.032	71.1	2.00	9.34	2.54	0.786
150	2.3364	1.012	103.4	4.426	13.8	5.84	0.758
200	1.7458	1.007	132.5	7.590	18.1	10.3	0.737
250	1.3947	1.006	159.6	11.44	22.3	15.9	0.720
300	1.1614	1.007	184.6	15.89	26.3	22.5	0.707
350	0.9950	1.009	208.2	20.92	30.0	29.9	0.700
400	0.8711	1.014	230.1	26.41	33.8	38.3	0.690
450	0.7740	1.021	250.7	32.39	37.3	47.2	0.686
500	0.6964	1.030	270.1	38.79	40.7	56.7	0.684
550	0.6329	1.040	288.4	45.57	43.9	66.7	0.683
600	0.5804	1.051	305.8	52.69	46.9	76.9	0.685
650	0.5356	1.063	322.5	60.21	49.7	87.3	0.690
700	0.4975	1.075	338.8	68.10	52.4	98.0	0.695
750	0.4643	1.087	354.6	76.37	54.9	109	0.702
800	0.4354	1.099	369.8	84.93	57.3	120	0.709
850	0.4097	1.110	384.3	93.80	59.6	131	0.716
900	0.3868	1.121	398.1	102.9	62.0	143	0.720
950	0.3666	1.131	411.3	112.2	64.3	155	0.723
1000	0.3482	1.141	424.4	121.9	66.7	168	0.726
1100	0.3166	1.159	449.0	141.8	71.5	195	0.728
1200	0.2902	1.175	473.0	162.9	76.3	224	0.728
1300	0.2679	1.189	496.0	185.1	82	238	0.719
1400	0.2488	1.207	530	213	91	303	0.703
1500	0.2322	1.230	557	240	100	350	0.685
1600	0.2177	1.248	584	268	106	390	0.688
1700	0.2049	1.267	611	298	113	435	0.685
1800	0.1935	1.286	637	329	120	482	0.683
1900	0.1833	1.307	663	362	128	534	0.677
2000	0.1741	1.337	689	396	137	589	0.672
2100	0.1658	1.372	715	431	147	646	0.667
2200	0.1582	1.417	740	468	160	714	0.655
2300	0.1513	1.478	766	506	175	783	0.647
2400	0.1448	1.558	792	547	196	869	0.630
2500	0.1389	1.665	818	589	222	960	0.613
3000	0.1135	2.726	955	841	486	1570	0.536
Ammonia (NH_3)							
300	0.6894	2.158	101.5	14.7	24.7	16.6	0.887
320	0.6448	2.170	109	16.9	27.2	19.4	0.870
340	0.6059	2.192	116.5	19.2	29.3	22.1	0.872
360	0.5716	2.221	124	21.7	31.6	24.9	0.872
380	0.5410	2.254	131	24.2	34.0	27.9	0.869

(continued)

Table A.4 ● Continued

T (K)	ρ (kg/m³)	C_p (kJ/kg K)	μ × 10⁷ (N s/m²)	ν × 10⁶ (m²/s)	k × 10³ (W/m K)	α × 10⁶ (m²/s)	Pr
Ammonia (NH₃) (continued)							
400	0.5136	2.287	138	26.9	37.0	31.5	0.853
420	0.4888	2.322	145	29.7	40.4	35.6	0.833
440	0.4664	2.357	152.5	32.7	43.5	39.6	0.826
460	0.4460	2.393	159	35.7	46.3	43.4	0.822
480	0.4273	2.430	166.5	39.0	49.2	47.4	0.822
500	0.4101	2.467	173	42.2	52.5	51.9	0.813
520	0.3942	2.504	180	45.7	54.5	55.2	0.827
540	0.3795	2.540	186.5	49.1	57.5	59.7	0.824
560	0.3708	2.577	193	52.0	60.6	63.4	0.827
580	0.3533	2.613	199.5	56.5	63.8	69.1	0.817
Carbon dioxide (CO₂)							
280	1.9022	0.830	140	7.36	15.20	9.63	0.765
300	1.7730	0.851	149	8.40	16.55	11.0	0.766
320	1.6609	0.872	156	9.39	18.05	12.5	0.754
340	1.5618	0.891	165	10.6	19.70	14.2	0.746
360	1.4743	0.908	173	11.7	21.2	15.8	0.741
380	1.3961	0.926	181	13.0	22.75	17.6	0.737
400	1.3257	0.942	190	14.3	24.3	19.5	0.737
450	1.1782	0.981	210	17.8	28.3	24.5	0.728
500	1.0594	1.02	231	21.8	32.5	30.1	0.725
550	0.9625	1.05	251	26.1	36.6	36.2	0.721
600	0.8826	1.08	270	30.6	40.7	42.7	0.717
650	0.8143	1.10	288	35.4	44.5	49.7	0.712
700	0.7564	1.13	305	40.3	48.1	56.3	0.717
750	0.7057	1.15	321	45.5	51.7	63.7	0.714
800	0.6614	1.17	337	51.0	55.1	71.2	0.716
Carbon monoxide (CO)							
200	1.6888	1.045	127	7.52	17.0	9.63	0.781
220	1.5341	1.044	137	8.93	19.0	11.9	0.753
240	1.4055	1.043	147	10.5	20.6	14.1	0.744
260	1.2967	1.043	157	12.1	22.1	16.3	0.741
280	1.2038	1.042	166	13.8	23.6	18.8	0.733
300	1.1233	1.043	175	15.6	25.0	21.3	0.730
320	1.0529	1.043	184	17.5	26.3	23.9	0.730
340	0.9909	1.044	193	19.5	27.8	26.9	0.725
360	0.9357	1.045	202	21.6	29.1	29.8	0.725
380	0.8864	1.047	210	23.7	30.5	32.9	0.729
400	0.8421	1.049	218	25.9	31.8	36.0	0.719
450	0.7483	1.055	237	31.7	35.0	44.3	0.714
500	0.673 52	1.065	254	37.7	38.1	53.1	0.710
550	0.612 26	1.076	271	44.3	41.1	62.4	0.710
600	0.561 26	1.088	286	51.0	44.0	72.1	0.707

(continued)

Table A.4 ● Continued

T (K)	ρ (kg/m³)	Cₚ (kJ/kg K)	μ × 10⁷ (N s/m²)	ν × 10⁶ (m²/s)	k × 10³ (W/m K)	α × 10⁶ (m²/s)	Pr
Carbon monoxide (CO) (continued)							
650	0.518 06	1.101	301	58.1	47.0	82.4	0.705
700	0.481 02	1.114	315	65.5	50.0	93.3	0.702
750	0.448 99	1.127	329	73.3	52.8	104	0.702
800	0.420 95	1.140	343	81.5	55.5	116	0.705
Helium (He)							
100	0.4871	5.193	96.3	19.8	73.0	28.9	0.686
120	0.4060	5.193	107	26.4	81.9	38.8	0.679
140	0.3481	5.193	118	33.9	90.7	50.2	0.676
160	–	5.193	129	–	99.2	–	–
180	0.2708	5.193	139	51.3	107.2	76.2	0.673
200	–	5.193	150	–	115.1	–	–
220	0.2216	5.193	160	72.2	123.1	107	0.675
240	–	5.193	170	–	130	–	–
260	0.1875	5.193	180	96.0	137	141	0.682
280	–	5.193	190	–	145	–	–
300	0.1625	5.193	199	122	152	180	0.680
350	–	5.193	221	–	170	–	–
400	0.1219	5.193	243	199	187	295	0.675
450	–	5.193	263	–	204	–	– .
500	0.097 54	5.193	283	290	220	434	0.668
550	–	5.193	–	–	–	–	–
600	–	5.193	320	–	252	–	–
650	–	5.193	332	–	264	–	–
700	0.069 69	5.193	350	502	278	768	0.654
750	–	5.193	364	–	291	–	–
800	–	5.193	382	–	304	–	–
900	–	5.193	414	–	330	–	–
1000	0.048 79	5.193	446	914	354	1400	0.654
Hydrogen (H₂)							
100	0.242 55	11.23	42.1	17.4	67.0	24.6	0.707
150	0.161 56	12.60	56.0	34.7	101	49.6	0.699
200	0.121 15	13.54	68.1	56.2	131	79.9	0.704
250	0.096 93	14.06	78.9	81.4	157	115	0.707
300	0.080 78	14.31	89.6	111	183	158	0.701
350	0.069 24	14.43	98.8	143	204	204	0.700
400	0.060 59	14.48	108.2	179	226	258	0.695
450	0.053 86	14.50	117.2	218	247	316	0.689
500	0.048 48	14.52	126.4	261	266	378	0.691
550	0.044 07	14.53	134.3	305	285	445	0.685

(continued)

Table A.4 ● Continued

T (K)	ρ (kg/m³)	Cₚ (kJ/kg K)	μ × 10⁷ (N s/m²)	ν × 10⁶ (m²/s)	k × 10³ (W/m K)	α × 10⁶ (m²/s)	Pr
Hydrogen (H₂) (continued)							
600	0.040 40	14.55	142.4	352	305	519	0.678
700	0.034 63	14.61	157.8	456	342	676	0.675
800	0.030 30	14.70	172.4	569	378	849	0.670
900	0.026 94	14.83	186.5	692	412	1030	0.671
1000	0.024 24	14.99	201.3	830	448	1230	0.673
1100	0.022 04	15.17	213.0	966	488	1460	0.662
1200	0.020 20	15.37	226.2	1120	528	1700	0.659
1300	0.018 65	15.59	238.5	1279	568	1955	0.655
1400	0.017 32	15.81	250.7	1447	610	2230	0.650
1500	0.016 16	16.02	262.7	1626	655	2530	0.643
1600	0.0152	16.28	273.7	1801	697	2815	0.639
1700	0.0143	16.58	284.9	1992	742	3130	0.637
1800	0.0135	16.96	296.1	2193	786	3435	0.639
1900	0.0128	17.49	307.2	2400	835	3730	0.643
2000	0.0121	18.25	318.2	2630	878	3975	0.661
Nitrogen (N₂)							
100	3.4388	1.070	68.8	2.00	9.58	2.60	0.768
150	2.2594	1.050	100.6	4.45	13.9	5.86	0.759
200	1.6883	1.043	129.2	7.65	18.3	10.4	0.736
250	1.3488	1.042	154.9	11.48	22.2	15.8	0.727
300	1.1233	1.041	178.2	15.86	25.9	22.1	0.716
350	0.9625	1.042	200.0	20.78	29.3	29.2	0.711
400	0.8425	1.045	220.4	26.16	32.7	37.1	0.704
450	0.7485	1.050	239.6	32.01	35.8	45.6	0.703
500	0.6739	1.056	257.7	38.24	38.9	54.7	0.700
550	0.6124	1.065	274.7	44.86	41.7	63.9	0.702
600	0.5615	1.075	290.8	51.79	44.6	73.9	0.701
700	0.4812	1.098	321.0	66.71	49.9	94.4	0.706
800	0.4211	1.22	349.1	82.90	54.8	116	0.715
900	0.3743	1.146	375.3	100.3	59.7	139	0.721
1000	0.3368	1.167	399.9	118.7	64.7	165	0.721
1100	0.3062	1.187	423.2	138.2	70.0	193	0.718
1200	0.2807	1.204	445.3	158.6	75.8	224	0.707
1300	0.2591	1.219	466.2	179.9	81.0	256	0.701
Oxygen (O₂)							
100	3.945	0.962	76.4	1.94	9.25	2.44	0.796
150	2.585	0.921	114.8	4.44	13.8	5.80	0.766
200	1.930	0.915	147.5	7.64	18.3	10.4	0.737
250	1.542	0.915	178.6	11.58	22.6	16.0	0.723
300	1.284	0.920	207.2	16.14	26.8	22.7	0.711

(continued)

Table A.4 ● Continued

T (K)	ρ (kg/m³)	C_p (kJ/kg K)	$\mu \times 10^7$ (N s/m²)	$\nu \times 10^6$ (m²/s)	$k \times 10^3$ (W/m K)	$\alpha \times 10^6$ (m²/s)	Pr
Oxygen (O₂) (continued)							
350	1.100	0.929	233.5	21.23	29.6	29.0	0.733
400	0.9620	0.942	258.2	26.84	33.0	36.4	0.737
450	0.8554	0.956	281.4	32.90	36.3	44.4	0.741
500	0.7698	0.972	303.3	39.40	41.2	55.1	0.716
550	0.6998	0.988	324.0	46.30	44.1	63.8	0.726
600	0.6414	1.003	343.7	53.59	47.3	73.5	0.729
700	0.5498	1.031	380.8	69.26	52.8	93.1	0.744
800	0.4810	1.054	415.2	86.32	58.9	116	0.743
900	0.4275	1.074	447.2	104.6	64.9	141	0.740
1000	0.3848	1.090	477.0	124.0	71.0	169	0.733
1100	0.3498	1.103	505.5	144.5	75.8	196	0.736
1200	0.3206	1.115	532.5	166.1	81.9	229	0.725
1300	0.2960	1.125	588.4	188.6	87.1	262	0.721
Water vapour (steam)							
380	0.5863	2.060	127.1	21.68	24.6	20.4	1.06
400	0.5542	2.014	134.4	24.25	26.1	23.4	1.04
450	0.4902	1.980	152.5	31.11	29.9	30.8	1.01
500	0.4405	1.985	170.4	36.68	33.9	38.8	0.998
550	0.4005	1.997	188.4	47.04	37.9	47.4	0.993
600	0.3652	2.026	206.7	56.60	42.2	57.0	0.993
650	0.3380	2.056	224.7	66.48	46.4	66.8	0.996
700	0.3140	2.085	242.6	77.26	50.5	77.1	1.00
750	0.2931	2.119	260.4	88.84	54.9	88.4	1.00
800	0.2739	2.152	278.6	101.7	59.2	100	1.01
850	0.2579	2.186	296.9	115.1	63.7	113	1.02

Source: Adapted from References 8, 14 and 15

Table A.5 ● Thermophysical properties of saturated liquids

T (K)	ρ (kg/m³)	C_p (kJ/kg K)	$\mu \times 10^2$ (N s/m²)	$\nu \times 10^6$ (m²/s)	$k \times 10^3$ (W/m K)	$\alpha \times 10^7$ (m²/s)	Pr	$\beta \times 10^3$ (K⁻¹)
Engine oil (unused)								
273	899.1	1.796	385	4280	147	0.910	47 000	0.70
280	895.3	1.827	217	2430	144	0.880	27 500	0.70
290	890.0	1.868	99.9	1120	145	0.872	12 900	0.70
300	884.1	1.909	48.6	550	145	0.859	6400	0.70
310	877.9	1.951	25.3	288	145	0.847	3400	0.70
320	871.8	1.993	14.1	161	143	0.823	1965	0.70
330	865.8	2.035	8.36	96.6	141	0.800	1205	0.70
340	859.9	2.076	5.31	61.7	139	0.779	793	0.70
350	853.9	2.118	3.56	41.7	138	0.763	546	0.70
360	847.8	2.161	2.52	29.7	138	0.753	395	0.70
370	841.8	2.206	1.86	22.0	137	0.738	300	0.70
380	836.0	2.250	1.41	16.9	136	0.723	233	0.70
390	830.6	2.294	1.10	13.3	135	0.709	187	0.70
400	825.1	2.337	0.874	10.6	134	0.695	152	0.70
410	818.9	2.381	0.698	8.52	133	0.682	125	0.70
420	812.1	2.427	0.564	6.94	133	0.675	103	0.70
430	806.5	2.471	0.470	5.83	132	0.662	88	0.70
Ethylene glycol [(C₂H₄(OH)₂]								
273	1130.8	2.294	6.51	57.6	242	0.933	617	0.65
280	1125.8	2.323	4.20	37.3	244	0.933	400	0.65
290	1118.8	2.368	2.47	22.1	248	0.936	236	0.65
300	1114.4	2.415	1.57	14.1	252	0.939	151	0.65
310	1103.7	2.460	1.07	9.65	255	0.939	103	0.65
320	1096.2	2.505	0.757	6.91	258	0.940	73.5	0.65
330	1089.5	2.549	0.561	5.15	260	0.936	55.0	0.65
340	1083.8	2.592	0.431	3.98	261	0.929	42.8	0.65
350	1079.0	2.637	0.342	3.17	261	0.917	34.6	0.65
360	1074.0	2.682	0.278	2.59	261	0.906	28.6	0.65
370	1066.7	2.728	0.228	2.14	262	0.900	23.7	0.65
373	1058.5	2.742	0.215	2.03	263	0.906	22.4	0.65

(continued)

Table A.5 ● Continued

T (K)	ρ (kg/m³)	C_p (kJ/kg K)	$\mu \times 10^2$ (N s/m²)	$\nu \times 10^6$ (m²/s)	$k \times 10^3$ (W/m K)	$\alpha \times 10^7$ (m²/s)	Pr	$\beta \times 10^3$ (K⁻¹)
Glycerin [$C_3H_5(OH)_3$]								
273	1276.0	2.261	1060	8310	282	0.977	85 000	0.47
280	1271.9	2.298	534	4200	284	0.972	43 200	0.47
290	1265.8	2.367	185	1460	286	0.955	15 300	0.48
300	1259.9	2.427	79.9	634	286	0.935	6 780	0.48
310	1253.9	2.490	35.2	281	286	0.916	3 060	0.49
320	1247.2	2.564	21.0	168	287	0.897	1 870	0.50
Freon (Refrigerant-12) (CCl_2F_2)								
230	1528.4	0.8816	0.0457	0.299	68	0.505	5.9	1.85
240	1498.0	0.8923	0.0385	0.257	69	0.516	5.0	1.90
250	1469.5	0.9037	0.0354	0.241	70	0.527	4.6	2.00
260	1439.0	0.9163	0.0322	0.224	73	0.554	4.0	2.10
270	1407.2	0.9301	0.0304	0.216	73	0.558	3.9	2.25
280	1374.4	0.9450	0.0283	0.206	73	0.562	3.7	2.35
290	1340.5	0.9609	0.0265	0.198	73	0.567	3.5	2.55
300	1305.8	0.9781	0.0254	0.195	72	0.564	3.5	2.75
310	1268.9	0.9963	0.0244	0.192	69	0.546	3.4	3.05
320	1228.6	1.0155	0.0233	0.190	68	0.545	3.5	3.5
Mercury (Hg)								
273	13 528	0.1404	0.1688	0.1240	8 180	42.85	0.0290	0.181
300	13 529	0.1393	0.1523	0.1125	8 540	45.30	0.0248	0.181
350	13 407	0.1377	0.1309	0.0976	9 180	49.75	0.0196	0.181
400	13 287	0.1365	0.1171	0.0882	9 800	54.05	0.0163	0.181
450	13 167	0.1357	0.1075	0.0816	10 400	58.10	0.0140	0.181
500	13 048	0.1353	0.1007	0.0771	10 950	61.90	0.0125	0.182
550	12 929	0.1352	0.0953	0.0737	11 450	65.55	0.0112	0.184
600	12 809	0.1355	0.0911	0.0711	11 950	68.80	0.0103	0.187

Source: Adapted from References 15 and 16

Table A.6 ● Thermophysical properties of saturated liquid-vapour, 1 atm

Fluid	T_{sat} (K)	h_{fg} (kJ/kg)	ρ_L (kg/m^3)	ρ_V (kg/m^3)	$\sigma \times 10^3$ (N/m)
Ethanol	351	846	757	1.44	17.7
Ethylene glycol	470	812	1 111[a]	–	32.7
Glycerin	563	974	1 260[a]	–	63.0[a]
Mercury	630	301	12 740	3.90	417
Refrigerant R-12	243	165	1 488	6.32	15.8
Refrigerant R-113	321	147	1 511	7.38	15.9

[a] Property value corresponding to 300 K
Adapted from References 8, 17 and 18

Table A.7 ● Thermophysical properties of saturated water

Temperature, T (K)	Pressure, p (bar)[a]	Specific volume (m³/kg)		Heat of vaporisation, h_fg (kJ/kg)	Specific heat (kJ/kg K)		Viscosity (N s/m²)	
		$V_L \times 10^3$	V_v		$C_{p,L}$	$C_{p,v}$	$\mu_L \times 10^6$	$\mu_v \times 10^6$
273.15	0.006 11	1.000	206.3	2502	4.217	1.854	1750	8.02
275	0.006 97	1.000	181.7	2497	4.211	1.855	1652	8.09
280	0.009 90	1.000	130.4	2485	4.198	1.858	1422	8.29
285	0.013 87	1.000	99.4	2473	4.189	1.861	1225	8.49
290	0.019 17	1.001	69.7	2461	4.184	1.864	1080	8.69
295	0.026 17	1.002	51.94	2449	4.181	1.868	959	8.89
300	0.035 31	1.003	39.13	2438	4.179	1.872	855	9.09
305	0.047 12	1.005	29.74	2426	4.178	1.877	769	9.29
310	0.062 21	1.007	22.93	2414	4.178	1.882	695	9.49
315	0.081 32	1.009	17.82	2402	4.179	1.888	631	9.69
320	0.1053	1.011	13.98	2390	4.180	1.895	577	9.89
325	0.1351	1.013	11.06	2378	4.182	1.903	528	10.09
330	0.1719	1.016	8.82	2366	4.184	1.911	489	10.29
335	0.2167	1.018	7.09	2354	4.186	1.920	453	10.49
340	0.2713	1.021	5.74	2342	4.188	1.930	420	10.69
345	0.3372	1.024	4.683	2329	4.191	1.941	389	10.89
350	0.4163	1.027	3.846	2317	4.195	1.954	365	11.09
355	0.5100	1.030	3.180	2304	4.199	1.968	343	11.29
360	0.6209	1.034	2.645	2291	4.203	1.983	324	11.49
365	0.7514	1.038	2.212	2278	4.209	1.999	306	11.69
370	0.9040	1.041	1.861	2265	4.214	2.017	289	11.89
373.15	1.0133	1.044	1.679	2257	4.217	2.029	279	12.02
375	1.0815	1.045	1.574	2252	4.220	2.036	274	12.09
380	1.2869	1.049	1.337	2239	4.226	2.057	260	12.29
385	1.5233	1.053	1.142	2225	4.232	2.080	248	12.49
390	1.794	1.058	0.980	2212	4.239	2.104	237	12.69
400	2.455	1.067	0.731	2183	4.256	2.158	217	13.05
410	3.302	1.077	0.553	2153	4.278	2.221	200	13.42
420	4.370	1.088	0.425	2123	4.302	2.291	185	13.79
430	5.699	1.099	0.331	2091	4.331	2.369	173	14.14

(continued)

Table A.7 ● Continued

Temperature, T (K)	Pressure, p (bar)[a]	Specific volume (m³/kg)		Heat of vaporisation, h_fg (kJ/kg)	Specific heat (kJ/kg K)		Viscosity (N s/m²)	
		$V_L \times 10^3$	V_v		$C_{p,L}$	$C_{p,v}$	$\mu_L \times 10^6$	$\mu_v \times 10^6$
440	7.333	1.110	0.261	2059	4.36	2.46	162	14.50
450	9.319	1.123	0.208	2024	4.40	2.56	152	14.85
460	11.71	1.137	0.167	1989	4.44	2.68	143	15.19
470	14.55	1.152	0.136	1951	4.48	2.79	136	15.54
480	17.90	1.167	0.111	1912	4.53	2.94	129	15.88
490	21.83	1.184	0.0922	1870	4.59	3.10	124	16.23
500	26.40	1.203	0.0766	1825	4.66	3.27	118	16.59
510	31.66	1.222	0.0631	1779	4.74	3.47	113	16.95
520	37.70	1.244	0.0525	1730	4.84	3.70	108	17.33
530	44.58	1.268	0.0445	1679	4.95	3.96	104	17.72
540	52.38	1.294	0.0375	1622	5.08	4.27	101	18.1
550	61.19	1.323	0.0317	1564	5.24	4.64	97	18.6
560	71.08	1.355	0.0269	1499	5.43	5.09	94	19.1
570	82.16	1.392	0.0228	1429	5.68	5.67	91	19.7
580	94.51	1.433	0.0193	1353	6.00	6.40	88	20.4
590	108.3	1.482	0.0163	1274	6.41	7.35	84	21.5
600	123.5	1.541	0.0137	1176	7.00	8.75	81	22.7
610	137.3	1.612	0.0115	1068	7.85	11.1	77	24.1
620	159.1	1.705	0.0094	941	9.35	15.4	72	25.9
625	169.1	1.778	0.0085	858	10.6	18.3	70	27.0
630	179.7	1.856	0.0075	781	12.6	22.1	67	28.0
635	190.9	1.935	0.0066	683	16.4	27.6	64	30.0
640	202.7	2.075	0.0057	560	26	42	59	32.0
645	215.2	2.351	0.0045	361	90	–	54	37.0
647.3[b]	221.2	3.170	0.0032	0	∞	∞	45	45.0

(continued)

Table A.7 ● Continued

Temperature, T (K)	Thermal conductivity (W/m K)		Prandtl number		Surface tension, $\sigma_L \times 10^3$ (N/m)	Expansion coefficient, $\beta \times 10^6$ (K^{-1})
	$k_L \times 10^3$	$k_V \times 10^3$	Pr_L	Pr_V		
273.15	569	18.2	12.99	0.815	75.5	− 68.05
275	574	18.3	12.22	0.817	75.3	− 32.74
280	582	18.6	10.26	0.825	74.8	46.04
285	590	18.9	8.81	0.833	74.3	114.1
290	598	19.3	7.56	0.841	73.7	174.0
295	606	19.5	6.62	0.849	72.7	227.5
300	613	19.6	5.83	0.857	71.7	276.1
305	620	20.1	5.20	0.865	70.9	320.6
310	628	20.4	4.62	0.873	70.0	361.9
315	634	20.7	4.16	0.883	69.2	400.4
320	640	21.0	3.77	0.894	68.3	436.7
325	645	21.3	3.42	0.901	67.5	471.2
330	650	21.7	3.15	0.908	66.6	504.0
335	656	22.0	2.88	0.916	65.8	535.5
340	660	22.3	2.66	0.925	64.9	566.0
345	668	22.6	2.45	0.933	64.1	595.4
350	668	23.0	2.29	0.942	63.2	624.2
355	671	23.3	2.14	0.951	62.3	652.3
360	674	23.7	2.02	0.960	61.4	697.9
365	677	24.1	1.91	0.969	60.5	707.1
370	679	24.5	1.80	0.978	59.5	728.7
373.15	680	24.8	1.76	0.984	58.9	750.1
375	681	24.9	1.70	0.987	58.6	761
380	683	25.4	1.61	0.999	57.6	788
385	685	25.8	1.53	1.004	56.6	814
390	686	26.3	1.47	1.013	55.6	841
400	688	27.2	1.34	1.033	53.6	896
410	688	28.2	1.24	1.054	51.5	952
420	688	29.8	1.16	1.075	49.4	1010
430	685	30.4	1.09	1.10	47.2	

(continued)

Table A.7 ● Continued

Temperature, T (K)	Thermal conductivity (W/m K)		Prandtl number		Surface tension, $\sigma_L \times 10^3$ (N/m)	Expansion coefficient, $\beta \times 10^6$ (K^{-1})
	$k_L \times 10^3$	$k_V \times 10^3$	Pr_L	Pr_V		
440	682	31.7	1.04	1.12	45.1	
450	678	33.1	0.99	1.14	42.9	
460	673	34.6	0.95	1.17	40.7	
470	667	36.3	0.92	1.20	38.5	
480	660	38.1	0.89	1.23	36.2	
490	651	40.1	0.87	1.25	33.9	—
500	642	42.3	0.86	1.28	31.6	—
510	631	44.7	0.85	1.31	29.3	—
520	621	47.5	0.84	1.35	26.9	—
530	608	50.6	0.85	1.39	24.5	—
540	594	54.0	0.86	1.43	22.1	—
550	580	58.3	0.87	1.47	19.7	—
560	563	63.7	0.90	1.52	17.3	—
570	548	76.7	0.94	1.59	15.0	—
580	528	76.7	0.99	1.68	12.8	—
590	513	84.1	1.05	1.84	10.5	—
600	497	92.9	1.14	2.15	8.4	—
610	467	103	1.30	2.60	6.3	—
620	444	114	1.52	3.46	4.5	—
625	430	121	1.65	4.20	3.5	—
630	412	130	2.0	4.8	2.6	—
635	392	141	2.7	6.0	1.5	—
640	367	155	4.2	9.6	0.8	—
645	331	178	12	26	0.1	—
647.3[b]	238	238	∞	∞	0.0	—

[a] 1 bar = 10^5 N/m^2
[b] Critical temperature
Source: Adapted from Reference 19

Table A.8 ● Thermophysical properties of liquid metals

Composition	Melting point (K)	T (K)	ρ (kg/m³)	Cp (kJ/kg K)	ν × 10⁷ (m²/s)	k (W/m K)	α × 10⁵ (m²/s)	Pr
Bismuth	544	589	10011	0.1444	1.617	16.4	0.138	0.0142
		811	9739	0.1545	1.133	15.6	1.035	0.0110
		1033	9467	0.1645	0.8343	15.6	1.001	0.0083
Lead	600	644	10540	0.159	2.276	16.1	1.084	0.024
		755	10412	0.155	1.849	15.6	1.223	0.017
		977	10140	–	1.347	14.9	–	–
Potassium	337	422	807.3	0.80	4.608	45.0	6.99	0.0066
		700	741.7	0.75	2.397	39.5	7.07	0.0034
		977	674.4	0.75	1.905	33.1	6.55	0.0029
Sodium	371	366	929.1	1.38	7.516	86.2	6.71	0.011
		644	860.2	1.30	3.270	72.3	6.48	0.0051
		977	778.5	1.26	2.285	59.7	6.12	0.0037
NaK, (45%/55%)	292	366	887.4	1.130	6.522	25.6	2.552	0.026
		644	821.7	1.055	2.871	27.5	3.17	0.0091
		977	740.1	1.043	2.174	28.9	3.74	0.0058
NaK, (22%/78%)	262	366	849.0	0.946	5.797	24.4	3.05	0.019
		672	775.3	0.879	2.666	26.7	3.92	0.0068
		1033	690.4	0.883	2.118	–	–	–
PbBi, (44.5%/55.5%)	398	422	10524	0.147	–	9.05	0.586	–
		644	10236	0.147	1.496	11.86	0.790	0.189
		922	9835	–	1.171	–	–	–
Mercury	234				See Table A.5			

Source: Adapted from *Liquid Materials Handbook* 23rd edn Washington, DC: Atomic Energy Commission, Department of the Navy (1952)

Table A.9 ● Total, normal (n) or hemispherical (h) emissivity of selected surfaces

Metallic solids and their oxides[a]

Description/composition		Emissivity, ϵ_n or ϵ_h, at various temperatures (K)										
		100	200	300	400	600	800	1000	1200	1500	2000	2500
Aluminum												
Highly polished, film	(h)	0.02	0.03	0.04	0.05	0.06						
Foil, bright	(h)	0.06	0.06	0.07								
Anodised	(h)			0.82	0.76							
Chromium												
Polished or plated	(n)	0.05	0.07	0.10	0.12	0.14						
Copper												
Highly polished	(h)			0.03	0.03	0.04	0.04	0.04				
Stably oxidized	(h)					0.50	0.58	0.80				
Gold												
Highly polished or film	(h)	0.01	0.02	0.03	0.03	0.04	0.05	0.06				
Foil, bright	(h)	0.06	0.07	0.07								
Molybdenum												
Polished	(h)					0.06	0.08	0.10	0.12	0.15	0.21	0.26
Shot-blasted, rough	(h)					0.25	0.28	0.31	0.35	0.42		
Stably oxidised	(h)					0.80	0.82					
Nickel												
Polished	(h)					0.09	0.11	0.14	0.17			
Stably oxidised	(h)					0.40	0.49	0.57				
Platinum												
Polished	(h)						0.10	0.13	0.15	0.18		
Silver												
Polished	(h)			0.02	0.02	0.03	0.05	0.08				
Stainless steels												
Typical, polished	(n)			0.17	0.17	0.19	0.23	0.30				
Typical, cleaned	(n)			0.22	0.22	0.24	0.28	0.35				
Typical, lightly oxidised	(n)						0.33	0.40				

(continued)

Table A.9 ● Continued

Metallic solids and their oxides[a] (continued)

Description/composition		Emissivity, ϵ_n or ϵ_h, at various temperatures (K)												
		100	200	300	400	600	800	1000	1200	1500	2000	2500
Stainless steels (continued)												
Typical, highly oxidised	(n)						0.67	0.70	0.76			
AISI 347, stably oxidised	(n)					0.87	0.88	0.89	0.90			
Tantalum												
Polished	(h)								0.11	0.17	0.23	0.28
Tungsten												
Polished	(h)							0.10	0.13	0.18	0.25	0.29

[a] Source: Adapted from Reference 1

Table A.9 ● Continued

Nonmetallic substances[b]

Description/composition		Temperature (K)	Emiisivity ϵ
Aluminum oxide	(n)	600	0.69
		1000	0.55
		1500	0.41
Asphalt pavement	(h)	300	0.85–0.93
Building materials			
Asbestos sheet	(h)	300	0.93–0.96
Brick, red	(h)	300	0.93–0.96
Gypsum or plaster board	(h)	300	0.90–0.92
Wood	(h)	300	0.82–0.92
Cloth	(h)	300	0.75–0.90
Concrete	(h)	300	0.88–0.93
Glass, window	(h)	300	0.90–0.95
Ice	(h)	273	0.95–0.98
Paints			
Black (Parsons)	(h)	300	0.98
White, acrylic	(h)	300	0.90
White, zinc oxide	(h)	300	0.92
Paper, white	(h)	300	0.92–0.97
Pyrex	(n)	300	0.82
		600	0.80
		1000	0.71
		1200	0.62
Pyroceram	(n)	300	0.85
		600	0.78
		1000	0.69
		1500	0.57
Refractories (furnace liners)			
Alumina brick	(n)	800	0.40
		1000	0.33
		1400	0.28
		1600	0.33
Magnesia brick	(n)	800	0.45
		1000	0.36
		1400	0.31
		1600	0.40
Kaolin insulating brick	(n)	800	0.70
		1200	0.57
		1400	0.47
		1600	0.53
Sand	(h)	300	0.90

(continued)

Table A.9 ● Continued

Nonmetallic substances[b] (continued)

Description/composition		Temperature (K)	Emiisivity ϵ
Silicon carbide	(n)	600	0.87
		1000	0.87
		1500	0.85
Skin	(h)	300	0.95
Snow	(h)	273	0.82–0.90
Soil	(h)	300	0.93–0.96
Rocks	(h)	300	0.88–0.95
Teflon	(h)	300	0.85
		400	0.87
		500	0.92
Vegetation	(h)	300	0.92–0.96
Water	(h)	300	0.96

[b] Source: adapted from References 1, 9, 24 and 25

Table A.10 ● Solar radiative properties for selected materials

Description/composition	α_s	ϵ^a	α_s/ϵ	τ_s
Aluminum				
Polished	0.09	0.03	3.0	
Anodised	0.14	0.84	0.17	
Quartz overcoated	0.11	0.37	0.30	
Foil	0.15	0.05	3.0	
Brick, red (Purdue)	0.63	0.93	0.68	
Concrete	0.60	0.88	0.68	
Galvanised sheet metal				
Clean, new	0.65	0.13	5.0	
Oxidised, weathered	0.80	0.28	2.9	
Glass, 3.2 mm thickness				
Float or tempered				0.79
Low iron oxide type				0.88
Metal, plated				
Black sulphide	0.92	0.10	9.2	
Black cobalt oxide	0.93	0.30	3.1	
Black nickel oxide	0.92	0.08	11	
Black chrome	0.87	0.09	9.7	
Mylar, 0.13 mm thickness				0.87
Paints				
Black (Parsons)	0.98	0.98	1.0	
White, acrylic	0.26	0.90	0.29	
White, zinc oxide	0.16	0.93	0.17	
Plexiglas, 3.2 mm thickness				0.90
Snow				
Fine particles, fresh	0.13	0.82	0.16	
Ice granules	0.33	0.89	0.37	
Tedlar, 0.10 mm thickness				0.92
Teflon, 0.13 mm thickness				0.92

[a] The emissivity values in this table correspond to a surface temperature of approximately 300 K
Source: Adapted with permission from Reference 25

A.4 ● **References to Appendix A**

1. Touloukian, Y.S. and Ho, C.Y. eds. (1972). *Thermophysical Properties of Matter.* Vol. 1: *Thermal Conductivity of Metallic Solids;* Vol. 2: *Thermal Conductivity of Nonmetallic Solids;* Vol. 4: *Specific Heat of Metallic Solids;* Vol. 5: *Specific Heat of Nonmetallic Solids;* Vol. 7: *Thermal Radiative Properties of Metallic Solids;* Vol. 8: *Thermal Radiative Properties of Nonmetallic Solids;* Vol. 9: *Thermal Radiative Properties of Coatings.* New York: Plenum Press

2. Touloukian, Y.S. and Ho, C.Y. eds. (1976). *Thermophysical Properties of Selected Aerospace Materials.* Part 1: Thermal Radiative Properties; Part II: Thermophysical Properties of Seven Materials. Thermophysical and Electronic Properties Information Analysis Center, CINDAS, Purdue University, West Lafayette, IN

3. Ho, C.Y., Powell, R.W. and Liley, P.E. (1974). *J. Phys. Chem. Ref. Data,* **3**, Supplement 1

4. Desai, P.D., Chu, T.K., Bogaard, R.H., Ackermann, M.W. and Ho, C.Y. (1976). Part I: Thermophysical Properties of Carbon Steels, Part II: Thermophysical Properties of Low Chromium Steels, Part III: Thermophysical Properties of Nickel Steels, Part IV: Thermophysical Properties of Stainless Steels, *CINDAS Special Report.* Purdue University, West Lafayette, IN

5. American Society for Metals (1961). *Metals Handbook.* Vol. 1: *Properties and Selection of Metals* 8th edn. Metals Park, OH: ASM

6. Hultgren, R., Desai, P.D., Hawkins, D.T., Gleiser, M., Kelley, K.K. and Wagman, D.D. (1973). *Selected Values of the Thermodynamic Properties of the Elements.* Metals Park, OH: American Society of Metals

7. Hultgren, R., Desai, P.D., Hawkins, D.T., Gleiser, M. and Kelley, K.K. (1973). *Selected Values of the Thermodynamic Properties of Binary Alloys.* Metals Park, OH: American Society of Metals

8. American Society of Heating, Refrigerating and Air Conditioning Engineers (1981). *ASHRAE Handbook of Fundamentals.* New York: ASHRAE

9. Mallory, J.F. (1969) *Thermal Insulation.* New York: Van Nostrand Reinhold

10. Hanley, E.J., DeWitt, D.P. and Taylor, R.E. (1977). The thermal transport properties at normal and elevated temperature of eight representative rocks. *Proceedings of the Seventh Symposium on Thermophysical Properties.* New York: American Society of Mechanical Engineers

11. Sweat, V.E. (1976). A Miniature Thermal Conductivity Probe for Foods. American Society of Mechanical Engineers, *Paper 76-HT-60,* August

12. Kothandaraman, C.P. and Subramanyan, S. (1975). *Heat and Mass Transfer Data Book.* New York: Halsted Press/Wiley

13. Chapman, A.J. (1984). *Heat Transfer* 4th edn. New York: Macmillan

14. Vargaftik, N.B. (1975). *Tables of Thermophysical Properties of Liquids and Gases* 2nd edn. New York: Hemisphere

15. Eckert, E.R.G. and Drake, R.M. (1972). *Analysis of Heat and Mass Transfer.* New York: McGraw-Hill

16. Vukalovich, M.P., Ivanov, A.I., Fokin, L.R. and Yakovelev, A.T. (1971). *Thermophysical Properties of Mercury,* State Committee on Standards, State Service for Standards and Handbook Data, Monograph Series No. 9, Izd. Standartov, Moscow

17. Bolz, R.E. and Tuve, G.L. eds. (1979) *CRC Handbook of Tables for Applied Engineering Science* 2nd edn. Boca Raton, FL: CRC Press

18. Liley, P.E. (1984) Private communication, School of Mechanical Engineering, Purdue University, West Lafayette, IN

19. Liley, P.E. (1984). Steam tables in SI units. Private communication, School of Mechanical Engineering, Purdue University, West Lafayette, IN

20. Perry, J.H. ed. (1963). *Chemical Engineer's Handbook* 4th edn. New York: McGraw-Hill

21. Geankoplis, C.J. (1972). *Mass Transport Phenomena*. New York: Holt, Rinehart & Winston
22. Barrer, R.M. (1941). *Diffusion In and Through Solids*. New York: Macmillan
23. Spalding, D.B. (1963). *Convective Mass Transfer*. New York: McGraw-Hill
24. Gubareff, G.G., Janssen, J.E. and Torborg, R.H. (1960). *Thermal Radiation Properties Survey*. Minneapolis, MN: Minneapolis-Honeywell Regulator Company
25 Kreith, F. and Kreider, J.F. (1978). *Principles of Solar Energy*. New York: Hemisphere

Mathematical relations and functions

B.1 ● Hyperbolic functions[a]

x	sinh x	cosh x	tanh x	x	sinh x	cosh x	tanh x
0.00	0.0000	1.0000	0.000 00	2.00	3.6269	3.7622	0.964 03
0.10	0.1002	1.0050	0.099 67	2.10	4.0219	4.1443	0.970 45
0.20	0.2013	1.0201	0.197 38	2.20	4.4571	4.5679	0.975 74
0.30	0.3045	1.0453	0.291 31	2.30	4.9370	5.0372	0.980 10
0.40	0.4108	1.0811	0.379 95	2.40	5.4662	5.5569	0.983 67
0.50	0.5211	1.1276	0.462 12	2.50	6.0502	6.1323	0.986 61
0.60	0.6367	1.1855	0.537 05	2.60	6.6947	6.7690	0.989 03
0.70	0.7586	1.2552	0.604 37	2.70	7.4063	7.4735	0.991 01
0.80	0.8881	1.3374	0.664 04	2.80	8.1919	8.2527	0.992 63
0.90	1.0265	1.4331	0.716 30	2.90	9.0596	9.1146	0.993 96
1.00	1.1752	1.5431	0.761 59	3.00	10.018	10.068	0.995 05
1.10	1.3356	1.6685	0.800 50	3.50	16.543	16.573	0.998 18
1.20	1.5095	1.8107	0.833 65	4.00	27.290	27.308	0.999 33
1.30	1.6984	1.9709	0.861 72	4.50	45.003	45.014	0.999 75
1.40	1.9043	2.1509	0.885 35	5.00	74.203	74.210	0.999 91
1.50	2.1293	2.3524	0.905 15	6.00	201.71	201.72	0.999 99
1.60	2.3756	2.5775	0.921 67	7.00	548.32	548.32	1.0000
1.70	2.6456	2.8283	0.935 41	8.00	1 490.5	1 490.5	1.0000
1.80	2.9422	3.1075	0.946 81	9.00	4 051.5	4 051.5	1.0000
1.90	3.2682	3.4177	0.956 24	10.00	11 013	11 013	1.0000

[a] The hyperbolic functions are defined as

$$\sinh x = \tfrac{1}{2}(e^x - e^{-x}) \qquad \cosh x = \tfrac{1}{2}(e^x + e^{-x}) \qquad \tanh x = \frac{e^x - e^{-x}}{e^x + e^{-x}} = \frac{\sinh x}{\cosh x}$$

The derivatives of the hyperbolic functions of the variable u are given as

$$\frac{d}{dx}(\sinh u) = (\cosh u)\frac{du}{dx} \qquad \frac{d}{dx}(\cosh u) = (\sinh u)\frac{du}{dx} \qquad \frac{d}{dx}(\tanh u) = \left(\frac{1}{\cosh^2 u}\right)\frac{du}{dx}$$

B.2 ● Gaussian error function[a]

w	erf w	w	erf w	w	erf w
0.00	0.000 00	0.36	0.389 33	1.04	0.858 65
0.02	0.022 56	0.38	0.409 01	1.08	0.873 33
0.04	0.045 11	0.40	0.428 39	1.12	0.886 79
0.06	0.067 62	0.44	0.466 22	1.16	0.899 10
0.08	0.090 08	0.48	0.502 75	1.20	0.910 31
0.10	0.112 46	0.52	0.537 90	1.30	0.934 01
0.12	0.134 76	0.56	0.571 62	1.40	0.952 28
0.14	0.156 95	0.60	0.603 86	1.50	0.966 11
0.16	0.179 01	0.64	0.634 59	1.60	0.976 35
0.18	0.200 94	0.68	0.663 78	1.70	0.983 79
0.20	0.222 70	0.72	0.691 43	1.80	0.989 09
0.22	0.244 30	0.76	0.717 54	1.90	0.992 79
0.24	0.265 70	0.80	0.742 10	2.00	0.995 32
0.26	0.286 90	0.84	0.765 14	2.20	0.998 14
0.28	0.307 88	0.88	0.786 69	2.40	0.999 31
0.30	0.328 63	0.92	0.806 77	2.60	0.999 76
0.32	0.349 13	0.96	0.825 42	2.80	0.999 92
0.34	0.369 36	1.00	0.842 70	3.00	0.999 98

[a] The Gaussian error function is defined as

$$\text{erf } w = \frac{2}{\sqrt{\pi}} \int_0^w \exp(-v^2) \, dv$$

The complementary error function is defined as

$$\text{erfc } w = 1 - \text{erf } w$$

Appendix C

Program listings

C.1 ● Onedxt.bas (see pp 51–5)

```
1000   REM Analytical Solution to 1-D Transient
Conduction Equation

1010   DIM xpos(100), theta(100), pn(10), a(10),
eterm(10), xterm(10), sum(10)

1013   OPEN "onedxt.RES" FOR OUTPUT AS #1
1014   OPEN "onedxt.cgd" FOR OUTPUT AS #2
1700   nroots = 10
1800   npos = 100
1900   rho = 7850#
1910   spht = 450#
1920   cond = 16#
1930   htc = 3200#
1940   thick = .015#
1950   tinit = 600#
1960   tfluid = 20#
1970   alpha = cond / (rho * spht)
2000   PRINT "Enter time (s)"
2001   INPUT tsec
2010   fo = alpha * tsec / (thick * thick)
2020   biot = htc * thick / cond
2021   FOR ix = 1 TO npos
2022   xpos (ix) = (ix - 1) / (npos - 1)
2023   NEXT .ix
2030   GOSUB 5000
2035   GOSUB 6000
2040   FOR ix = 1 TO npos
2050   x = xpos(ix)
2060   GOSUB 7000
2070   theta(ix) = result
2080   NEXT ix
2081   GOSUB 8000
2082   qdot = result * cond * (tfluid - tinit) / thick
2083   GOSUB 9000
2084   qxfer = result
```

```
2085    FOR ix = 1 TO npos
2090    PRINT #1, ix; " "; xpos(ix); " "; theta(ix)
2095    NEXT ix
2096    PRINT #1, qdot; " "; qxfer
3000    STOP

5000    REM -----------------------------------------------
5001    REM Find roots of the equation p tan(p) = Biot
5010    REM Using Newton-Raphson Iteration Method
5020    REM <beta> is an underrelaxation factor.
5030    REM -----------------------------------------------
5050    tol = .00001                            'Tolerance
5070    pi = 3.1415927#
5150    beta = 1!
5160    IF biot > 2# THEN beta = 1# / biot
5200    FOR ipi = 1 TO nroots
5300    p0 = (ipi * pi) - (.75# * pi)
5350    p = p0
5360    f = (p * (SIN(p) / COS(p))) - biot
5370    fdash = (p / (COS(p) * COS(p))) + (SIN(p) /
        COS(p))
5500    pnew = p - ((f / fdash) * beta)
5600    diff = ABS(pnew - p)
5700    IF diff > tol THEN p = pnew: GOTO 5360
5800    pn(ipi) = p
5850    NEXT ipi
5870    RETURN
5890    END

6000    REM ---------------------------------------------
6001    REM Evaluate constants
6100    FOR i = 1 TO nroots
6300    a(i) = 4# * SIN(pn(i)) / ((2# * pn(i)) + SIN(2#
*           pn(i)))
6400    eterm(i) = EXP(-1# * pn(i) * pn(i) * fo)
6500    NEXT i
6600    RETURN
6700    END

7000    REM ---------------------------------------------
7100    REM Evaluate Series Solution
7150    sum = 0#: result = 0#
7200    FOR i = 1 TO nroots
7500    xterm(i) = COS(pn(i) * x)
7600    sum(i) = xterm(i) * eterm(i) * a(i)
7700    NEXT i
7800    FOR i = 1 TO nroots
```

```
7900   result = sum(i) + result
7950   NEXT i
7960   RETURN
7970   END

8000   REM ----------------------------------------------
8100   REM Evaluate surface heat transfer rate
8150   sum = 0#: result = 0#
8200   FOR i = 1 TO nroots
8300   sum = a(i) * pn(i) * SIN(pn(i)) * eterm(i)
8400   result = sum + result
8500   NEXT i
8600   RETURN
8700   END

9000   REM ----------------------------------------------
9100   REM Evaluate total heat transfer
9150   sum = 0#: result = 0#
9200   FOR i = 1 TO nroots
9300   termi = SIN(pn(i)) * SIN(pn(i)) * eterm(i) * 2#
       / pn(i)
9400   term2 = pn(i) + (SIN(pn(i)) * COS(pn(i)))
9500   sum = term1 / term2
9600   result = sum + result
9700   NEXT i
9800   RETURN
9900   END
```

C.2 ● Erf.bas (see pp 58–9, 320)

```
6000   REM Calculate error function
6050   term = 0#
6070   pi = 3.1415927#
6080   count = 0
6100   p = .3#                          'argument of erf
6200   FOR i = 3 TO 11 STEP 2
6300   count = count + 1
6400   fac = 1#
6500   FOR j = 1 TO count
6600   fac = fac * j
6700   NEXT j
6810   term = (((-1#) ^ count) * (p ^ i) / (i * fac)
       + term
6900   NEXT i
6910   erfp = (2# / (pi ^ .5)) * (p + term)
6920   PRINT erfp
```

```
7000   STOP
8000   END
```

C.3 ● Pn(nr).bas (see p52)

```
5000   REM Find roots of the equation p tan(p) = Biot
5010   REM Using Newton-Raphson Iteration Method
5020   REM <alpha> is an underrelaxation factor.
5030   REM ----------------------------------------------
5050   tol = .00001                          'Tolerance
5070   pi = 3.1415927#
5100   biot = .01#
5150   alpha = 1!
5160   IF biot > 2# THEN alpha = 1# / biot
5200   FOR ipi = 1 TO 6
5300   p0 = (ipi * pi) - (.75# * pi)
5350   p = p0
5400   GOSUB 6000
5500   pnew = p - ((f / fdash) * alpha)
5600   diff = ABS(pnew - p)
5700   IF diff > tol THEN p = pnew: GOTO 5400
5800   PRINT ipi; "   "; p
5850   NEXT ipi
5870   STOP

6000   REM Calculate function (f) and derivative
       (fdash)
6100   f = (p * (SIN(p) / COS(p))) - biot
6200   fdash = (p / (COS(p) * COS(p))) + (SIN(p) /
       COS(p))
6300   RETURN
6400   END
```

C.4 ● SS2D(NR).bas (see pp 47–50)

```
100    REM Does 2-D Analytical Solution for htc boundary
       condition
110    REM Chris Long, July 16th 1998
120    REM
130    DIM term(100), a(100)
150    nterm = 20
160    l = .015: x = .01
170    w = .01: y = .0075
180    cond = .3
```

```
190   htc = 10!
999   OPEN "ss3d.res" FOR OUTPUT AS #1

1000  GOSUB 5000
1500  FOR iterm = 1 TO nterm
1600  a1 = COS(a(iterm) * x)
1650  arg = a(iterm) * (w - y)
1700  a2 = ((EXP(arg)) + (1! / EXP(arg))) / 2!
1800  a3 = (((((a(iterm) ^ 2!) + ((htc / cond) ^ 2!)))
      * 1) + (htc / cond)
1900  arg = a(iterm) * w
2000  a4 = ((EXP(arg)) + (1! / EXP(arg))) / 2!
2050  a5 = COS(a(iterm) * 1)
2100  term(iterm) = (a1 * a2) / (a3 * a4 * a5)
2150  PRINT #1, iterm; a1; a2; a3; a4; a5; term(iterm)
2200  NEXT iterm
2999  sum = 0!
3000  FOR iterm = 1 TO nterm
3050  sum = term(iterm) + sum
3100  REM PRINT iterm; "   "; term(iterm)
3600  NEXT iterm
3800  theta = sum * 2! * htc / cond
3900  PRINT #1, theta
4000  STOP

5000  REM Find roots of the equation p tan(p) = Biot
5010  REM Using Newton-Raphson Iteration Method
5020  <alpha> is an underrelaxation factor.
5030  REM -------------------------------------------
5050  tol = .00001                            'Tolerance
5070  pi = 3.1415927#
5100  biot = htc * 1 / cond
5150  alpha = 1!
5160  IF biot > 2# THEN alpha = 1# / biot
5200  FOR ipi = 1 TO nterm
5300  p0 = (ipi * pi) - (.75# * pi)
5350  p = p0
5400  GOSUB 6000
5500  pnew = p - ((f / fdash) * alpha)
5600  diff = ABS(pnew - p)
5700  IF diff > tol THEN p = pnew: GOTO 5400
5750  a(ipi) = p / 1
5800  PRINT #1, ipi; "   "; a(ipi)
5850  NEXT ipi
5870  RETURN

6000  REM Calculate function (f) and derivative
      (fdash)
```

```
6100   f = (p * (SIN(p) / COS(P))) - biot
6200   fdash = (p / (COS(p) * COS(p)) + (SIN(p) /
       COS(p))
6300   RETURN
6400   END
```

C.5 ● FDEXPXT.BAS (see pp 80–9)

```
999    REM 'FDEXPXT.BAS' Does Example 3.2
1000   REM Program to do EXPLICIT finite difference of
       1-D Transient Eqn.
1001   DIM anp(80), asp(80), bnew(80), c(80, 80),
       b(80), t(80), tnew(80), diff(80)
1002   OPEN "fdexpxt.COEF" FOR OUTPUT AS #1
1003   OPEN "fdexpxt.RES" FOR OUTPUT AS #2
1004   OPEN "fdexpxt.cgd" FOR OUTPUT AS #3

1027   itrace = 1
1040   cond = .72: spht = 835!: rho = 1920!
1050   tinit = 30!
1060   tair = 20!
1070   qdot = 10000!
1080   htc = 10!
1100   wide = .2
1200   nz = 41: npts = nz
1400   dz = wide / (nz - 1)
1410   alpha = cond / (rho * spht)
1420   dt = 5!
1430   nsteps = 1000

1450   GOSUB 9000
1500   GOSUB 2000
1600   GOSUB 3000
1650   GOSUB 6000
1690   FOR istep = 1 TO nsteps
1695   atim = istep * dt
1700   GOSUB 5000
1705   GOSUB 7000
1710   GOSUB 8000

1720   NEXT istep

1810   CLOSE #1
1820   CLOSE #2
1830   CLOSE #3
1900   STOP
```

```
2000    REM Calculate coefficients
2050    delfo = alpha * dt / (dz * dz)
2250    cp = (1! - (2! * delfo))
2300    cep = delfo
2350    cwp = delfo
2950    RETURN

3000    REM Place coefficients in matrix
3200    FOR iax = 1 TO nz
3300    ipoint = iax

3320    REM 4100 Convective boundary at z = 0: 4200
        Constant heat flux at z = L
3400    IF ipoint = 1 THEN 4100
3410    IF ipoint = nz THEN 4200

3800    REM Interior points by default
3840    c(ipoint, ipoint) = cp
3850    c(ipoint, ipoint + 1) = cwp
3860    c(ipoint, ipoint - 1) = cep
3870    GOTO 4990

4100    c(ipoint, ipoint) = cp - (cwp * 2! * dz * htc /
        cond)
4104    c(ipoint, ipoint + 1) = cep + cwp
4105    b(ipoint) = (2! * dz * htc * cwp * tair / cond)
4106    GOTO 4990

4200    c(ipoint, ipoint) = cp
4203    c(ipoint, ipoint - 1) = (cwp + cep)
4204    b(ipoint) = ((2! * qdot * dz / cond) * cep)

4211    GOTO 4990

4990    NEXT iax
4999    RETURN

5000    REM Sum elements of C to get R.H.S. vector, B.
5100    FOR iax = 1 TO npts
5200    bnew(iax) = b(iax) + (t(iax) * c(iax, iax)) +
        (t(iax - 1) * c(iax, iax - 1)) + (t(iax + 1) *
        c(iax, iax + 1))
5300    NEXT iax
5400    RETURN

6000    REM Print out
6100    FOR i = 1 TO npts
6200    FOR j = 1 TO npts
6300    PRINT #1, i; j; c(i, j)
6400    NEXT j: NEXT i
```

```
6500    FOR i = 1 TO npts
6600    PRINT #1, i; bnew(i)
6700    NEXT i
6900    RETURN

7000    REM Direct (explicit) solution
7100    FOR iax = 1 TO npts
7200    tnew(iax) = bnew(iax)
7250    t(iax) = tnew(iax)
7300    NEXT i
7400    RETURN

8000    REM PRINT out solution
8010    PRINT #3, istep; CHR$(9); atim; CHR$(9);
        tnew(itrace)
8050    PRINT #2, "Solution for time step No.  "; istep
8060    PRINT , "Solution for time step No.  "; istep
8100    FOR ipoint = 1 TO npts
8200    PRINT #2, ipoint; tnew(ipoint)
8250    NEXT ipoint
8300    RETURN

9000    REM Initialise Temperature Field
9100    FOR ipoint = 1 TO npts
9320    t(ipoint) = tinit
9335    NEXT ipoint
9500    RETURN
```

C.6 ● FDRZ.Bas (see pp 71–80)

```
999     REM 'FDRŻ.BAS' does Example 3.1
1000    REM Program to do finite differece of Steady
        State r-z Eqn.
1001    DIM anp(16), asp(16), htc(16), a(91, 91), b(91),
        t(91), tnew(91), diff(91)
1002    OPEN "fdrz.COEF" FOR OUTPUT AS #1
1003    OPEN "fdrz.RES" FOR OUTPUT AS #2
1004    OPEN "fdrz.ggd" FOR OUTPUT AS #3
1007    REM Line 7500 for iteration scheme
1008    omega = 1.8
1027    niters = 200: itrace = 35
1028    cond = 20!
1029    trim = 600
1030    tair = 450
1100    rin = .15: rout = .3: wide = .015
1200    nz = 7: nr = 13: npts = nr * nz
```

```
1300   dr = (rout - rin) / (nr - 1)
1400   dz = wide / (nz - 1)
1500   GOSUB 2000
1600   GOSUB 3000
1700   GOSUB 6000
1800   GOSUB 7000
1810   CLOSE #1
1820   CLOSE #2
1830   CLOSE #3
1900   STOP

2000   REM Calculate coefficients
2100   aep = (1! / (dz * dz))
2200   awp = (1! / (dz * dz))
2300   FOR jrad = 1 TO nr
2400   radius = rin + ((jrad - 1) * dr)
2500   htc(jrad) = 133! * (radius ^ .6)
2600   anp(jrad) = ((1! / (dr * dr)) + (1! / (2! *
       radius * dr)))
2700   asp(jrad) = ((1! / (dr * dr)) - (1! / (2! *
       radius * dr)))
2800   NEXT
2900   ap = -((2! / (dr * dr)) + (2! / (dz * dz)))
2950   RETURN

3000   REM Place coefficients in matrix
3100   FOR jrad = 1 TO nr
3200   FOR iax = 1 TO nz
3300   ipoint = iax + (nz * (jrad - 1))
3310   t(ipoint) = 0!: b(ipoint) = 0!: a(ipoint,
       ipoint) = 0!
3400   IF ipoint = 1 THEN 4000
3410   IF ipoint = nz THEN 4100
3420   IF ipoint = npts - nz + 1 THEN 4200
3430   IF ipoint = npts THEN 4300
3440   IF ipoint = 1 + (nz * (jrad - 1)) THEN 4400
3450   IF ipoint = nz + (nz * (jrad - 1)) THEN 4500
3460   IF ipoint > (npts - nz + 1) AND ipoint < npts
       THEN 4600
3470   IF ipoint > 1 AND ipoint < nz THEN 4700
3800   REM Interior points by default
3810   a(ipoint, ipoint) = ap
3820   a(ipoint, ipoint + 1) = awp
3830   a(ipoint, ipoint - 1) = aep
3840   a(ipoint, ipoint + nz) = anp(jrad)
3850   a(ipoint, ipoint - nz) = asp(jrad)
3860   GOTO 4999
```

```
4000   a(ipoint, ipoint) = ap
4001   a(ipoint, ipoint + 1) = aep + awp
4002   a(ipoint, ipoint + nz) = anp(jrad) + asp(jrad)
4003   GOTO 4999

4100   a(ipoint, ipoint) = ap - (awp * 2! * dz *
       htc (jrad) / cond)
4101   a(ipoint, ipoint - 1) = aep + awp
4102   a(ipoint, ipoint + nz) = anp(jrad) + asp(jrad)
4103   b(ipoint) = -2! * dz * htc(jrad) * awp *
       tair / cond
4104   GOTO 4999

4200   a(ipoint, ipoint) = 1!
4201   t(ipoint) = trim
4202   b(ipoint) = trim
4203   GOTO 4999

4300   a(ipoint, ipoint) = 1!
4301   t(ipoint) = trim
4302   b(ipoint) = trim
4303   GOTO 4999

4400   a(ipoint, ipoint) = ap
4401   a(ipoint, ipoint + nz) = anp(jrad)
4402   a(ipoint, ipoint - nz) = asp(jrad)
4403   a(ipoint, ipoint + 1) = awp + aep
4404   GOTO 4999

4500   a(ipoint, ipoint) = ap - (awp * 2! * dz *
       htc (jrad) / cond)
4501   a(ipoint, ipoint - 1) = aep + awp
4502   a(ipoint, ipoint - nz) = asp(jrad)
4503   a(ipoint, ipoint + nz) = anp(jrad)
4504   b(ipoint) = -2! * awp * dz * htc(jrad) *
       tair / cond
4505   GOTO 4999

4600   a(ipoint, ipoint) =1!
4601   t(ipoint) = trim
4602   b(ipoint) = trim
4603   GOTO 4999

4700   a(ipoint, ipoint) = ap
4701   a(ipoint, ipoint + nz) = anp(jrad) + asp(jrad)
4702   a(ipoint, ipoint - 1) = aep
4703   a(ipoint, ipoint + 1) = awp
4704   GOTO 4999

4999   NEXT iax: NEXT jrad
5000   RETURN
```

```
6000    REM Print out
6100    FOR i = 1 TO npts
6200    FOR j = 1 TO npts
6300    PRINT #1, i; j; a(i, j)
6400    NEXT j: NEXT i
6500    FOR i = 1 TO npts
6600    PRINT #1, i; b(i)
6700    NEXT i
6900    RETURN

7000    REM Iterative solution
7100    FOR ipoint = 1 TO npts
7320    IF ipoint >= (npts - nz + 1) AND ipoint <= npts
        THEN tnew(ipoint) = t(ipoint)
7330    IF ipoint < (npts - nz + 1) THEN t(ipoint) =
        (trim + tair) / 2!
7335    NEXT ipoint

7339    FOR itno = 1 TO niters
7340    FOR ipoint = 1 TO npts
7400    rowsum = 0!
7410    FOR isum = 1 TO npts
7412    tval = t(isum)
7413    IF isum < ipoint THEN tval = tnew(isum)
7415    acoef = a(ipoint, isum)
7416    REM IF ipoint=isum THEN acoef = 0!
7420    rowsum = rowsum + (acoef * tval)
7430    NEXT isum

7500    tnew(ipoint) = t(ipoint) + (omega * (b(ipoint) -
        rowsum) / a(ipoint, ipoint))
7600    diff(ipoint) = (tnew(ipoint) - t(ipoint)) /
        (trim - tair)

7700    NEXT ipoint
7750    GOSUB 8000
7760    FOR ipoint = 1 TO npts: t(ipoint) =
        tnew(ipoint): NEXT ipoint
7800    NEXT itno
7900    RETURN

8000    REM PRINT out solution
8010    PRINT #3, itno; CHR$(9); tnew(itrace)
8050    PRINT #2, "Solution for iteration No.  "; itno
8100    FOR ipoint = 1 TO npts
8200    PRINT #2, ipoint; tnew(ipoint); t(ipoint);
        diff(ipoint)
8250    NEXT ipoint
8300    RETURN
```

Index

Topics covered within worked examples are referenced here using italic typeface for the page number. Topics covered within an end of chapter question are indicated with the question number enclosed in parentheses.

e.g. Black body, enclosure 249, *250–1*, (8.2), (8.4), (8.10) indicates that the subject is described on p. 249; it appears either as the main theme or as a sub-topic in the worked example on pp. 250–1; and it is addressed in Questions 8.2, 8.4 and 8.10 at the end of Chapter 8.

absorptivity, 11, 239, 240, 242
acceleration parameter, 78, 79
adiabatic boundary condition, numerical
    implementation, 69, 73
adiabatic wall temperature, 114
analogies between heat and momentum 121–3
annular film boiling, 226, 227
area ratio, heat exchanger, 296, 297
average, *see specific entries*, e.g. Nusselt number

Benard cells, 195, 196
Bernoulli's equation, 109, 124, 296
binomial theorem, 258
Biot number, 21, 57, 317
black body radiation, 11, *(1.5)*, *(8.1)*
black body, 233, 241
    enclosure, 249, *250–1*, *(8.2)*, *(8.4)*, *(8.10)*
black body radiation, 236, 237, 238, 249, *250, 251*
Blasius formula, 145, *148*, 155, 156
body force, 97, 102, 112, 125, 129
boiling, forced convection, *see* flow boiling
boiling, free convection, 218
boiling 217–27
boiling curve, 217
Boltzman constant, 235
boundary conditions, 64
    1st, 2nd and 3rd kinds, 18, 19
    numerical implementation 69–70, *73–4, 84–5*,
        (3.3–5)

boundary layer
    free convection, 175
    mixed (laminar and turbulent), *158, 159*, *(5.8)*, *(5.10)*
    pipe, 139
    pressure in, 108
    transition in, 115, *(4.10)*
    velocity distribution, 125, *(4.10)*
boundary layer equations, 109, 123, 136
boundary layer theory, 106–9, 130
boundary layer thickness, 111, 126, *182*
    laminar flat plate, 152, 153, 154
    laminar free convection, 178
    from scale analysis, 127, *(4.7)*
    thermal, 107, 109
    turbulent flat plate, 157
    velocity, 107, 109
Boussinesq approximation, 112, 190
building heat loss, *23, 24, 25, 26, 27*, *(2.12)*, *(6.2)*
bulk average velocity, 140, 156
bulk mean temperature, 114, 141
buoyancy force, 175

calibration of thermocouples, 307, 308, *(10.1–2)*
caloric, 1
cartesian coordinates, 18, 80
central difference, 66, 81
characteristic length, 111
colburn analogy, 123, *(4.12)*, 144, 155, 180, *(5.12)*,
        (6.12)

cold junction, thermocouple, 306, 307
cold junction compensation, 308, 309, *310*, 312, (*10.4*)
Colebrook formula, 145, *148*
combined convection and conduction, 9, (*1.6*), (*2.2–4*)
combined radiation and convection (*1.2*), (*1.4*), (*1.6–7*), (*5.12*), *182, 184*, (*6.6*), *223, 238, 256, 261, 263*, 264, (*8.3*), (*8.5*), (*8.9–11*), 314, *317*, (*10.5*), (*10.6*), (*10.9–10*)
compatibility condition, 151
compensation cable, thermocouple, 308, 309, (*10.3*)
compressor blade, 56, 59
compressor disc *71–5, 161*
computer methods, 3
condensation, 205–17
   forced flow, 216, 217, (*7.3*)
   rate of formation, 208
conduction, 3, 4, *5, 6*, (*1.3*), 14–63
   composite walls, *25–7*, (*2.2*)
   dimensionless groups, 20
   in fluids, 14, 192, *195*
   numerical, steady-state, 2-D, 67–70, *71–6*, (*3.2–3*), (*3.5*)
   numerical, 64–93
   numerical transient, 80–3, *83–9, (3.1*), (*3.4*)
   plane walls, 22–5, *23–5*, (*2.4*)
conduction equation, 313
   constant properties, 17, (*2.1*)
   general, 15, 16, 17
conduction solution, 314–7, (*10.6*)
constant heat flux boundary condition, numerical implementation, 70, 84
continuity equation, 95, 96, 97, 127
   cylindrical coordinates, 97, (*4.8*), 139
   time averaged, 118
contraction coefficient, heat exchanger, 296
convecting boundary condition, numerical implementation, 69, 74, 85
convection, 6, 7, 8, *9*, 10, 94–134
   dimensionless groups 109–11
   during a change of phase, 205
   free and forced, *198*, (*6.2–3*), (*6.5*)
convection regimes, 94, 130
   film condensation, 210
   flow boiling, 225, 226
   internal free convection, 192
   pool boiling, 218, *223*
convergence, 77, 78, 87
corner points, *see* boundary conditions
Couette flow, 135, 136, *137, 138*, (*5.1*)
counter flow heat exchanger, 269, 270, 283, (*9.7*), (*9.11*)
   effectiveness, 292, 293
   NTU, 292, 293

Crank–Nicolson scheme, 82, *87, 88*
critical insulation radius, 33, (*2.8*)
critical Rayleigh number, 195
cross flow heat exchanger, 269, 270, 283, (*9.10*)
   correction factor, 283, 284
cylinder
   film condensation, 214, *215, 216*, (*7.1*)
   free convection, 180, 186
   nucleate pool boiling heat transfer, 220
cylinder in crossflow, 166, *167, 168*, (*5.2*), (*5.7*)
cylindrical coordinates, 18, (*2.1*), 81, 105

diffuse emitter, 233
dimensional analysis 109–11, (*4.11*)
direct contact heat exchanger, 268
direct solution, 77
directional dependence, 233
discretisation, 64
discretisation error, 66, 76
dissipation function, 103
Dittus–Boelter equation, 145, (*7.2*), *277*, (*9.3*)
double pipe heat exchanger, 271, *276, 284, 285*
droplet condensation, 206
dryout, 226

Eckert number, 111, 138
eddy diffusivity, 119, 120, 121, 122
electric motor, (*2.6*), (*5.1*)
effectiveness, heat exchanger, 289
effectiveness, NTU analysis, 289–92, *292–5*, (*9.9*)
electromagnetic spectrum, 232
electronics heat sink, (*1.7*), *41, 42*
emission, 239
emissivity, 11, 240, 241, 242, 313
enclosure
   free convection 192–200
   inclined, 196, 197
   heated from below, 195, *197, 198, 199*
   heated from the side 192–5, (*6.5*)
energy equation 100–5
   boundary layer, 109, 119, 127
   cylindrical coordinates, 105, 139
   time averaged, 119
energy integral equation, 125
enhancement of heat transfer, 36
entrance loss, heat exchanger, 296, *299*
entropy, 104
entry length
   hydrodynamic, 140, 144
   thermal, 141, 144
   pipe, 139

environmental radiation, 259, 260, (*8.11*)
error function, 58, 320
errors, *see specific entries*, e.g truncation error
excess temperature, in pool boiling, 217, 218
exit loss, heat exchanger, 296, *299*
expansion coefficient, heat exchanger, 296
experimental measurement 304–23
explicit scheme, 81, 82, *85*
extended surface, *see* fin
external convection, 94, 114, 121

fictitious point, 69, 70, *74*
film boiling, 219, 222, 223, *224*, (*7.5*)
film condensation
  inclined surface, 214
  laminar, 207–10, *212*, *215*, *216*, (*7.1*)
  turbulent, 210, 211, *213*, (*7.2*)
  vertical tube, *211–14*
film Reynolds number, 210, *213*
fin, 36–44, (*2.5*), (*2.6*), (*3.2*), 278, (*9.2*)
  adiabatic tip, 39
  as a parasitic loss, 36, (*2.12*)
  assumptions, 37
  complex geometry, 41
  convecting tip, 39, *42*, *43*
  effectiveness, 40, 41, *43*
  efficiency, 40, 41, *43*, *182*, *279*, *280*
  equàtion, 38
  infinite, 40, 41, *42*, *43*, 327
  known tip temperature, 40
  performance, 40
  uniform 38–40
finite difference
  approximations, *see* central/forward difference
    etc.
  coefficients, *73*
  equations, 65–70, *73*
  grid, 67, *72*
  interior point, 68
  nodes, 67, *72*
  numerical accuracy, 90
  solution methods, 76, *86*
finite elements, 65
first law of thermodynamics, 1, 100, 208
flat plate
  laminar 149–55
  turbulent 155–9
flow boiling, 225–7, (*7.6*)
flow boiling regimes, 225, 226
fluctuating components, 116, 117
forced convection, 8, 121, 126, 135–72, (*4.5–6*),
  (*5.2*)

cylinder, 166, (*5.2*)
dimensionless groups, 111
external flow, 149–168
flat plate, 149–59, (*5.6*), (*5.10*), (*5.12*)
heat transfer coefficient, 110
impinging jet, 163–5, (*5.4*)
parallel flow, 135–8
pipe flow, 139–43, (*5.7*)
rotating disc, 159–62, (*5.8–9*)
scale analysis, 127, 128
sphere, 168, 325
forced convection boiling, *see* flow boiling
form drag, 107
forward difference, 81
fouling resistance, 275, *278*, (*9.1*)
  table of representative values, 276
Fourier coefficient, 46
Fourier number, 21, 53, 57, 59, 317
Fourier step length, 81
Fourier's equation, 17, 50, 315
Fourier's law, 4, 5, 14, 101, 138, 208
free convection, 8, 126, (*4.6*), (*4.11*), 173–204
  boiling, 217
  channel flow, 182
  cylinder, (*5.2*), 186
  dimensionless groups, 112
  external flow, 175–89
  heat transfer coefficient, 113
  horizontal plate, 182–6, (*6.3*), (*8.11*), (*10.8*)
  inclined surface, 182–3
  internal flow, 192–200
  lumped mass approximation, 56
  scale analysis, 128, 129
  sphere, 186, 187
  uniform heat flux, 180
  vertical channel flow, 189, 190, 191
  vertical cylinder, 180
  vertical plate 175–82
freestream, 106, 122, 124
friction factor, 141
friction velocity, 156
frictional heating, 114
fully developed pipe flow, 139, 140

gas turbine engines, 50, 159
gauge sensitivity, 323
Gauss–Seidel, 78, *87*, (*3.2*)
Gaussian elimination, 77
Gaussian error function, 58
glass, transmissivity, 240, 313
Grashof number, 112, 113, 116, •129, (*4.3*), 174
greenhouse effect, 240

grey body, 241
   enclosure, 251, 252, *253, 254, 255, 256, (8.6), (8.8)*
grid, *see* finite difference
grid dependence, *89*

heat capacity rate, 283
heat exchanger, 144, 166, *(5.7), (6.11)*, 268–303
   classification, 269, 270, 272
   compactness, 272
   design, 269, *284, 285, 286, (9.7), (9.10–11)*
   examples of application, 269
   maximum heat transfer rate, 289, 290
   mixed/unmixed streams, 269, 270
   performance, *286–8, 292, 294, 295, (9.6), (9.8),*
     (9.9), (9.11)
   pressure loss 295–7, 297–300, *(9.11)*
heat flux, 5, *6*, 14, 119
   average value, 113, *158*
   in nucleate boiling, 219
   measurement *(1.6)*, 313–23, *(10.6), (10.8–10)*
heat loss from an insulated pipe, *31*
heat sink, *181*
heat transfer
   boiling and condensation, 8
   definition of, 1
   modes of, 2, *(1.1)*
heat transfer augmentation in heat exchangers, 269,
     278, 279, 280, 281, (9.4), (9.5)
heat transfer coefficient, 6, 22, 95, 110, 130
   average value, 113
   controlling, 275, *278, (9.1–2)*
   in film boiling, 222, 223, *224, (7.5)*
   film condensation
      horizontal cylinder, 214
      inclined surface, 214
      sphere, 214
   laminar film condensation, 209, *212, 215, 216*
   laminar flat plate, 154, 155
   in nucleate boiling, *221*
   order of magnitude values, 8
   pipe, 141
   solar panel, 262
   turbulent film condensation, 211, *213*
heat transfer model, 2, *(6.3)*
heat transfer rate, 5, 14
Heisler chart, plane wall, 53
horizontal surface, nucleate pool boiling heat
     transfer, 220, *221*
hot junction, thermocouple, 306, 307
hydraulic diameter, 143, 210, *277*
hydrophobic coatings, 206
hydrostatic pressure gradient, 112, 190

hyperbolic cosine, 39
hyperbolic sine, 39, 46
hypothetical surface, in radiation analysis, 252

impinging jets, 163, 164, *164, 165, (5.4)*
implicit scheme, 82
inclined surface, free convection 182–3
incompressible flow, 97
incompressible fluid, 105
indirect solution, 77
inertial terms, 99, 122, 127
infrared radiation, *see* radiation
initial conditions, 18
instrumentation 304–32
insulation, effect on heat loss, *26, 27*
integral equations 123–6, 130, 176–8, *(6.12)*
integral method, laminar flat plate 149–55, *(5.11)*
internal convection, 94, 114
internal energy, 101
internal heat generation, 15, *(2.4)*
international practical temperature scale, 305
inverse square law, 260
inviscid core, pipe flow, 139
irradiation, 239, 240
isothermal box, 309, 311

Jacobi, 78
Jakob number, 209, *212*
journal bearing, *137*

Kapton, 322
kinematic viscosity, 150
kinetic theory of gases, 3
Kirchhoff's law of radiation, 242, 252, 259, 261

Lambert's cosine law, 236
laminar flow, 115, 125
Laplace's equation, 17, 44, 105
latent heat of vaporisation, 206, 209
   water, 220
Leidenfrost temperature, 219
liquid crystals, 312
liquid in glass thermometer, 306
liquid metals, 127, 146
logarithmic mean temperature difference 281–4,
     284–9, (9.6–9.8) (9.10–11)
low Biot number approximation, *see* lumped mass
lumped capacitance, *see* lumped mass
lumped mass, validity, 57
lumped mass approximation, 55–7, *56–7, (2.9–11),*
     (3.1), 165, (5.4), 313, 317, *318, 319*, 324, *(10.8)*

matrix equations
  finite difference, 75
  grey body radiation analysis, 253
maximum heat flux, in boiling, 218, 220, *221*
Maxwell's relations for a perfect gas, 104
measurement error
  heat transfer coefficient, 315
  temperature, (*2.5*), 315, 323, 324, *325*, *326*, 327
minimum heat flux, in boiling, 219, 222
mist flow, 226
mixed boundary layer, *see* boundary layer
mixed convection, 94, 174, 188
mixing length, 120, 121
momentum equations, 97–100, 108
  boundary layer, 109, 127
  cylindrical coordinates, 100, (*4.9*), 139, 189
  time averaged, 118
momentum integral equation, 125, 157
monochromatic radiation, 239
Moody chart, 146, *149*, 296
multi-dimensional conduction, 15, 16, 22

natural convection, *see* free convection
Navier–Stoke's equations, 99, 105
Newton–Raphson iteration, *185*, *186*, *199*, (*6.1*), (*8.9–11*)
non–intrusive measurement, 312, 313
normal stress, 98
no–slip condition, 7, 95, 119, 159
NTU, 289
nucleate flow boiling, 226, 227, (*7.6*)
nucleate pool boiling, 218, (*7.4*)
nucleation sites, 217
number of transfer units, 289
numerical errors, *76*
numerical instability, *82*, *89*
Nusselt number, 111, 113, (*4.4*), (*10.6*)
  average value, 114, (*4.5*), 155, 157, 158, *161*, (*5.7–8*), (*5.10*), (*5.12*), 179, *181*, (*6.11*)
  correlations for boiling and condensation, 229
  correlations for forced convection, 169
  correlations for free convection, 201
  Couette flow, *138*
  cylinder in crossflow, 166, *167*, *168*, (*5.2*)
  definition of average value, 113
  enclosure heated from below, 195, 196, *197*, *198*, *199*
  enclosure heated from the side, 192, 193, *194*, (*6.5*)
  forced convection, 128, (*4.10*), (*10.10*)
  free convection, 129
    horizontal plate, 184, *184*, *185*, (*6.3*)
    rotating disc, 188, *188*, *189*

    vertical plate, 179, 180, *181*, *182*, (*6.4*), (*6.9–10*), (*6.12*)
  from analogy methods, 123, (*4.12*), (*5.12*), (*6.12*)
  from scale analysis, 126
  horizontal cylinder, free convection, 186, *187*
  impinging jet, *164*, *165*, (*5.4*)
  inclined enclosure, 196, 197
  laminar flat plate, 155
  laminar pipe flow, 143
  mixed boundary layer, *158*, *159*, (*5.8*), (*5.10*)
  rotating disc, 160, *161*, *162*
  sphere, 168
    free convection, 186, *187*, (*6.6*)
  turbulent flat plate, 157
  turbulent pipe flow, 144, 145, *146*, *148*
  vertical channel flow free convection, 191, (*6.7–8*), (*6.10*)
Nusselt's analysis, 207, 208, 209, 210

one seventh power law, 125, 156
overall heat transfer coefficient, *25*, *32*, *33*, *195*, *272–6*, *276–81*, (*9.1–3*), (*9.11*)
  representative values in heat exchangers, 274

parabolic velocity distribution, 125, 151, 152
parallel flow, 135–8
parallel flow heat exchanger, 269, 270, 282, 283, (*9.6–9*)
  effectiveness, 291, 292, 293
  NTU, 292, 293
partial differentiation, 7
passage loss, heat exchanger, 296, *298*
Peltier effect, 306
penetration depth, 321
perfect gas, 105
pipe flow
  heat transfer, *147*, (*5.3*), (*5.7*)
  laminar 139–43
  turbulent (*4.12*), 143–9,
Planck distribution, 234
Planck's constant, 235
plate fin heat exchanger, 271
platinum resistance thermometer, 307, 311, (*10.5*)
plug gauge, 321, 322
Poiseuille flow, 135
Poisson's equation, 17
pool boiling, 217–25, (*7.4*)
power law, velocity profile, 156
power law variation, 113, 160, *161*
Pr $\simeq$ 1 (1 assumption, 109, 154
Prandtl number, 109, 111, 112, 113, 127, 128, 129, (*4.1*), 154, 174
Prandtl number effect in free convection, 179

radial conduction, 28–33, 272, 273
  composite wall *31–3*
  current carrying wire, *29, 30*
  heat flow, 28
  pipe insulation, *31*, *(2.3)*
  temperature distribution, 28
radiation, 10, 11, *(1.2)*, *12*, 232–67
  between two grey bodies, 257, 258, 259
  in boiling, 219, 223, *225*
  from gas flames, 259
  in gases, 240
radiation configuration factor, *see* view factor
radiation heat transfer coefficient, 11, *33*, 223, 224
radiation pyrometer, 312, 313
radiation shield, 325
radiative power, 235, 236
radiosity, 239, 240, 252, *255*
Rayleigh number, 113, 129, 180
Rayleigh number, for a uniform heat flux, 180
reciprocity rule, 244, *246*, 249, *251*, 252, *254*
recovery factor, 114
recuperative heat exchanger, 268
reference junction, thermocouple, 308, 309, *310*,
  (10.3)
reference temperature, 114
reflectivity, 11, 239, 240
regenerative heat exchanger, 268
residual, 78
resistance thermometer, 311
Reynolds analogy, 123, 144, 295
Reynolds number, 111, 113, 115, 127, *(4.2)*, 152, 174
Reynolds stress, 118
Rohsenow equation, 219, *221*, *(7.4)*
rotating disc, 159, 160, *161, 162, (5.8–9)*, 188, *188,
  189, 318, 319*
rotational Reynolds number, 111, 160, *162*
round-off error, *76*

saturated nucleate boiling, 218
saturation temperature, 205, 217
scale analysis, 21, 130, 107–8, 126–9, 153, 155
scattering, atmospheric, 260
second law of thermodynamics, 1, 290
second order backward difference, *88, 89, (3.2), (3.5),*
  316, *317, (10.6)*
Seebeck effect, 306
Seider–Tate equation, 145
semi infinite solid, 57, 58, *59*, 313, 320, 321, 327
  validity, 59, 321
sensible heat, 209
separation of variables 45–6, 51, *(2.7)*
series solution 46–7, *48–50*, 52–3, 54–5, *(2.7), (2.10)*

shallow enclosure, 192
shape factor, *see* view factor
shear stress, 98, 118, 125
  turbulent flat plate, 156
shell and tube heat exchanger, 271, *297, 298, 299*
similarity variables, 178
Simpson's rule, *(5.5), (5.6)*
skin friction coefficient, 122, *(4.12)*, 144, 152, 157,
  (5.12)
sky temperature, 260, *261*
solar absorptivity, *184*, 262
solar constant, 260
solar heating panels, 262, *263*
solar radiation, 10, *184*, 236, 242, 259, 260, *(8.9)*
solid angle, 235
solid state temperature sensors, 312
specific heat at constant pressure, 104
spectral dependence, 233
spectral emissive power, 236
spectral intensity, 234
sphere
  film condensation, 214
  free convection, 186, 187
  forced convection, 168
stagnation point, 166
steady flow, 96
steady state conduction, 15
  1-D, 22–44, 313, *314, 315*
  2-D, 44–8, *48–50, (2.7)*
Stefan–Boltzmann constant, 11, 237, 238
steradian, 235
Stoke's hypothesis, 99
sub-cooled boiling, 218
subcooling, 208
successive over relaxation, 78
summation rule, 244, *246*, 249, *254*
surface finish, effect on boiling heat transfer, 219,
  220, *222*
surface force, 97
surface roughness, 145, *148, (5.7)*
surface tension, 219
surface tension, water, 220
symmetry boundary condition, *48*, 69, 73

Taylor series, 16, 65
temperatue coefficient of resistance, 311
temperature difference, average, 113
temperature distribution, laminar free convection,
  177
temperature measurement, *(8.5)*, 305–13
  dynamic error, 323, 324
  embedding error, 327

lead wire conduction error, 326, 327
  radiation error, 324, *325*, *(10.5)*
temperature scale, 305
thermal boundary layer, 107
thermal conductivity, 15
  anisotropy, 15
  of solids liquids and gases, 3
  table of common materials, 4
  temperature dependent, 34, *35*
thermal contact resistance, 34
thermal diffusivity, 17, 51, 150
thermal disturbance error, 327
thermal paint, 312
thermal radiation, *see* radiation
thermal resistance, *10, 25*, 274
thermistor, 311
thermocouple, 306, 307, 308, *(10.1–4)*
  ideal circuit, 307
  materials, 306
thermography, 313
thermopile, 322, *(10.7)*
time-averaged equations, 121
time-averaged equations, 116–19
time dependent, *see* transient
total emission, 233
total emissive power, 237
transcendental equation, 52
transient conduction, 15, 50–3, *54–5*, 55–6, *56–7*,
  57–8, 59, (2.9–10)
  2D, 315
  plane wall, 51, *83*
transition boiling, 219
transition criteria
  cylinder in crossflow, 166
  external forced, 155
  external free convection, horizontal plate, 184,
    184, 185
  film condensation, 210
  forced convection, 168, *(10.10)*
  free convection, 179
  pipe flow, 140
  rotating disc, 160, 162, *(5.8)*

transmissivity, 11, 239, 240
truncation error, 76
tube fin heat exchanger, 271
turbulence, 115–23, 125
turbulence
  effect on heat transfer, 116
  free convection, 179, 180, *(6.12)*
  transition criteria, 115, 116
turbulence modelling, 120
turbulent eddies, 116, 118
turbulent heat flux, 119
turbulent Prandtl number, 122
turbulent shear stress, 118, 119

unheated starting length, 155, 157, *(5.6)*

vapour quality, 226, 227
velocity boundary layer, 106, 107
velocity distribution
  laminar film condensation, 208
  laminar flat plate, 151, 152
  laminar free convection, 177
  laminar pipe flow, 140
  turbulent flat plate, 156
  turbulent pipe flow, 156
vertical plate, free convection 175–182
view factor, 11, 243, 244, *246, 249*
  coaxial discs, 248, *(8.5)*
  concentric cylinders, 248
  opposed rectangles, 247
  perpendicular rectangles, 247
  table of formulae for some 2–D geometries, 245
  table of formulae for some 3–D geometries, 246
view factor algebra, 244, 249, *254, (8.4–5), (8.7)*
viscosity, 107
viscous dissipation, 138, *(5.1)*
viscous stress, 99
viscous sub-layer, 122
volume expansion coefficient, 104, 112

Wien's displacement law, 237, *237*
work done on a fluid, 101, 102